AF505583

Plant Biotechnology

Agnès Ricroch · Surinder Chopra · Marcel Kuntz
Editors

Plant Biotechnology

Experience and Future Prospects

Second Edition

 Springer

Editors
Agnès Ricroch
AgroParisTech
Paris, France

Marcel Kuntz
Laboratory of Plant and Cellular Physiology
University Grenoble-Alpes
CNRS, INRAE and CEA
Grenoble, France

Surinder Chopra
Department of Plant Science
The Pennsylvania State University
University Park, State College
PA, USA

ISBN 978-3-030-68344-3   ISBN 978-3-030-68345-0 (eBook)
https://doi.org/10.1007/978-3-030-68345-0

Foreword

Reflections on Innovation and Progress

An examination of the meaning of words reveals that *innovation* and *progress* are strikingly inequivalent Examining their Latin derivations, *innovation* means entering into novelty by introducing something new into a preexisting device. *Progress* means going forward (*progressus*) and increasing (*progressio*). In the contemporary uses, *innovation* is a descriptive term, whereas *progress* includes an often positive value judgment. However, terminology may be a factor of progress or regression. Terminological conservatism (e.g., GMO) is an obstacle to scientific progress in some sectors. We cannot underestimate the power of words in social life as they condense real questions: How do we make scientific and technical innovation a progress for everybody? Is progress in science equivalent to progress in society?

The latter of these questions was addressed in a conference of the European Academy of Sciences in Brussels in 2016. The Standing Secretary of the French Academy of Sciences Catherine Bréchignac stressed that society embraces technology rather than science. A member of the French Academy of Technology, François Guinot, declared that technological innovation induces science to hide behind technology. Moreover, the exponential character of technological development creates a phase lag affecting many layers of society and resulting in political problems.

Innovation is a key factor of economic growth by producing new variations of preexisting devices, which provide consistency and the basis for innovation itself. Genome editing is a prominent example of true technological innovation based on a whole preexisting scientific story in fields such as bacterial genetics and virology. For example, the CRISPR-Cas9 system represents striking progress in genome cleavage thanks to its unprecedented precision. This system was significantly improved and simplified by Emmanuelle Charpentier and Jennifer Doudna awarded the Nobel Prize in Chemistry 2020.

Genome editing became a routine practice not only in biological and medical research, but also as a commonly used tool in agriculture and more applied in medicine. For example, success in the treatment of childhood leukemia was recently reported. However, the European Union decided to consider genome editing as GMO technology. This label carried the consequence that it could not be used for agricultural production in Europe. Regarding medicine, eugenic practice on embryonic cells, which is different from editing somatic, differentiated cells in organs and tissues, was recently performed by a Chinese physician who was able to induce a gene mutation and prevent possible HIV infection in the baby. This practice was widely condemned. The French National Academy of Medicine and the French Academy of Sciences issued in 2018 a joint, balanced declaration, stating essentially: (1) in the present state of knowledge, it is not advisable to give birth to such embryonic-modified babies; (2) in case such a procedure could be started in the future, it must undergo academic and ethical approval and in-depth public debate; (3) responsible research using DNA-modifying technologies, including at the embryonic level, is important to human beings. Consequently, both Academies support such research.

At this point, we meet the question of the relationship between fundamental and applied research. Indeed, this relationship is bidirectional in its essence. "Translational medicine" consists of fostering applied, clinical testing in order to provide useful results for basic research. In this context, the question of risk evaluation arises, together with the famous aphorism "absence of evidence is not evidence of absence"—a quite ambiguous story with dubious logical foundations. Most of the time, the aphorism is used, according to the spirit of the *precautionary principle*, to prevent further research on supposedly toxic substances. Whether or not this use turns out to be justified, generally this aphorism should be used to foster research rather than to hinder it.

How can we ensure that scientific and technological innovation creates progress in society? Progress in society means applying several kinds of human values, not only knowledge, to the same effect. While progress is impossible without innovation, it is unimaginable if it does not pertain to society as a whole.

Claude Debru
Emeritus professor of philosophy of science at the Ecole normale supérieure, Paris, France
Member of the French Academy of Sciences and of the French Academy of Agriculture

Paris, France

Claude Debru
claude.debru@ens.fr

Claude Debru, Dr. phil., Ph.D. is the author of books and collections on the history and philosophy of life sciences and medicine, including such fields as protein biochemistry, physiology, neurosciences, biotechnologies, and hematology. A former professor at the universities Louis-Pasteur of Strasbourg, Paris-Diderot and at the Ecole normale supérieure, he is a member of the French Academy of sciences, of the French Academy of agriculture, of the German National Academy of sciences Leopoldina, and of several other international scientific societies.

Contents

Part I
The Tools for Engineering Plants

Chapter 1
The Evolution of Agriculture and Tools for Plant Innovation

Agnès Ricroch

Genetically engineered crops are playing an increasingly important role in world agriculture, enabling scientists to reach across genera for useful genes to enhance tolerance to drought, heat, cold, and waterlogging, all likely consequences of global warming. I believe biotechnology will be essential to meet future food, feed, fiber, and biofuel demand. The battle to ensure food security for hundreds of millions of miserably poor people is far from won. We must increase world food supplies but also recognize the links between population growth, food production, and environmental sustainability. Without a better balance, efforts to halt global poverty will grind to a halt.
Norman Borlaug—Science, 318: 359, 19 October 2007

Abstract Plants such as cereals and legumes on which humans depend on today were domesticated gradually and independently by ancient farmers in many different parts of the world over a few thousand years. Over time, ancient farmers converted hundreds wild species into cultivated crops (some of the world's most important crops). In the transition from foraging to farming 10,000 years ago, the wild forms of these plants mutated and were selected to result into new, domesticated species that were easier to harvest. This process continues today. Since the beginning of the twentieth century, innovation in plant genetic technologies has accelerated and produced better crops through increased resistance to pests and diseases, tolerance to drought and flooding, and biofortification. Together with the advancement of whole genome sequencing technologies dramatic and rapid progress has been made in our understanding and ability to alter gene expression in plants and in techniques for the identification, isolation and transfer of genes of interest. In many cases, this progress has been facilitated by the availability of efficient gene transfer methods, New breeding techniques (NBT) have rapidly emerged in the 2000s. Compared to the early versions of gene editing tools, such as ODM (oligonucleotide-directed mutagenesis), meganucleases (MNs), zinc fingers nucleases (ZFNs), and transcription activator-like effector nucleases (TALENs), clustered regularly interspaced short

A. Ricroch (✉)
AgroParisTech and Paris-Saclay University, Paris and Sceaux, France
e-mail: agnes.ricroch@agroparistech.fr

palindromic repeat (CRISPR) system is capable of altering a genome more efficiently and with high accuracy. Most recently, new CRISPR systems, including base editors and prime editors, confer reduced off-target activity with improved DNA specificity and an expanded targeting scope. Geneticists use a wide variety of gene transfer methods to introduce foreign DNA (from microorganisms, plants, animals) into plants. Plant genetic improvement offer an effective approach to increase food production and food security in order to support the world's growing population, especially in inhospitable climates. Plant innovations can also improve production of medicines for all.

Keywords Griculture · Crop · Domestication · Breeding · NBT · CRISPR · Biotechnologies

1.1 Multiple Origins of Agriculture

Did you know that the oldest bee pollinators from 100 million-years ago were found in pieces of amber? Scientists also found evidence that human ancestors used fire one million years ago. According to fossils of starch grains from grinding stones and cooking pots found in archaeological sites, archaeologists stated that the history of plant breeding and cultivation of major cereals started about 10,000 years ago.

1.1.1 Emergence of Agriculture

The adaptation of crop plants to human needs and cultivation is a slow process evolving on a time scale of millennia. Wild cereals could have been cultivated for over one millennium before the emergence of domesticated landraces (Tanno and Willcox 2006). Human domestication of plants can be divided into three stages: "gathering," in which people gathered plants from wild stands; "cultivation," in which wild plants were systematically sown in fields of choice; and "domestication," in which mutant plants with desirable traits were raised (Weiss et al. 2006).

Based on recent DNA studies and radiocarbon dating of archaeobotanical remains, farming arose several times in several locations once the Ice Age had ended and climatic and environmental conditions were favourable for farming. Soon after humans adopted a sedentary existence agriculture arose (Tanno and Willcox 2006). These discoveries show the greatest revolution in human history: the transition from gathering foods from the wild to producing them on farms.

Foremost among the creations of ancient plant breeders are the cereals—rice, wheat, and maize, provide more than 50% of the calories consumed by humans today (Ross-Ibarra et al. 2007). However, 70% of the calories consumed by humans come from only 15 crops, which were domesticated in different countries worldwide. The Neolithic transition, which broadly describes the shift from foraging to farming, is one of the most important events in human history. Agriculture happened first in the early villages of the Near East in the Fertile Crescent, a region from the Mediterranean Sea to Iran including modern-day Israel, Syria, Jordan northeastern Iraq and southeastern Turkey and subsequently occurred in different parts of the world including China, Mesoamerica and the Andes, Near Oceania, sub-Saharan Africa, and eastern North America (Riehl et al. 2013; Meyer and Purugganan 2013). As early as 13,000 years ago, hunter-gatherers first began to gather and plant seeds from wild cereals and legumes, such as wheat, barley, and lentils and began their cultivating more than 11,500 years ago. Plants were domesticated gradually and independently by people in many different parts of the world. Japonica rice, a subspecies of *Oryza sativa*, was bred about 10,000 years ago in the upstream region of the Yangtze River in China. Key crops such as rice and soybean originated in eastern Asia. This region is also the original home of several minor crops, such as certain types of millet. Maize eaten today by over 1 billion people was domesticated approximately 10,000 years ago in southwestern Mexico. For further information refer to the book «1491» by Charles C. Mann. Starting from 12,000 years ago in the Middle East, the Neolithic lifestyle spreads across Europe via separate continental and Mediterranean routes (Rivollat et al. 2020).

1.1.2 The 'Domestication Syndrome'

The dawn of agriculture, as well as of crop domestication, was a process of trials and errors. During domestication, humans subjected several key events to selection that make up the 'domestication syndrome'. During this process, ancient farmers, either consciously or unconsciously, saved seeds from plants with favoured characters to be sown the next year. The 'domestication syndrome' defined as phenotypic traits associated with the genetic change to a domesticated form of an organism from a wild progenitor form include loss of seed falling (shattering), decreased dispersal, loss of seed dormancy, increased number of seeds, change in seed shape, compact growth habit (reduced branching, reduced plant size, dwarfism), increased size of fruits, adaptation of flowering time to local areas, and reduced content of toxic compounds (safer food). Humans have also selected crops for disease-resistance.

The cereals—botanically a grass, from which the fruit which is called a caryopsis (grain) is harvested—, and most other crops, share a feature—a character or trait—central to domestication: their grains remain attached to the plant for harvest by humans rather than falling from the plant, as required by wild species to produce their next generation. For example, domestication of maize involved a plant architecture transformation from the wild ancestor (progenitor), *Zea mays* ssp. *parviglumis* resulting into an unbranched plant with seed attached to a cob, thereby making maize dependent on humans for cultivation. Subsequent to domestication, maize has been subject to intensive improvement efforts, culminating in the development of hybrid maize lines that are highly adapted to modern agricultural practices.Understanding the origins and domestication of crops is of evolutionary interest. Understanding crop origins also allows the identification of useful genetic resources for crop improvement. Thus, domesticated plants provide a model system for studying adaptation of plants to their environment (the concept of adaptation is central in Darwin's work). Domestication shapes the genetic variation that is available to modern breeders as it influences diversity at the DNA level. Indeed, scientists today can follow how domestication proceeded at the level of DNA sequence change, from wild ancestors (progenitors) to cultivated crops. Insights into the domestication process reveal useful DNA information (at the gene level) for future crop breeding.

1.2 The Toolbox of Crop Improvement: Hybrids and First Biotechnologies

To accomplish the objectives for crop improvement plant breeders develop various tools and methods to broaden the possibilities for breeding new plant varieties: conventional breeding such as hybridization and mutation breeding, to advanced breeding techniques such as genetic modification.

The work of Charles Darwin (1859) and Georg Mendel (1866) created the scientific foundation for plant breeding (Fedoroff 2004). The Austrian monk Gregory Mendel showed the importance of statistics in breeding experiments and the predictability in selective breeding. In 1866 he formulated the laws of inheritance on garden peas and discovered of unit factors (later defined as genes). Previously, the French family of the Vilmorins, who established the first seed company in 1727 in France (today part of the Limagrain Cooperative), introduced the pedigree method of breeding in 1830 (based on selected individual plants). The first seed company in North America was established by David Landreth in 1784. He published a catalog of vegetable seeds in 1799. The twentieth century efforts were devoted to improving the productivity, reliability, and nutrition of crops: maize (George Beadle and Paul Mangelsdorf), fruits, vegetables, and ornamental flowers (Luther Burbank) to cite some. Indeed since the beginning of the 20th-century the plant breeder's toolbox has been developed to cause specific and permanent changes (genetic modifications): from first-generation hybrids (of maize and many other crops), wide-species crosses, mutation breeding, to genetic engineering. The new tools and methods are more and more rapid in their ability to create varieties with new and interesting traits.

1.2.1 Hybridization (Crosses Between Plants or Species)

The transfer of traits between genetically distant or closely related species is not a new technique. Hybridization which is a cross between two parental plants which carry interesting traits has been achieved in numerous crops. It takes almost 15–20 years to create a new hybrid variety such as in sunflower, maize, oilseed rape, or wheat. In these wide crosses thousands of genes are affected while in transgenic plants one to six genes can be added (for the moment).

In 1919 in Connecticut Donald F. Jones developed the double-cross method in maize, which involved a cross between two single crosses (four inbred lines generated from the mating of parents who The domestication syndrome can be defined as the characteristic collection of phenotypic traits associated with the genetic change to a domesticated form of an organism from a wild progenitor form.are closely related genetically are used). This technique made the commercial production of hybrid maize seed economically-viable. In 1923 in Iowa, Henry C. Wallace developed the first commercial hybrid maize. In 1926, he then founded the Hi-Bred Maize Company (today Pioneer Hi-Bred, a DuPont Company).

Hybrid seed technology generates heterozygous plants with improved yield and disease resistance by adding traits from two different parents. Average maize yields over the past 40 years have doubled in the USA, but this did not occur everywhere in the world.

1.2.2 Chemical- and Radiation-Induced Mutagenesis

Chemical- and radiation-induced mutagenesis (using Gamma-rays and X-rays since ca.1920) increases the frequency of genetic variations which can be used to create new mutant varieties. A mutant is a plant/organism in which a base-pair sequence change occurs within the DNA of a gene or chromosome resulting in the creation of a new character or trait. These mutations can be interesting for crop improvement, such as reducing the height of the plant, changing seed colour, or providing tolerance or resistance to abiotic (e.g. salinity and drought) and biotic (e.g. pests and diseases) stresses. In the UK, much of the beer was produced using a mutant variety of barley (the 'Golden Promise' variety, salt-tolerant spring barley with semi-dwarfness in stature). Wheat varieties developed through mutation breeding technique are used today for bread and pasta (e.g. induced mutability for yield). Many physiological and morphological mutants have been obtained (in banana, cassava, cotton, date palm, grapefruit, pea, peanut, pear, peppermint, rice, sesame, sorghum, and sunflower … and also horticultural plants, see https://mvd.iaea.org/.) Over 3332 crop and legume varieties developed through chemical or radiation induced mutagenesis have been released worldwide in more in 73 countries: Seeds of tomato variety Bintomato-7 irradiated with gamma ray (370 Gy) were released in 2018 for cultivation in winter season (November-February) of Bangladesh, The mutant variety of wheat with low amylose was developed by treatment with chemical mutagen sodium azide (NaN3) in Japan; the first mutant semi-dwarf table rice 'Calrose 76' released in the US, a mutated indica rice stain developed by irradiation of seeds with gamma rays (250 Gy) with short stature (95 cm against 120 cm of Calrose), shortening of all internodes. In organic agriculture farmers use the 'Calrose 76' strain of brown rice, also developed through mutagenesis. Lewis J. Stadler of the University of Missouri was the first to use X-rays on barley seeds in 1920s and ultraviolet radiation on maize pollen in 1936.

Different kinds of mutagens are used in plant breeding, such as chemical mutagens like EMS (ethyl methanesulfonate) to generate mutants.

It takes more than ten years to create a variety with such mutations, which will be then crossed with an elite variety adapted to local agronomical and climatic conditions. Such varieties carry a huge number of genes affected. The random results of this genetic technique illustrate how spontaneous mutations create the genetic diversity that drives evolution (one of the Darwin's concept), and the material upon which selective breeding can operate.

1.2.3 Other Techniques: In Vitro Techniques, Genome Sequencing and Gene Mapping

Other breeding techniques using in vitro tissue culture—micropropagation, and embryo rescue—permits the crossing of incompatible plants and allows the production of uniform plants.

Thanks to the knowledge at molecular (DNA) level and bioinformatics the latest step of innovation in plant breeding, dating from the 1980s, came from biotechnologies. Molecular marker–assisted selection (MAS) is now widely used to localize characters or traits on the genetic map of the crop and select commercially important characters or traits. In MAS for example, a DNA marker closely linked to a disease resistance locus can be used to predict whether a plant is likely to be resistant to that disease (Tester and Langridge 2010).

In 1944, DNA as the genetic material was discovered in pseudococcus by Oswald Avery, Colin MacLeod, and Maclyn McCarty, from the Rockefeller Institute in the USA. Then in 1953 James Watson, Francis Crick, Rosalind Franklin, and Maurice Wilkins determined the structure of DNA. Since the 50s DNA sequencing has seen rapid progress. The first sequencing of a natural gene from yeast was made in 1965 and took 2.5 years. In 1976, the first genome was sequenced (a bacteriophage). In 2008, the first human genome (6 billion base pairs of DNA of James Watson's genome) was sequenced in four months and cost less than US\$ 1.5 million. The price is dropping rapidly due to new DNA sequencing technologies. According to the National Human Genome Research Institute (USA) today the cost to generate a whole-exome sequence is generally below US \$1,000.

A complete genome sequence is available for several crops since the late 1990s: bread wheat, rice, maize, papaya, grape, apple, soybean, potato, sorghum, strawberry, date palm, cassava, cacao, foxtail millet, cotton, banana… The latest sequenced genomes of 2013 are of chickpea, peach, sweet orange, and wild rice.

1.2.4 The Green Revolution

Since 1940, foundations such as the Ford, the Rockefeller, the Howard Buffet or the Bill and Melinda Gates Foundations have played a major role in collaboration with governments for breeding of crops. The Green Revolution started in 1943 when the Mexican government and the Rockefeller Foundation co-sponsored a project, the Mexican Agricultural Program, to increase food production in Mexico, in particular wheat production. Using a double-concept (interdisciplinary approach and international team effort), the scientific team headed by an American wheat breeder at the Rockefeller Foundation, Norman E. Borlaug, started to assemble genetic resources (germplasm) of wheat from all over the world. The life and legacy of the father of the Green Revolution, Borlaug, who received the Nobel Peace Prize in 1970, is celebrated in 2014 for the 100th anniversary of his birth.

After the famine of 1961 in India, Borlaug advanced the development of high-yielding varieties such as IR8—a semi-dwarf rice variety, along with expansion of irrigation infrastructure, and modernization of management techniques, distribution of hybrid seeds, fertilizers, and pesticides to farmers.

Today almost two billion people suffer from chronic hunger and malnutrition in developing countries. This makes agricultural development in developing countries a pressing need as they have the fastest population growth rate and they are also more at risk from resource shortages and the effects of climate change. Increasing food supply without deforestation or a net change in land use means increasing production. This makes agricultural development through crop improvement a pressing need. As deplored by Paarlberg (2009), modern agriculture—including biotechnology—has recently been kept out of Africa).

1.3 Advanced Breeding Techniques: Genetic Modification Technologies

In 1946 J. Lederberg and E. L. Tatum were the first to discover that DNA naturally transfers between organisms. Genetic engineering, also known as genetic modification (GM), exploits recombinant DNA technology as new tool for plant breeders. As a technique that is faster and able to deliver genetic changes that would never occur through conventional methods, GM is uniquely useful in the plant breeder's toolbox.

Conventional breeding today encompasses all plant breeding methods that do not fall under current regulations for GMOs. For example in Europe, the European legal framework defines GMOs and specifies various breeding techniques that are excluded from the GMO regulations (the European Directive 2001/18/EC on the deliberate release of GMOs into the environment). Excluded from this GMO Directive (and thus may be viewed as conventional breeding) are hybridization (cross breeding), in vitro fertilization, polyploidy induction, mutagenesis and fusion of protoplasts from sexually compatible plants. In the USA transgenic (GM) plants are deregulated and not labeled as GMOs (except in some States). Edited plants are deregulated in the USA. The case of Europe is examined in the Joachim Schiemann's chapter.

1.3.1 Genetic Engineering Technologies

Transgenic techniques provide genetic modification or genetic engineering of a recipient plant with one or more foreign genes. These foreign genes can come from plant or non-plant organisms. Transgenic plants are used for precise crop improvement because of transfer of limited genetic material as oppose to conventional breeding in which one half of the genome from each parental line is combined after hybridization. Genetic engineering also makes possible genetic changes, including between animals and plants, which would be highly unlikely or would never occur using mutagenesis or other conventional breeding techniques.

Advances in molecular biology in the 1970s made it possible to identify the specific gene responsible for a trait, isolate it, and transfer it, from any type of organism, to plant cells. Instead of making tens of thousands of genetic changes (cross or mutation breeding), with transgenesis a gene with a known single beneficial trait is inserted into the plant genome. Plant breeders embraced transgenesis because it offered this precision and a quicker way of obtaining a desired trait in a plant.

Ethical questions on growing GM crops were addressed by scientists involved molecular biology research. The first GM experiment, published in 1972, described the insertion of bacteriophage genes into an animal viral DNA. Consequently scientists raised questions about potential risks of recombinant DNA to human health and organized the Asilomar Conference in 1975 in California in the USA, attended by scientists, lawyers and government officials to discuss the technology. They concluded that experiments could proceed under strict guidelines drawn up by the US National Institutes of Health (Berg et al. 1975).

There are several vectors to genetically engineer plants: (i) infecting plant tissue by recombinant *Agrobacterium tumefaciens* carrying a gene of interest will lead to integration of this gene in the plant DNA, a mechanism of genetic engineering discovered by Marc Van Montagu and Jeff Schell (in Belgium) and Mary-Dell Chilton (in the USA) in 1977, or (ii) shooting plant tissue with a 'particle gun' carrying tungsten or gold particles coated with the gene to be transfered (also called as biolistic particle delivery system; developed in 1984 by John Sanford, Edward Wolf, and Nelson Allen in the USA).

Introduced genes fall randomly amid the DNA strands. Plant mutation breeding (discussed above, 2.2) may induce more changes than transgene insertions through genetic engineering. Regeneration of a genetically engineered plant is a rather fast process, however, since such a variety need to be crossed with elite varieties adapted to specific agronomical and climatic conditions, it takes a few years to create a variety with added transgenes.

A special feature of genetic modification is that it allows the transfer into crop plants of one or a few genes from unrelated organisms (microorganisms such as bacteria, animal or human). Conventional breeding (hybridization between very distinct plants even from different genus) cannot form plants with genes coming from different kingdoms. Additional techniques of modern plant breeding are discussed in the second chapter by Surinder Chopra.

1.3.2 Traits Expressed by the Genetic Engineering Technologies

The first GM plant produced was an antibiotic-resistant tobacco plant in 1982. The first commercialized GM crop was the FlavrSavr® tomato in 1994 in the USA. It contained a trait that suppressed early ripening in tomato to maintain flavor and taste. In the UK, a concentrated tomato paste using these GM tomatoes went on sale in 1996 (by Zeneca). It received an award in France for the best innovation. The earliest crops produced by transgenesis (insect-resistant and herbicide-tolerant varieties) have been commercially cultivated since 1995. A GM variety of maize developed to express a protein from *Bacillus thuringiensis,* ('*Bt* maize') protects maize against the European maize borer and some other lepidopteran insects. *Bt,* originally discovered in 1911 in the province of Thuringia in Germany, has been used as a spray by organic farmers. The *Bt* genes produce insecticidal CRY proteins which are an alternative to chemical pesticides. These are introduced in more than a thousand elite varieties of maize, but also in cotton, cowpea, soybean and sugarcane as examples.

The global area cultivated with GM varieties was over 191.7 million hectares in twenty-six countries (21 developing and 5 industrialized countries) in 2018. A total of 26 countries adopted GM crops through cultivation and 44 additional countries imported. Crops grown commercially today contain traits for mainly herbicide tolerance, insect resistance, or both. These have been developed for commodity crops such as soybean, cotton, maize, oilseed rape and alfafa. It is estimated that, for example, 88% of the cotton grown in India is now GM due to its greater resistance to pests. The cultivation of GM insect-resistant crops, particularly varieties of cotton, in India and China, is also reducing the exposure of farmers to harmful organo-phosphate insecticides. There are a lot of products from GM crops in the food chain. In Europe it is estimated that 90% of some animal feed (maize and soybean) is derived from GM varieties because of their low cost and large amount available.

The list of approved GM crop varieties modified by transgenesis (gene transfer or silencing using RNAi) is long: alfalfa, Argentine canola, apple, bean, canola, carnation, creeping bentgrass, cotton, cowpea, eucalyptus, flax, maize, melon, miscanthus, papaya, petunia, plum, Polish canola, potato, rice, rose, squash, safflower, sorghum, sugar beet, sugarcane, sweet pepper, soybean, tobacco, tomato, wheat (for updated data visit https://www.isaaa.org/gmapprovaldatabase/default.asp). The list of edited crop varieties which are deregulated includes e.g. alfafa, bahiagrass, camelina, citrus, chrysanthemum, flax, maize, pennycress, Petunia, potato, rice, setaria (wild millet), soybean, tobacco, tomato or wheat.

Many genes of interest have been discovered including pest and disease (fungi, virus, bacterial) resistance genes, and new ones are being discovered at a rapid rate. Some of these genes have been incorporated into commercial varieties to breed for specialty traits and these include heat and drought tolerance, nitrogen use efficiency, modified alpha amylase, male sterility, modified amino acid, modified flower color (in dianthus), modified oil/fatty acid, and virus resistance. In Pamela Ronald's laboratory in UC Davis (USA) the discovery of the gene XA21 confers resistance to a bacterial disease, and the discovery of a gene of submergence tolerance of rice allows drowning weeds without drowning the rice, providing a method for weed management without relying on a herbicide (Ronald and Adamchak 2008).

Radical innovations concern nutritional benefits. Healthier vegetable oils with fewer trans-fats are being developed. Bio-fortifying key crops including cassava in Africa or rice in Asia illustrate the potential of genetic engineering to fight malnutrition. In developing countries, especially in Asia, vitamin-A deficiency causes childhood blindness. The most famous attempt to combat this deficiency is the development of 'Golden rice' by Ingo Potrykus in Switzerland and his colleagues (Zeigler 2014). They genetically transformed rice plants with carotenoid biosynthetic genes that result in more vitamin-A precursors. Today, geneticists are also trying to reduce allergens in foods using genetic engineering. Technologies such as genomic selection, genome editing and the role of bioinformatics could be galvanized by using speed breeding to enable plant breeders to keep pace with a changing climate and environment in plant adaptation to environmental and biotic contraints,

The ability to manipulate plant genes to produce certain human enzymes is not new. Interest in deriving pharmaceuticals from plants (known as 'bio-pharming'), first took off in the 1990s after scientists showed that monoclonal antibodies could be produced in tobacco plants. Plant-derived biologic treatments have proven successful in drugs given to animals in recent years and today in human patients suffering from Gaucher disease or development of vaccines against COVID-19 (discussed in Kathleen Hefferson's chapter). This led to genetic engineering of plants to produce vaccines, antibodies and proteins for therapeutics.

1.3.3 Development of New Breeding Techniques

In the past two decades, additional applications of biotech and molecular biology in plants have emerged, with the potential to further enlarge the plant breeder's toolbox. Making precise changes in the genomes of organisms is challenging for most techniques. Several recently described genome editing techniques allow for site-directed mutagenesis of plant genes (to knock out or modify gene functions) and the targeted deletion or insertion of genes into plant genomes. New breeding techniques have rapidly emerged in the 2000s. Compared to the early versions of gene editing tools, such as ODM (oligonucleotide directed mutagenesis), meganucleases (MNs), zinc finger nucleases (ZFNs), and transcription activator-like effector nucleases (TALENs), clustered regularly interspaced short palindromic repeat (CRISPR) system is capable of altering a genome more efficiently and with high accuracy. In 2012, researchers transformed a bacterial immune system (CRISPR system) into a fast and versatile tool for genome editing. The Royal Swedish Academy of Sciences has decided to award the Nobel Prize in Chemistry 2020 to Emmanuelle Charpentier (Max Planck Unit for the Science of Pathogens, Berlin, Germany) and Jennifer A. Doudna (University of California, Berkeley, USA) for the development of a method for genome editing. Since 2014 plants were edited with CRISPR-cas9 and notably the hexaploid wheat (Ricroch 2017). Regarding plant species and countries in which the research is performed, one can note the importance of rice, mainly in China, which is in accordance with the Chinese research and economic contexts, while the application of CRISPR/Cas systems in maize is more prevalently studied in the USA (Ricroch et al. 2017). China is now taking the lead in the industrial and agricultural applied sectors and in the total number of patents per year (Martin-Laffon et al. 2019). Another innovative trend is the use of transgenes solely as a tool to facilitate the breeding process. In this application, transgenes are used in intermediate breeding steps and then removed during subsequent crosses, eliminating them from the final commercial variety (null segregants). Among new tools are accelerated breeding techniques, where genes that promote early flowering are used to speed up breeding, and reverse breeding, a technique that produces homozygous parental lines from heterozygous elite plants (Lusser et al. 2012). New tools also concern three techniques: cisgenesis, intragenesis, and the zinc finger nuclease-3 technique (ZFN-3). Cisgenesis is the genetic modification of a recipient organism with a gene from a crossable–sexually compatible organism (same species or closely related species). Intragenesis is a genetic modification of a recipient organism that leads to a combination of different gene fragments from donor organism(s) of the same or a sexually compatible species as the recipient. ZFN-3 allows the integration of gene(s) in a predefined insertion site in the genome of the recipient species. In 2012, the researchers transformed a bacterial immune system into the fast and versatile tool for genome editing (CRISPR system).

A search-and-replace method, also known as prime editing, was developed that can introduce user-defined sequence into a target site without requiring double-stranded breaks (DSBs) or repair templates (Anzalone et al. 2019). For precision breeding of crops this genome engineering using prime editing system was developed in rice (Hua et al. 2020) and wheat (Lin et al. 2020). China and the USA lead scientific research in crop editing while Nigeria being headquarters to numerous research consortia mainly using transgenesis (Ricroch 2019).

1.4 How to Meet 70% More Food by 2050?

Global population has risen from 2.6 billion in 1950 to around 7.8 billion in 2020, and is predicted to rise to a world population of near 10 billion people by 2050. According to the Food and Agricultural Organization of the United Nations, the demand for food could rise by 70% by 2050. To meet this goal an average annual increase in production of 44 million metric tons per year is required, representing a 38% increase over historical increases in production, to be sustained for 40 years.

This accomplishment will be particularly challenging in the face of global environmental change. The challenge for major changes in the global food system is that agriculture must meet the double challenge of feeding a growing population, with rising demand for meat and high-calorie diets, while simultaneously minimizing its global environmental impacts (Seufert et al. 2012).

Today farmers will have to hit targets for reducing greenhouse gas emissions, improving water use efficiency and meeting the demands of consumers for healthful food and high-value ingredients. In this context, new plant breeding techniques are needed to contribute to improvements in crop productivity and sustainability in a climate-smart agriculture framework.

New technologies must be developed to accelerate breeding through improved DNA methods and by increasing the available genetic diversity in breeding germplasm (collection of wild types and varieties). Scientists underline the importance of conserving and exploring traditional germplasm. Introgression of characters or traits (pest and disease resistances or adaptation to salinity, cold or heat temperatures for example) into locally adapted varieties is expected to considerably enhance productivity in protecting crops from new pests and diseases due to climate change variability and under abiotic stress conditions (e.g. drought). The most gain will come from delivering these technologies in developing countries, but the technologies will have to be economically accessible and readily disseminated.

With governments, the private sector, foundations, and development agencies faced with feeding a growing and hungry world, research to increase agricultural productivity and access to affordable and safe medicines is needed including against COVID-19. The rush to develop a vaccine for COVID-19 the disease caused by the novel coronavirus SARS-COV-2 has extended to public and private laboratories, where scientists are using the tools of genetic engineering to develop edible vaccines in plants. The challenges of intellectual property rights and genetic resources preservation that play major roles in the plant breeding enterprise. The twenty-first century will witness radical plant innovations.

References

Anzalone AV, Randolph PB, Davis JR et al (2019) Search-and-replace genome editing without double-strand breaks or donor DNA. Nature 576:149–157. https://doi.org/10.1038/s41586-019-1711-4

Berg P, Baltimore D, Brenner S, Roblin RO, Singer MF (1975) Summary statement of the Asilomar conference on recombinant DNA molecules. Proc Natl Acad Sci U S A 72(6):1981

Fedoroff N (2004) Mendel in the Kitchen: a scientist's view of genetically modified food. National Academies Press

Hua K, Jiang Y, Tao X, Zhu J-K (2020) Precision genome engineering in rice using prime editing system. Plant Biotech J 1–3. https://doi.org/10.1111/pbi.13395

Lin Q, Zong Y, Xue C, Wang S, Jin S, Zhu Z, Wang Y et al (2020) Prime genome editing in rice and wheat. Nat Biotechnol 38:582–658

Lusser M, Parisi C, Plan D, Rodríguez-Cerezo E (2012) Deployment of new biotechnologies in plant breeding. Nat Biotech 30(3):231–239

Mann CC (2005) 1491. Knopf, New revelations of the americas before Columbus

Martin-Laffon J, Kuntz M and A Ricroch (2019) Worldwide CRISPR patent landscape shows strong geographical biases. Nat Biotechnol 37:601–621

Meyer RS, Purugganan MD (2013) Evolution of crop species: genetics of domestication and diversification. Nat Rev Genet 14:840–852

Paarlberg R (2009) Starved for science. In: How biotechnology is being kept out of Africa. Forewords by Norman Borlaug and Jimmy Carter. Harvard University Press

Ricroch AE (2017) What will be the benefits of biotech wheat for European agriculture?. Wheat Biotechnology, pp 25–35

Ricroch A, Clairand P, W Harwood (2017) Use of CRISPR systems in plant genome editing: towards new opportunities in agriculture. Emerg Topics Life Sci 1:169–182 (Special issue)

Ricroch A (2019) Keynote at the OECD conference, 28th May 2018. Global developments of genome editing in agriculture. Transgenic Res 133. https://doi.org/10.1007/s11248-019-00133-6

Riehl S, Zeidi M, Conard NJ (2013) Emergence of agriculture in the foothills of the Zagros Mountains of Iran. Science 341(6141):65–67

Rivollat et al (2020) Ancient genome-wide DNA from France highlights the complexity of interactions between Mesolithic hunter-gatherers and Neolithic farmers. Sci Adv 6:eaaz5344

Ronald PC, Adamchak RW (2008) Tomorrow's table: organic farming, genetics, and the future of food. Oxford University Press

Ross-Ibarra J, Morrell PL, Gaut BS (2007) Plant domestication, a unique opportunity to identify the genetic basis of adaptation. Proc Natl Acad Sci U S A 104:8641–8648

Seufert V, Ramankutty N, Foley JA (2012) Comparing the yields of organic and conventional agriculture. Nature 485:229–232

Tanno K, Willcox G (2006) How fast was wild wheat domesticated? Science 311:1886–1886

Tester M, Langridge P (2010) Breeding technologies to increase crop production in a changing world. Science 327:818

Weiss E, Mordechai EK, Hartmann A (2006) Autonomous cultivation before domestication. Science 312:1608–1610

Zeigler RS (2014) Biofortification: vitamin a deficiency and the case for Golden Rice. In Plant Biotechnology. Springer, Cham, pp 245–262

Agnès Ricroch (Ph.D., HDR) is associate professor in evolutionary genetics and plant breeding at AgroParisTech (Paris, France) and adjunct professor at Pennsylvania State University (USA). She was visiting professor at the University of Melbourne (Australia). She has strong experience in benefits-risk assessment of green biotechnology at Paris-Saclay University. She is editor of six books on plant biotechnologies. She is the Secretary of the Life Sciences section at the Academy of Agriculture of France since 2016. She was awarded the Limagrain Foundation Prize of the Academy of Agriculture of France in 2012. She is a Knight of the National Order of the Legion of Honour of France in 2019.

Chapter 2
Techniques and Tools of Modern Plant Breeding

Dinakaran Elango, Germán Sandoya, and Surinder Chopra

Abstract Field and vegetable crops are primary source of food. Field crops are also a rich source of cellulosic biomass and carbohydrates for biofuels. One of the major challenges facing agriculture today is improving the productivity of crops in an environmentally sustainable manner. Annual climate variation causes temperature extremes, floods, and droughts which all exacerbate the vulnerability of crops to pests and diseases. Conventional plant breeding has evolved and molecular and modern breeding methods have enhanced the pace of crop improvement work. Plant breeders now use molecular and genetic techniques to selectively identify phenotypes and genotypes that are associated with traits of interest. Such functional genomics studies help plant breeders efficiently utilize the germplasm. Cutting edge molecular tools are now available in economically important crops as well as model plant systems. Gene expression techniques have been combined with forward and reverse genetic methods for isolation and introgression of desirable alleles into breeding populations that are used to develop hybrid crops. This chapter focuses on modern techniques and resources that field and vegetable crop scientists use to generate genetic information and efficient breeding strategies.

Keywords Association mapping · Genetics · Genomics · Germplasm · Marker assisted selection

2.1 Plant Breeding and Plant Ideotypes

Plants are the primary source of food, feed and energy and without them life on earth cannot be imagined. With a tremendous increase in human population, dramatic variability in the climatic patterns from year to year-enhanced efforts are needed to breed efficient plants. Plant breeding specifies plant improvement which can be attained

D. Elango · S. Chopra (✉)
The Pennsylvania State University, University Park, PA 16802, USA
e-mail: sic3@psu.edu

G. Sandoya
University of FL—EREC, Belle Glade, FL 33430, USA

either via sexual or asexual methods of propagation. Plant breeding process may involve domestication of a plant species from its wild native environment, developing pure lines, single or double cross breeding, and finally developing hybrids. The science of plant breeding relies on the principles of genetics, information on chemistry and physiology of metabolic pathways, and growth and development of the plant. On the other hand, plant breeder has a special art or a skill and an eye for selecting plants with morphological traits (phenotypes) or features that conform to a preconceived ideotype. A plant breeder also focuses on development of pest, disease and stress tolerant varieties. Thus, depending upon the plant organ to be harvested and climatic conditions for growing that particular variety, the definition of an ideotype can be developed. Ideotype development is dictated by how efficiently a plant utilizes natural resources. Modern plant breeders have several genetic tools and techniques available, which can be used to enhance the process of final product development.

2.2   Plant Breeding Exploits Phenotype and Genotype

Phenotype is the result of interaction of genes of a plant among themselves and with the environment in which the plant is growing. The biological processes involved in plant growth and development are complex and influenced by individual genes as well as a combination of several genes. Traits that are controlled by single genes give rise to qualitative variation while multigenic traits produce quantitative variation. Such quantitative traits exhibit complexity and are highly influenced by environmental conditions. Plant domestication is one of the examples of phenotypic selection in which by growing a wild form the ancient farmers have selected modern types. This is best exemplified by the domestication of teosinte into our modern maize. One can find landraces of field crops as well as horticultural and ornamental plants, which represent selections made by farmers and breeders in a specific climatic condition or in a geographical region. A plant breeder can now make use of sophisticated phenotyping tools to precisely measure the phenotypic effect of a trait.

Mendel's laws provided the genetic basis of segregation of traits (genes) and as the science of plant breeding evolved, a successful program combines traits from different germplasm sources by hybridizations (crossing). After hybridization a plant breeder then grows the subsequent generations to select the best combinations. Cultivars were developed by the use of breeding methods like pure line selection in a self-pollinated crop like wheat. One drawback of pure line breeding has been the genetic homogeneity, which caused instability especially during the growing season when a new race of a disease appeared. In open pollinated crops however, random mating of plants within a population followed by selection provided some advantage by selecting a population that performed better. Thus, phenotypic selection used by conventional plant breeders and success of this art of selection has been well documented in the form of release of high yielding inbreds, hybrids and varieties.

2.3 Molecular Markers and Plant Breeding

In traditional plant breeding, genetic composition of the populations and progenies was not known. However, with the availability of DNA sequences for several plant genomes it is now possible to develop molecular markers. Examples of commonly used include SSR (Simple Sequence Repeat) also known as satellite markers, and SNP (Single Nucleotide Polymorphic) markers. These reliable markers are based on PCR (Polymerase Chain Reaction) methodology. First and foremost, markers are used to enhance the process of breeding and this method is called marker assisted selection (MAS) based on the polymorphisms between the two parental lines used in a cross. The segregating progenies from F_2 (second filial) generation onwards are then screened using the markers that are genetically linked with specific traits in one or the other parental line. Plant breeders also use molecular markers in mapping of genes. In rice, for example, the *Sub1* locus which provides tolerance to submergence, was introgressed from a landrace of *Oryza sativa* into rice cultivars by the use of a method known as marker-assisted backcrossing (MAB) (Septiningsih et al. 2009). In lettuce, several single genes were mapped for disease resistance and MAS can be developed but pathogens are in constant evolution making these markers ineffective. By the use marker assisted breeding, genes for salinity tolerance have been introgressed in wheat and rice. Simple traits that are controlled by single genes can be mapped with relative ease by the use of molecular markers through backcross breeding. DNA-gel-blot-based restriction fragment length polymorphic (RFLP) markers have been used previously in maize, sorghum, and barley to identify quantitative trait loci (QTL) of complex traits conferring tolerance to drought and diseases. In addition to gene and QTL mapping, molecular markers are used in association mapping studies at the single candidate gene level. There are several examples of association of candidate gene and QTL with a given trait. In maize, genome wide association studies (GWAS) allowed the identification of loci associated with leaf length, width and angle. Flowering time variation analysis in maize has led to the association of markers in *dwarf8* gene, provitamin A and molecular markers in the *IcyE* gene in maize. Several agronomic traits in rice have been associated with markers based on single nucleotide polymorphism (SNP). The same applies to vegetable like lettuce with less studied cases; the significant relationship between a marker and a trait was identified for shelf-life (Kandel et al. 2020).

Technological innovations have led to modern genotyping platforms that evolved from laborious gel-based methods. These innovations have also reduced the cost of DNA sequencing which in turn has improved the efficiency of generating new markers to assist with MAS. These genotyping methods have been employed for field crops like rice, maize, barley, and wheat. The re-sequencing efforts of diverse rice germplasm through the Rice SNP Consortium (https://www.ricesnp.org) have provided valuable information on millions of SNP markers. Vegetable crop species as lettuce with larger genomes have been sequenced within the *Compositae* family, one of the largest in the plant kingdom.

2.4 Recombinant Inbred Lines for Plant Breeding

Plant breeders rely on natural variability in order to exploit genetic diversity available within a plant species. Molecular markers that are associated with specific traits are then used to identify diverse germplasm of economically important crops like maize, wheat, sorghum and soybeans. Plant scientists have developed resources to tap genetic diversity and these are used for GWA mapping studies. For example, in maize, nested association mapping (NAM) RIL (Recombinant Inbred Line) populations have been developed (https://www.panzea.org) by crossing twenty-five diverse parental lines with B73, a common parental line (200 RILs per cross). These 5000 RILs capture approximately 136,000 recombination events. These and other plant breeding resources that capture natural variation or genetic diversity allow plant breeders to study the effect of different alleles that are present in diverse parents used for developing association mapping panels. In self-pollinating species populations such as Multi-parent Advanced Generation Inter-Cross (MAGIC) are developed by intercrossing multiple diverse parents to generate a mapping population with a wider genetic background to detect QTLs (Huang et al. 2015). Breeding lines can be also derived from individual families used for mapping.

2.5 Plant Breeding with Haploids

Haploid breeding allows to achieve homozygous condition and this property is very crucial for quick release and dissemination of plant cultivars (Dwivedi et al. 2015; Gilles et al. 2017). Haploids were first reported in Jimson weed (Blakeslee et al. 1922), and later in several crop species. The commercial exploitation of haploids in plant breeding was recognized only after the discovery of anther-culture in Datura. There are numerous ways of generating haploids in plants. One of the recent approach is CENH3 mediated chromosome elimination to generate haploids (Ravi and Chan 2010). Other notable methods to create haploids are anther culture, interspecific and intergenic hybridizations, agrobacterium-mediated transformations, and haploid inducer lines. Heterosis was realized in maize after the discovery of haploid inducer lines. Later the gene underpinning the genetic regulation of haploid induction was identified as NOT LIKE DAD (NLD)/MATRILINEAL (MTL)/ZmPHOSPHOLIPASE A1 (ZmPLA1) in maize.

2.6 Speed Breeding

One of the key bottlenecks for plant breeding is the long generation times of crops, which hinders the rapid development of new crop varieties. Scientists at the University of Queensland during 2003 coined the term 'speed breeding' for set of improved methods to hasten wheat breeding program (Hickey et al. 2019). Speed breeding reduces the crop cycle by extending the photoperiods using artificial lights with controlled temperatures in a growth chamber (Hickey et al. 2019). Speed breeding protocols were developed for important field crops and are being developed for other orphan as well as short-day crops like sorghum. Speed breeding accelerates the rate of genetic gain which helps in fast forward genomic selection, express genome editing. This technology can be adapted according to the crop needs as some of the plants are sensitive to constant light.

2.7 Genome Wide Association Mapping in Plants

GWA mapping studies have been extensively conducted to dissect the genetic causes of complex traits. GWAS is powerful because of high allelic diversity, recombination rates, and the availability of molecular markers densely distributed across genomes. Advent of high throuput genome sequencing technologies made it possible to sequence large number of germplasm collections with an affordable price. The rare alleles present in the wide collection of germplasm materials could be tapped in plant breeding using GWAS. Core, mini-core and association mapping panels have been developed for different field crops and successfully utilized in GWAS to identify promising candidate genes and QTL regions. Novel traits like epi-cuticular wax genes were mapped using such panels in sorghum (Elango et al. 2020).

2.8 Availability of Sequenced Genomes of Field Crops

With the advent of modern DNA sequencing technologies, several plant genomes have been sequenced and are publicly available (https://www.gramene.org/info/about/species.html). These genome sequences provide tremendous opportunities of efficient crop improvement. First of all, genome sequences are rich sources for developing molecular markers. As explained above, these markers can exploit polymorphisms among germplasm lines of that plant species. Plant breeders can then perform allele mining based on these sequence polymorphisms and use selected alleles in the breeding program. Secondly, plant breeders use these reference genome sequences to perform gene mapping of the traits of interest.

2.9 Plant Breeding and Gene Expression Techniques

Crick (1970) described the Central Dogma of molecular biology in which the genetic information from DNA is converted first into RNA, which is then translated into protein. Over the past four decades, the science of molecular biology has exploded because of innovations in technology as well as computational biology. Current focus of a crop improvement program is to develop strategies and decisions based on gene expression. These expression-based techniques help identify, validate, and use desirable genes in the breeding programs. Field crop scientists are now routinely using gene expression as a molecular marker to decide about the strength of an allele of the given gene. Gene expression technologies include expressed sequence tags (ESTs), which are short cDNA (complementary DNA) sequences that can provide information about the expression of genes. EST sequences available for different plant tissues can provide tissue-specific or tissue-preferred expression data. EST sequences are now being used to develop gene-specific markers of expressed genes that crop scientists use in MAS breeding projects. DNA microarray is a gene expression technique in which DNA of all the genes of a plant species is fixed on a slide or a support. These slides are then used to hybridize with RNA from the same tissue of different parental lines or different tissue of the same parental line. DNA microarrays thus provide RNA expression information (i.e. similarities and differences) among different breeding lines as well as tissue-specific changes of genes. RNA-seq is another gene expression analysis tool which generates large data sets from a high throughput sequencing platform. Bioinformatic techniques have been developed to statistically analyze large gene expression data sets. RNA-seq thus provides global gene expression from thousands of genes and this analysis can be extended to multiple breeding lines. The Illumina based sequencing platforms HiSeq and MiSeq can be used for multiplexing large number of samples and these innovations provide huge data on expression of thousands of gene for hundreds of parental lines. These high throughput sequencing (HTPS) techniques have provided gene expression data for important field crops and vegetables. Gene expression profiling has further revolutionized the characterization of complex traits, which are controlled by multiple genes and their effects have been mapped as QTL. The association between phenotype and genotype by the use of molecular markers is done during the identification of a QTL. Expression QTL (eQTL) utilizes the concept of traditional QTL mapping in concert with genotyping information from transcription profiling data. Agronomically important traits are complex traits and eQTL mapping offers an efficient breeding tool. These marker–trait associations have been further exploited by validating them in order to use the relationship across different related species. For example, a major QTL identified in maize has been employed in sorghum to achieve virus and downy mildew resistance. More than 5,000 eQTLs regulate the expression of 4,105 genes; of which 9 eQTLs associated with flavonoid biosynthesis in addition of 6 loci likely responsible for anthocyanin variation in lettuce leaves which gives the red characteristic to leafy lettuce.

2.10 Forward Genetics for Plant Breeding

The goal of forward genetics is to identify the genetic variation underlying a trait. Mutants or variants are either naturally existing or new mutants can be generated artificially. New natural mutations occur at low frequency because these are the direct result of the evolutionary processes. Naturally occurring mutations represent the types that have adapted to a certain environment or a disease or insect pressure. Since naturally occurring mutations are not found for all traits, especially for traits of agronomic importance, plant breeders use artificial methods of generating mutations. Mutation breeding involves use of chemical, physical and insertional mutagens to generate new mutations and then identify plant phenotypes. Commonly used chemical mutagens are methyl nitroso urea (MNU) and ethyl methane sulfonate (EMS). EMS causes single base pair changes and can produce large number of mutations per kilobase pair of the DNA. Dominant mutations are screened in the M_0 generation. However, selfed progeny of the M_0 plants give rise to M_1 where segregation of traits takes place and thus recessive mutations are then identified in the M_1 generation. Physical mutagens including gamma rays and fast neutrons have been used to generate deletions in chromosomes. In the forward breeding programs, mutations are selected and introgressed to improve crops. A highly efficient EMS mutant library was developed recently in sorghum. EMS has been used to mutate vegetable populations of lettuce for seed germination at high temperatures resulting in families with higher germination at high temperature and for tolerance to herbicides (Huo et al. 2016).

The third types of mutations are caused by insertion or transposable elements or jumping genes. Barbara McClintock was awarded a Nobel Prize for her discovery of transposable elements in maize. Transposable elements excise from one place in the genome and re-insert at another place randomly. When transposons jump into a region of the gene that encodes a protein, the function of that gene is disrupted giving rise to a mutation. Today, plant scientists use transposable elements as genetic tags that can be used to clone the flanking gene sequence for which the mutation resides. Similar to the transposons, T-DNA (transfer DNA) has been used to develop insertion libraries of mutants. T-DNA is a plasmid DNA present in the engineered *Agrobacterium tumefaciens* and which can be engineered to transfer genes of interest. T-DNA has been used as an insertion tool to carry certain marker and/or antibiotic or herbicide resistant genes into plant cell. Public resources of maize, sorghum, brachypodium and rice are available for crop scientists to screen their desired mutations. Once a mutation is identified, a co-segregation analysis is performed in segregating generations to associate the insertion allele with the novel phenotype. In Arabidopsis as well as in many of the vegetables to date sequenced genomes have been valuable source of information for researchers to screen for mutations.

In forward genetics, once a mutation is identified, a mapping population is developed by crossing the mutant line with another parental line with large number of polymorphisms. In addition, the second parental line is chosen based on the availability of other genetic and genomic resources developed from this particular parental line. For example, the maize inbred line B73 has been used to develop the reference genome sequence, transposon insertion databases, and availability of transcriptome, proteome and metabolome resources and databases.

2.11   Reverse Genetics Tools

Plant breeding efforts have been enhanced by the availability of genome sequences. One of the challenges is to ascribe a function to a putative gene sequence. Forward genetics can identify a limited number of phenotypic mutations followed by the mapping of the underlying genes. Reverse genetics utilizes the available genic sequence to identify its function by developing gain of function or loss of function mutants. Reverse genetics tools thus allow crop scientists to dissect the function of a putative gene sequence. Transposon and T-DNA insertion libraries are extensively used for reverse genetics in Arabidopsis. Several functional genomics techniques have been made available to perform reverse genetics in crop plants including maize, sorghum, rice and tomato (Char et al. 2020; Emmanuel and Levy 2002; Ram et al. 2019).

Chemical mutagenesis via EMS has been advanced to the isolation of mutants by the use of TILLING (Targeting Induced Local Lesions in Genomes; Till et al. 2006). TILLING has been successfully used for maize, rice, wheat, barley, and sorghum and vegetables as lettuce. Insertional mutagenesis using transposons and T-DNA transgenes has been one of the popular reverse genetics' techniques. Insertion elements are dispersed throughout the genome and such plant populations are then used to screen presence of insertion in the gene of interest. In general, insertion mutagenesis identifies loss of function mutations. However, there are examples in maize where 'gain of function' mutations have been identified as well.

RNA induced gene-silencing method also known as RNA interference (RNAi) is being used to study the function of a known sequence. In RNAi mutagenesis, plants are transformed with a vector which generates a double stranded RNA corresponding to the gene of interest. Synthesis of double stranded RNA in the plant cell triggers the cellular machinery to degrade the RNA produced from the gene of interest. Similar to RNAi, virus induced gene silencing (VIGS) has been used as a reverse genetics tool; but compared to RNAi, VIGS is relatively quick because it does not involve development of transgenic plants. Thus, VIGS produces transient phenotypes that are not heritable, while RNAi generated mutations are heritable and can be deleterious. The advantage of RNAi over KO (knock out) insertions is that if the KO mutation is lethal, RNAi will usually be not lethal and thus allow isolation of mutants with reduced expression level.

2.12 Targeted Genome Editing Technology

Genome editing can be successful in enhancing the plant breeding process. Crossing two diverse parents to transfer segments of DNA carrying genes often has drastic effects because of the linkage drag. Backcrossing for several generations with the adapted parental line traditionally rectifies these deleterious defects but this process of backcross breeding is time consuming and use lots of resources. New technologies like Zinc finger nucleases (ZFN), transcription activator-like effector nucleases (TALENS) and clustered regularly interspaced short palindromic repeats (CRISPR) and CRISPR associated protein 9 or 12 or 13 (Cas-9/12/13) have shown promise in genome editing. In these methods, sequence-specific nucleases cleave targeted loci enabling entire sequence replacement of an allele, insertion of new DNA, and formation of indels (insertions and deletions). While there are many mutations to be achieved using genome editing, several "editions" proved to be successfully at improving specific traits in many crops; lettuce for example is able to germinate at temperatures as high as 35 °C; but the crop is better suited to germinate in temperatures under 28 °C.

References

Blakeslee AF, Belling J, Farnham ME, Bergner AD (1922) A haploid mutant in the jimson weed, datura stramonium. Science 55(1433):646–647

Char SN, Lee H, Yang B (2020) Use of CRISPR/Cas9 for targeted mutagenesis in sorghum. Curr Protocols Plant Biol 5(2):1–20

Crick F (1970) Central dogma of molecular biology. Nature 227:561–563

Dwivedi SL, Britt AB, Tripathi L, Sharma S, Upadhyaya HD, Ortiz R (2015) Haploids: constraints and opportunities in plant breeding. Biotechnol Adv 33(6):812–829

Elango D, Xue W, Chopra S (2020) Genome-wide association mapping of epi-cuticular wax genes in Sorghum bicolor. Physiol Mol Biol Plants 26(8):1727–1737

Emmanuel E, Levy AA (2002) Tomato mutants as tools for functional genomics. Curr Opin Plant Biol 5(2):112–117

Gilles LM, Martinant JP, Rogowsky PM, Widiez T (2017) Haploid induction in plants. Curr Biol 27(20):R1095–R1097

Hickey LT, Hafeez AN, Robinson H, Jackson SA, Leal-Bertioli SCM, Tester M, Gao C, Godwin ID, Hayes BJ, Wulff BBH (2019) Breeding crops to feed 10 billion. Nat Biotechnol 37(7):744-754

Huang BE, Verbyla KL, Verbyla AP, Raghavan C, Singh VK, Gaur P, Leung H, Varshney RK, Cavanagh CR (2015) MAGIC populations in crops: current status and future prospects. Theor Appl Genetics 128(6):999–1017

Huo H, Henry IM, Coppoolse ER, Verhoef-Post M, Schut JW, de Rooij H, Vogelaar A et al (2016) Rapid identification of lettuce seed germination mutants by Bulked segregant analysis and whole genome sequencing. Plant J 88(3):345–360

Kandel S, Jinita HP, Hayes RJ, Mou B, Simko I (2020) Genome-wide association mapping reveals Loci for Shelf Life and developmental rate of lettuce. Theor Appl Genet 133(6):1947–1966

Ram H, Soni P, Salvi P, Gandass N, Sharma A, Kaur A, Sharma TR (2019) Insertional mutagenesis approaches and their use in rice for functional genomics. Plants 8(9):310

Ravi M, Chan SWL (2010) Haploid plants produced by centromere-mediated genome elimination. Nature 464(7288):615–618

Septiningsih EM, Pamplona AM, Sanchez DL, Neeraja CN, Vergara GV, Heuer S, Ismail AM, Mackill DJ (2009) Development of submergence-Tolerant rice cultivars: the Sub1 locus and beyond. Ann Bot 103(2):151–160
Till BJ, Zerr T, Comai L, Henikoff S (2006) A protocol for TILLING and Ecotilling in plants and animals. Nat Protoc 1(5):2465–2477

Dinakaran Elango Ph.D. is a plant breeder and geneticist with more than 10 years of research and development experience primarily focused on sorghum crop improvement. He worked with the International Crops Research Institute for the Semi-Arid Tropics, Pioneer Hi-Bred International, and International Potato Center in various capacities. He is a specialist in participatory plant breeding, seed system development, plant–microbe interactions, gender mainstreaming, conventional and modern plant breeding.

Germán Sandoya is a plant breeder and geneticist improving lettuce cultivars. He has 15 years of experience combined improving lettuce and maize for biotic and abiotic stresses on these crops. He has released several lettuce breeding lines with resistance to diseases. He serves as a graduate faculty in the Horticultural Sciences Department where he mentors undergraduate, graduate students and postdocs. His current breeding program focuses in abiotic and biotic stresses in lettuce and other leafy vegetables. He is also a statewide extension specialist focused on solving issues in leafy vegetables.

Surinder Chopra has a Ph.D. from Vrije University Brussels, Belgium and a post-doctoral fellowship with Thomas Peterson, Iowa State University. He served the ICRISAT, India and currently directs a cereal molecular genetics and epigenetics research program focusing on plant development, plant-pathogen and plant–insect interactions. These basic and applied research projects have been supported by NSF, USDA NIFA, and USAID grant programs. He mentors post-doctoral fellows, graduate and undergraduate students and teaches plant genetics and crop biotechnology courses.

Chapter 3
Genomic Methods for Improving Abiotic Stress Tolerance in Crops

Jai Singh Rohila, Ramanjulu Sunkar, Randeep Rakwal, Abhijit Sarkar, Dea-Wook Kim, and Ganesh Kumar Agrawal

Abstract A tremendous progress towards meeting food security, commonly known as "green revolution", was gained in the twentieth century by utilizing conventional plant breeding methods. However, in the twenty-first century we collectively face numerous challenges including alleviating hunger, a nutritious diet for the poorer populations, quality food for all, and the sustainable management of natural resources such as land and water. Additionally, natural climatic changes leading to extreme weather patterns and human-influenced changes world over, which include the ever-growing population, have been blamed for adversely affecting the agricultural environment, causing an enormous strain on the food production chains. In particular, the crops that feed and sustain human health and development are being hampered by the increasing incidences of diverse environmental stresses (abiotic and biotic). This warrants urgent and speedy actions from the plant scientific community to devise rational strategies and minimize the impact of these stresses on crop yields. In this chapter, authors have provided an update on the approaches, especially the genomic methods that are useful to researchers to improve field crops under diverse abiotic environmental stresses.

J. S. Rohila (✉)
United States Department of Agriculture—Agricultural Research Service, Dale Bumpers National Rice Research Center, Stuttgart, AR, USA

R. Sunkar
Department of Biochemistry and Molecular Biology, Oklahoma State University, Stillwater, OK, USA

R. Rakwal
Faculty of Health and Sport Sciences, University of Tsukuba, Tsukuba, Japan

A. Sarkar
Laboratory of Applied Stress Biology, Department of Botany, University of Gour Banga, Malda, West Bengal, India

D.-W. Kim
Rural Development Administration, National Institute of Crop Science, Wanju, Republic of Korea

R. Rakwal · G. K. Agrawal
Research Laboratory for Biotechnology and Biochemistry, Kathmandu, Nepal

© The Author(s), under exclusive license to Springer Nature Switzerland AG 2021

A. Ricroch et al. (eds.), *Plant Biotechnology*,
https://doi.org/10.1007/978-3-030-68345-0_3

Keywords Abiotic stress · Plant · Crop · Breeding · Gene · Omics

3.1 Introduction

The current global population is approaching eight billion people, and is estimated to increase to ten billion by 2050. Correspondingly, the global demand for agricultural products is projected to rise by at least 50% over the next two decades, but the current trends for yield gains in major global crops are insufficient to meet this challenge (Ray et al. 2013). Cereals are the most critical food sources for humans and animals. It is estimated that more than one billion tons of cereals will be required to feed the projected world population. The increased demand for food comes not only from the growing populations in developed countries, but also from developing countries such as China and India, where increased economic benefits have resulted in higher meat consumption. The demand for higher value foods such as meat has resulted in a shift of land use from growing food crops for people to cultivating feed crops for animals. Cultivation of crops for renewable energy sources is another compounding factor that puts pressure on land use. These biofuel crops are shifting the food and feed equation further out of equilibrium. For example, food crops like corn and sugarcane are now being cultivated for biofuel production, thus reducing the net food production.

Globally, limited amounts of natural resources, such as land and water, are available for crop cultivation. It is unwise to increase land availability by deforestation or increase water availability by withdrawing more water from underground aquifers than their recharging rates because of the urgent need to preserve natural resources for environmental sustainability. Furthermore, existing crop-production practices suffer from improper management techniques; this problem is compounded by the variability in irrigation and soil fertility. Moreover, the arable lands are often used for cultivating only one type of plant, such as cash crops (e.g., coffee) that are more profitable to the farmers than the food crops. Taking all these factors into account, a careful consideration to increase net arable land area could be the use of marginal lands for food crop production; however, this strategy brings its own challenges such as drought, salinity, temperature extremes, poor fertility, flooding, and even the threat of radioactive and heavy metal contaminants in the soil. Because of variable and unpredictable climatic conditions, the occurrence of most of the above-mentioned abiotic stresses will probably become more frequent in the future, even for the farm lands with currently ideal climates. We do not wish to dwell on the causes of climate change—'global warming', but we are instead focusing on the fact that rising air and sea temperatures affect weather patterns and that these changes drastically affect crop yields in unpredictable ways. Current crop simulation models predict that climate change has been, and continues to be, a threat to agricultural production worldwide, having the most severe effect in tropical and sub-tropical countries (Wheeler and von Braun 2013). Therefore, scientists must find effective solutions to lessen the damage from extreme climatic conditions that will bring stability in food production.

One effective strategy toward achieving this goal is to integrate the use of appropriate technologies for genetic improvements of crop plants that will enable them to withstand unpredictable environment in the field. In the twentieth century the green-revolution was brought on by conventional plant breeding, which was able to greatly increase the productivity of major cereal crops such as wheat and rice. Since then, it has been a challenge for breeders to achieve another dramatic increase in yields because the traditional breeding methodology has its own limitations. However, with the recent advancements in genomics, breeders can now access a huge amount of genomic information, commonly called big data. This wealth of genomic data is extremely useful for identifying novel genes or allelic combinations, and in turn for genetic improvement of existing crop varieties. Modern plant breeders aim to not only increase crop yield, but also enhance profitability of crop production by incorporating protection against a rapidly changing environment, cultural practices, and the stressors associated with the local growing environments. The availability and use of genomic tools provide the foundation of a new strategy to improve stress tolerance in crops that will result in our ability to meet the current and future demands for food production in a sustainable manner.

3.2 Limitations of Conventional Breeding Methods Based on Phenotypic Evaluations of Stress Tolerance

3.2.1 Difficulties in Improving Abiotic Stress Tolerance Trait

Abiotic stress tolerance in plants is a complex heritable trait that is determined by combinations of genetic, environmental, and crop management factors. Most traits for abiotic stress tolerance are associated with simultaneous expression of multiple genes and metabolites. Because each gene/metabolite contributes relatively smaller effects, these are called quantitative traits. A successful breeding program for improving abiotic stress tolerance requires bringing optimal combination of genes and alleles associated with the stress tolerance. In a conventional breeding process, breeders empirically select germplasm to cross with an elite cultivar and select new lines from the offspring based on their phenotypic performances. Because of the complex quantitative nature of abiotic stress tolerance and associated plasticity of plant organs and quality parameters in field crops, it is extremely difficult to improve stress tolerance using only the empirical methods of plant breeding.

In practice, after several years of breeding for improved yield and enhanced quality, most pools of modern crop varieties have been reduced to a narrow range of genetic diversity, implying that they are not highly variable in either their genotype or phenotype. For example, it was estimated that all parental germplasm in modern US rice varieties can be traced back to only 22 or 23 introductions (Dilday 1990). Despite this bottleneck, the good news is that most crop species have diverse germplasm collections with a high probability of harboring genes and/or allelic combinations which might be better suited to many abiotic stresses. As the first step in a breeding program for stress tolerance, it is essential to find a native/exotic germplasm that has superior gene/alleles associated with the tolerance of a specific stress or stress combinations (Fig. 3.1). In this case, wild relatives or land-race crop species could be a good choice of germplasm for breeding because they were evolved prior to or during domestication and have survived under diverse environments. Thus, gene banks [for example, Germplasm Resources Information Network (GRIN) Global, USDA; https://npgsweb.ars-grin.gov/gringlobal/search.aspx] coupled with the currently available vast genome sequence data (for example, 3000 genome sequence in rice; https://snp-seek.irri.org/) are great resources for the mining of superior genes/alleles to identify stress-tolerant accessions in this post-genomic era.

After a genetic cross is made, progenies with stress tolerance can be selected, which may also have agriculturally important traits such as high yield and desired quality parameters. Because in the initial filial generations the desired traits segregate among the progenies, repeated selection in each generation is needed to fix the traits of interest. However, if the breeders select germplasm and progenies based only on phenotypic evaluations, the process will be costly and time-consuming. Moreover, phenotypic evaluations in the field may not work due to reasons such as the absence of threshold levels of the stress. Thus, relying on empirical methods alone to achieve abiotic stress tolerance is unlikely to deliver a successful outcome with the optimal genetic makeup and at a speed that modern breeders require.

Therefore, genetic markers were developed to assist with phenotypic evaluations and to eliminate portions of breeding populations for more efficient selection of breeding materials. The markers should identify the genotypic differences between individual lines/progenies and between germplasm. A molecular marker is a fragment of DNA located in a known region of the genome. It is useful because its nature and function is not affected by either environmental or developmental factors/stages of the crop, or in some instances when a phenotype could be masked by the presence of other associated genes the molecular marker still works.

3.2.2 Some Basic Concepts of QTL Analysis and Marker-Assisted Selection Performed at Gene Level

In 1985, the first RFLP markers were introduced to breeders, and since then numerous other types of molecular markers, including RAPDs, AFLPs, and SSRs are being utilized by plant breeders. These markers, each with different advantages and disadvantages, have been used extensively in genetic studies and have been integrated into crop breeding programs. The availability of molecular markers has revolutionized the plant breeding programs. Since the sequence information of genes was first identified, molecular markers were employed to detect quantitative trait loci (QTLs), which are regions in the genome that control quantitative traits.

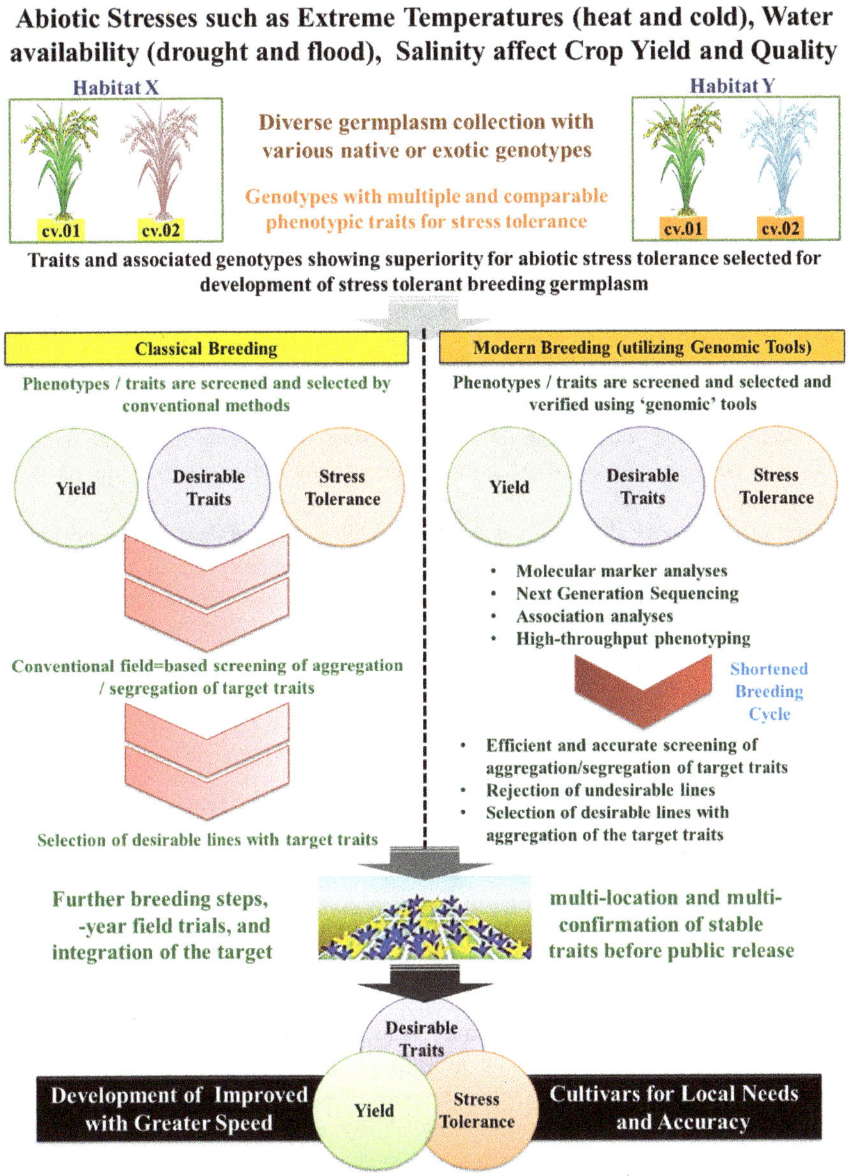

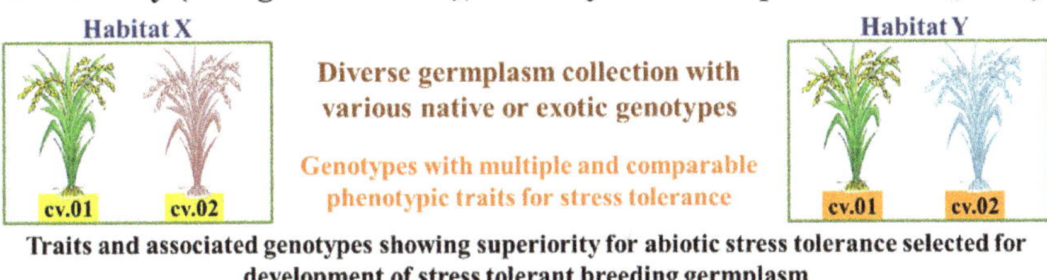

Fig. 3.1 Schematic illustration to highlight the comparison between the conventional breeding and the genomics-assisted breeding for abiotic stress tolerance. Integration of genomic methods increases breeding efficiency and reduces time and cost involved in the development of improved varieties

Breeders routinely handle a large amount of breeding materials during the screening of germplasm for crosses and selection of offspring from segregating populations. Therefore, selecting or rejecting a subset or individual plants based on reliable genetic markers that are tightly linked to QTLs may greatly enhance breeding efficiency. This diagnostic breeding tool, called marker-assisted selection (MAS), enables breeders to distinguish among progenies that have the desired traits based on the differences in their DNA sequences and almost eliminates the necessity for phenotypic evaluations. MAS can substitute for the time-consuming phenotypic evaluation of a large number of breeding materials, which can be complicated by environmental factors. Moreover, plants carrying the desired traits can be selected at the seedling stage in a greenhouse with specific markers prior to field tests and without phenotypic evaluations at initial stages. Currently, most of the molecular markers are derived from differences in regions located near (or linked to) a gene that controls the trait of interest, thus the feasibility of covering the whole genome remains to be seen. Thus, low selection accuracy may reduce the benefits of MAS for breeding.

Among the classical molecular marker technologies, simple sequence repeat (SSR) markers have been used extensively to discover genetic variation in breeding materials. In recent years, because of the progress in next generation sequencing (NGS) and high-throughput marker detection technology, the highly abundant single nucleotide polymorphism (SNPs) markers have become the markers of choice in many breeding projects.

3.3 Genomic Methods Are Available for Gene Discovery and Increasing Breeding Efficiency

3.3.1 Next Generation Sequencing (NGS)

The first DNA sequencing technology, called Sanger sequencing, was developed in the 1970s. This method was the primary sequencing tool for more than 30 years, and this time-consuming and costly technology was only applied to the genome sequencing of a few organisms such as humans, Arabidopsis, and rice. By 2007, an alternative type of sequencing technology, abbreviated NGS, had been developed. Using NGS, it became possible to rapidly analyze genome sequences of various organisms with a modest budget. Some may ask why one or the science needs the genome sequences of all crops or plants. The fully sequenced genomes of crop species facilitate the identification of the positional and functional information of genes that controls phenotypic expression. This genotyping tool accelerated genomics-based breeding for crop improvement. Since the first whole genome sequence of rice (*Oryza sativa* L.) was released, the genome sequences of other important food crops such as maize, soybean, and wheat have been completed, and genome sequences of other crop species and their wild relatives are expected to be unraveled in the coming years.

With the huge quantity of data generated by NGS and their analyses of the genomes, researchers need computational tools to effectively use the genomic information in breeding programs. With such computer algorithms, and software, genome sequence data can be adequately processed, assembled in order, and analyzed for their relevance to the biological properties of crops. The results of a bioinformatics analysis are stored as an easily accessible database that can be provided to the breeders or researchers for public access.

The identification of genes involved in abiotic stress tolerance will allow the rapid screening of the germplasm for alleles that confer tolerance and will enable these alleles to be introgressed into cultivars for improved yield using molecular markers (Voss-Fels and Snowdon 2016). SNP discovery at a large scale has become more appealing by sequencing the parental genotypes of a mapping population by the NGS. In the case of a crop for which the reference genome sequence data are available, the detection of large-scale SNPs is possible, and subsequently can be employed as functional markers for crop breeding. However, in the absence of a reference genome for the crop species, the NGS data can be aligned with the transcript data from the RNA-seq projects.

3.3.2 Association Analysis

Association analysis is used to detect allelic variation associated with traits of interest in a population of genetically unrelated individuals based on linkage disequilibrium (LD), the non-random occurrence of allelic combinations. In the conventional method for QTL detection, a population of segregating progeny is developed for linkage analysis. A population of related individuals, such as a recombinant inbred line population, is prepared over a few generations and results in a limited number of recombination events. By using the natural germplasm instead of preparing a segregating population, association analysis can provide a much wider range of genetic diversity. Hence, association analysis is a time-saving approach that offers a better analysis of natural genetic variation or the multiple genetic variants in QTLs. Currently, two approaches have been employed for association mapping: candidate gene association mapping and genome-wide association studies (GWAS). The statistical power of GWAS is significantly greater than traditional linkage analysis, but has its own limitations such as: population structure dependent analysis, difficulties in identification of multiple functional alleles within a gene, and mapping of rare stress-tolerant alleles in a population. To overcome such difficulties nested association mapping (NAM) and multi-parent advanced generation inter-cross (MAGIC) populations can be useful (Zhou and Huang 2019).

By sequencing candidate genes using NGS technology, polymorphisms can be identified that are associated with the desired phenotypes. For instance, in maize, the candidate genes *Dwarf8*, *Vgt1*, and *ZmRap2.7* were reported to be associated with flowering time (Pérez-de-Castro et al. 2012). In a GWAS, whole genome DNA sequence of a large set of natural population is compared between accessions showing stress tolerance and the accessions that do not show stress tolerance. In this process millions of genetic variants (a pool of DNA sequences of all accessions in the study) are compared and the DNA variants (i.e. allele) are identified which are more frequently associated with the stress tolerant phenotype. Recently, this approach was used to study a panel of 162 accessions from around the world to screen for salt tolerance of rice in the seedling stage (Rohila et al. 2019). This led to identification of six novel SNPs, and 16 candidate genes in their vicinity. This knowledge will be useful for improving salt tolerance in modern rice varieties, which may lead to increased profitability of rice production in salt-affected soils.

3.3.3 Genome-Wide Selection

Complex quantitative traits, such as abiotic stress tolerance, are often controlled by several genes that individually have small effects. Therefore, it is crucial to combine many genes with small effects into one plant for crop improvement, but this task is highly challenging. Due to the availability of genome-wide molecular markers and their genome-wide estimated breeding values, the genomic selection (GS) approach has great potential for its ability to combine multiple small-effect QTLs at the whole genome level (Juliana et al. 2019). Compared with traditional MAS, which requires the identification of markers associated with QTLs, GS can be conducted without this step. In the GS approach, a statistical model for determining genomic-estimated breeding values (GEBVs) is developed from the phenotype data and the genotype data, which are obtained using genome-wide markers, for the individuals in a reference population. Once the GEBV model is established, a plant (that has the desired traits) from the breeding population can be selected using the GEBV method with the genotypic data, and this process does not require a phenotypic evaluation. Despite having no requirement for previous marker selection, the information from this approach can be employed to detect QTLs that control the desired traits.

Nevertheless, even the identification of the QTLs responsible for a certain trait does not imply the identification of the specific gene(s) controlling the trait or an understanding of the mode of action. Models applied in genomic selection are useful to predict breeding values and, in some cases, to detect chromosomal regions associated with a trait; however, further work is necessary to identify the gene(s) responsible for the phenotypic variability observed. Therefore, we briefly introduce a collection of high-throughput technologies, called—omics (see below), which might be used to generate improved crop plants. The future exploitation of these strategies could facilitate the identification of candidate genes underlying the traits of interest and make MAS more meaningful and efficient.

3.3.4   Omics

In the twenty-first century, the most prominent tools of functional genomics are the -omics technologies. As explained above, unraveling the numerous plant species' (including crops) genomes has brought a paradigm shift in the approaches to plant biology and crop breeding. The goal is very clear: crop improvement to meet the future global food demand in a sustainable manner. In this context, the identification and cataloging of genes, proteins and metabolites via high-throughput technologies such as—transcriptomics (the expression of genes), proteomics (the expression of proteins), and metabolomics (the levels of metabolites) have become powerful approaches that can reveal the function of each gene in the genome (Weckwerth 2011). More recently, clustered regularly interspaced short palindromic repeats (CRISPR)—guided targeted gene editing technologies are emerging as efficient tools for functional analyses of individual genes (Arora and Narula 2017). The -omics can help in addressing fundamental biological questions, and the -omics tools will create novel expression data on the molecular factors that could confer stress tolerance. Their subsequent analysis through functional genomics could be used to create next generations of plants that can withstand the adverse climatic conditions. With the development and increased use of these -omics technologies, breeders can exploit the omics-driven discovery of new candidate genes, proteins, and metabolites in their breeding programs.

3.4   Conclusions

The above-mentioned approaches, namely NGS, GWAS, GS, and -omics, are some of the well-proven genomic methods that can be implemented with conventional breeding methods to transform them into modern breeding methods that are suitable for present and future needs. The optimized molecular markers will help the modern plant breeders to select, deploy and stack useful traits in their crop of interest, enable them to achieve the "ideotype" concept for optimization of food crops to grow profitably the in current changing climatic conditions and unpredictable environments. In particular, the high-throughput -omics technologies have the advantage of identifying prospective candidate genes, proteins, and metabolites that can complement the molecular markers identified using genomic methods. Thus far, the -omics-based biomarker discovery program has contributed relatively little to the overall genetic improvement of crops, and utility of this technology requires renewed efforts to screen larger sets of germplasm or selected germplasm resources. High throughput phenotypic evaluations of traits are also crucial for improving abiotic stress tolerance of crop plants. In order to reach the full potential of genomic methods, it is essential to combine genomic methods and precise high throughput phenotypic evaluations. Based on the above approaches the chances of obtaining suitable and usable biomarkers for improving the field crops can be increased, including the development of stress tolerant germplasm for further detailed genetic analyses. Other than the required large-scale screening, multiple abiotic (and biotic) stresses also must be investigated using all the above-mentioned approaches to link potential biomarkers to a particular stress, thereby increasing the likelihood that the newly generated information for designing stress-resilient varieties can be translated to commercial production by the farmers.

Acknowledgements The authors acknowledge Dr. David Gealy, Dr. Yulin Jia and Dr. Trevis D. Huggins at the USDA-ARS for constructive comments. Supports from NSF-EPSCoR award R2-Track 2 FEC#1826836 (for RS), University and Tsukuba (for RR), University Grant Commission, MHRD, GoI, New Delhi, India research grant [Reference No. F.30-393/2017(BSR)] (for AS) are thankfully acknowledged.

References

Arora L, Narula A (2017) Gene editing and crop improvement using CRISPR-Cas9 system. Front Plant Sci 8:1932. https://doi.org/10.3389/fpls.2017.01932

Dilday RH (1990) Contribution of ancestral lines in the development of new cultivars of rice. Crop Sci 30:905–911. https://doi.org/10.2135/cropsci1990.0011183X003000040030x

Juliana P et al (2019) Improving grain yield, stress resilience and quality of bread wheat using large-scale genomics. Nat Genet 51:1530–1539. https://doi.org/10.1038/s41588-019-0496-6

Pérez-de-Castro AM, Vilanova S, Cañizares J, Pascual L, Blanca JM, Díez MJ, Prohens J, Picó B (2012) Application of genomic tools in plant breeding. Curr Genomics 13:179–195. https://doi.org/10.2174/138920212800543084

Ray DK, Mueller ND, West PC, Foley JA (2013) Yield trends are insufficient to double global crop production by 2050. PLoS ONE 8(6):e66428. https://doi.org/10.1371/journal.pone.0066428

Rohila JS, Edwards JD, Tran GD, Jackson AK, McClung AM (2019) Identification of superior alleles for seedling stage salt tolerance in the USDA rice mini-core collection. Plants 8(11):472. https://doi.org/10.3390/plants8110472

Voss-Fels K, Snowdon RJ (2016) Understanding and utilizing crop genome diversity via high-resolution genotyping. Plant Biotechnol J 14:1086–1094. https://doi.org/10.1111/pbi.12456

Weckwerth W (2011) Green systems biology-from single genomes, proteomes and metabolomes to ecosystems research and biotechnology. J Proteomics 75:284–305. https://doi.org/10.1016/j.jprot.2011.07.010

Wheeler T, von Braun J (2013) Climate change impacts on global food security. Science 341:508–513. https://doi.org/10.1126/science.1239402

Zhou X, Huang X (2019) Genome-wide association studies in rice: how to solve the low power problems? Mol Plant 12:10–12. https://doi.org/10.1016/j.molp.2018.11.010

Jai Singh Rohila is a scientist at the USDA-ARS, Dale Bumpers National Rice Research Center, Stuttgart, Arkansas, USA. He also holds an adjunct Professor position at Oklahoma State University Stillwater, USA. Dr. Rohila received Faculty Excellence Award for Global Research at SDSU, and Excellence in Environmental Efforts at the USDA-ARS. Currently, Dr. Rohila's research focus is to understand the genetic basis of water stress tolerance and its underlying molecular mechanisms and apply them to improve US rice germplasm for robust physiology and agronomic traits that are significant for commercial and sustainable rice production.

Ramanjulu Sunkar is currently Professor at the Department of Biochemistry and Molecular Biology, OklahomaStateUniversityStillwater, USA. His laboratory works on molecular characterization of genes/microRNAs critical for plant stress tolerance. He was included in the list of Thomson Reuters Highly Cited Researchers. He edited three books and currently serving as an associate/handling editor for Journal of Experimental Botany.

Randeep Rakwal is a Professor at the University of Tsukuba (Japan), at the Faculty of Health and Sport Sciences. Being expert in multidiscipline, he continues his research work on plants, human health and animal-human systems using omics. His major research interests in plant environmental stress biology are—jasmonic acid, ozone, heat, radiations, plant pathogens using "omics" approaches with close collaborators in Japan, South Korea, Nepal, USA, Italy, Australia and India. He is the initiator and one of the founding members of the International Plant Proteomics Organization (INPPO).

Abhijit Sarkar is a Professor who recently joined the Department of Botany, University of Gour Banga, Malda, West Bengal, India. Prior to this position he has several years of strong research experience in studying plant stress tolerance mechanisms in numerous plant species. Based on his academic and research achievements during his Ph.D. program he was awarded Senior Research Fellowship by Council of Scientific and Industrial research (CSIR), Government of India.

Dea-Wook Kim is a scientist at National Institute of Crop Science (NICS), Republic of Korea. He holds Ph.D. in Applied Biochemistry from University of TsukubaTsukuba, Japan. Throughout his career he was continuously focused on crop improvement utilizing a combination of conventional plant breeding and cutting-edge technologies.

Ganesh Kumar Agrawal is the Associate Director of RLABB, a non-profit research organization focusing on biotechnology and biochemistry in Kathmandu, Nepal. Dr. Agrawal is a multidisciplinary scientist focused on food security and human nutrition using high-throughput and targeted omics techniques. He has edited a comprehensive "Plant Proteomics: Technologies, Strategies, and Applications" book (John Wiley & Sons, NY, USA). He earned his Ph.D. in Applied Biological Chemistry (Biochemistry & Biotechnology, Tokyo University of Agriculture and Technology, Japan). He is an initiator of the International Plant Proteomics Organization.

Chapter 4
New Technologies for Precision Plant Breeding

Shdema Filler-Hayut, Cathy Melamed-Bessudo, and Avraham A. Levy

Abstract Recent developments in high-throughput sequencing and genome editing enable breeders to use a new tool kit for precision breeding. DNA double strand breaks can be targeted to desired genomic loci by custom-designed nucleases such as Zinc-finger nucleases, TALEN and CRISPR-Cas9. The repair products of induced breaks vary depending on the repair mechanisms and the repair template. These may be targeted small insertions or deletions, gene conversions or crossovers as well as gene replacement. In this chapter we review the concepts, achievements, opportunities and challenges of these new technologies.

Keywords CRISPR-Cas9 · Targeted mutation · Gene targeting · Inter-homologs recombination

4.1 Plant Breeding, Then and Now

During more than 10,000 years of plants domestication and breeding, humans have collected plant variants with beneficial traits and eventually crossed them and selected for new varieties with desired phenotypes. Two main pillars of classical breeding are: (1) Biodiversity, namely a collection of plant variants. The more diverse the collection, the higher the chances to find genetic material with the desired traits. This reservoir consists of natural variation, such as wild plants, wild relatives or genetically distant varieties or of induced variation, such as mutations obtained through exposure to ionizing radiation or chemical mutagens. (2) Meiotic recombination—the exchange of genetic material between homologous chromosomes is an obligatory step that leads to the formation of diverse gametes (pollen or egg cells) containing new combinations of parental chromosomal segments. This process is at the core of breeding in sexually reproducing crops. Both mutagenesis and meiosis are random and therefore breeders invest a great amount of time and financial resources in growing and screening large plant populations.

S. Filler-Hayut · C. Melamed-Bessudo · A. A. Levy (✉)
Department of Plant and Environmental Sciences, Weizmann Institute of Science, 7610001 Rehovot, Israel
e-mail: avi.levy@weizmann.ac.il

© The Author(s), under exclusive license to Springer Nature Switzerland AG 2021
A. Ricroch et al. (eds.), *Plant Biotechnology*,
https://doi.org/10.1007/978-3-030-68345-0_4

Traditionally, plant breeding was completely based on phenotypes. The development of molecular biology towards the end of the twentieth century, enabled to identify DNA markers associated with complex phenotypes and to select plants based on the marker alone, thus bypassing a costly phenotyping, in particular for traits affected by the environment. The on-going Next-Generation Sequencing (NGS) revolution made whole genome sequencing cheaper and faster. Many crop plant genomes have been fully sequenced and annotated, providing a better resolution of genome structure, an unlimited amount of genetic markers and facilitating to identify genes and sequences that are associated with beneficial phenotypes for crop plants. This revolution opened the door for higher resolution breeding. For example, formerly elusive Quantitative Trait Loci (QTL) can now be narrowed down from big genomic segments of tens or hundreds of Kilo-base pairs, to a specific sequence element or even to one Single Nucleotide Polymorphism (SNP) that associates with the trait. Whole genome sequences thus enable to breed crops more accurately and faster. Moreover, it provides a long list of candidate genes and alleles that can be included or excluded from a variety by genetic recombination and segregation or that can be mutated, for mutagenesis-based breeding, or that can be modified using targeted editing approaches for crop improvement.

A large body of research on DNA Double Strand Breaks (DSBs) has shown that the DSB repair process can lead to mutations or homologous recombination (HR) at the DSB site. The ability to target a DSB to a specific genomic locus has therefore been the "holy grail" of precision breeding because it enables to obtain a desired mutation in a surgical manner compared to the traditional shotgun approaches of random mutagenesis. Moreover, it also enables to target HR to a specific locus as discussed below. Upon DNA DSB induction, the plant endogenous repair mechanisms are activated and the break may be repaired either by the error-prone Non-Homologous End Joining (NHEJ) repair machinery or by the precise HR machinery that requires a homologous template for repair e.g. a homologous chromosome/sister chromatid/homologous repeat/an extrachromosomal DNA fragment with homologous sequences to the sequence flanking the DSB site (See review [Schmidt et al. 2019]) on DSB repair and genome editing in plants). A DNA DSB, either induced by a custom-designed nuclease (Fig. 4.1a) or natural, shows different fates (Fig. 4.1b): The NHEJ repair products may form small insertions or deletions (indels) at the DSB site, generating a new mutation; HR repair products depend on the type of homologous template sequence. When an extrachromosomal sequence is delivered in the cell it can recombine with homologous sequences flanking the DSB site, leading to a targeted integration of the delivered DNA. This process is called gene targeting, or Homologous-dependent repair (HDR) or gene replacement. When the repair template is an endogenous homologous chromosome, HR-mediated repair can induce recombination between the two homologs, a process referred to as Inter-Homologs Recombination (IHR). In this chapter we review how IHR, gene targeting and targeted mutagenesis are becoming new emerging technologies for precision breeding. In addition, we describe the challenges that still lie ahead before these methods can be widely used.

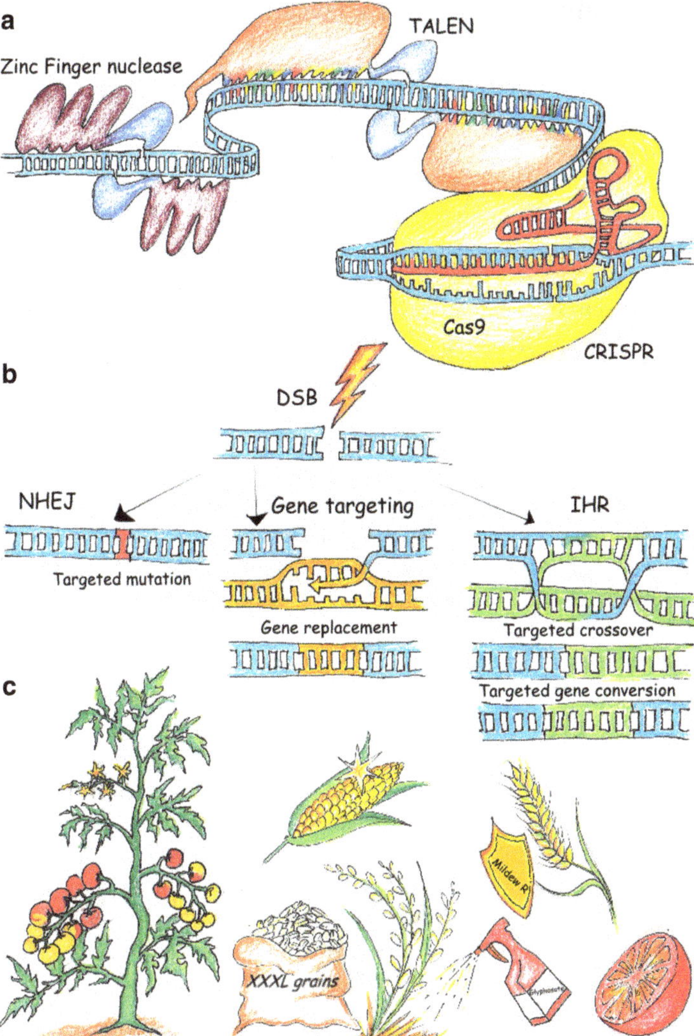

Fig. 4.1 Precision breeding: **a**. Custom-designed nucleases: zinc-finger nuclease, TALEN or CRISPR-Cas9, all can induce a targeted DSB (shown as a lightning). **b**. Double strand break (DSB) repair mechanisms and precision breeding. DSB repair product via error-prone NHEJ can lead to targeted mutations, such as small insertions or deletions (shown in red). In the presence of exogenous "donor DNA" (in orange) with sequence homology to the DSB site, homologous recombination between the "donor DNA" and the DSB-flanking region can lead to targeted gene replacement also known as gene targeting. Under recombination between homologous chromosomes (in blue and green) the repair product via IHR can be targeted crossover or targeted gene conversion. **c**. Examples of precision breeding products include mildew-resistant wheat, high yield-high amylopectin maize, optimization of fruit size, inflorescence architecture in tomato, high grain weight and herbicide resistance in rice

4.2 DNA Double-Strand Break (DSB) Induction

Efficient induction of targeted DSBs is the first step for precision breeding. The increase in both NHEJ and HR repair products under induction of DSB was first demonstrated in plants using expression of the site-specific and rare-cutting meganuclease I-SceI (Carroll 2014). Later on, custom designed nucleases (Fig. 4.1a) were developed for a broad range of targets, namely the Zinc-Finger Nucleases (ZFNs) and Transcription Activator-Like Effector Nucleases (TALENs). Although successful experiments with these nucleases were preformed, they were not suitable for wide usage in genome engineering due to low number of recognition sites on the genome (I-SceI, ZFN), the need for protein engineering for every single target (ZFN, TALEN) and often, a low efficiency of cleavage (for reviews on targetable nucleases see Carroll (2014)).

Decoding of bacterial and archaeal Clustered Regulatory Interspaced Short Palindromic Repeats (CRISPR) immune systems gave rise to a new breakthrough technology in the field of custom-designed nucleases that opened the prospect for routine and precise genome engineering. In 2012, *in vitro* induction of DNA double strand break (DSB) at a specific genomic sequence was achieved by using only three elements from the CRISPR immune system, two RNA molecules: The CRISPR RNA (crRNA) and the trans-activating CRISPR RNA (tracrRNA); and one protein: CRISPR associated protein (Cas9). Shortly after, the two RNA molecules were fused to one single guide RNA (gRNA) molecule that matches the target and recruits the Cas9 protein and in vivo activity was shown in human cells. Rapidly, successful genome editing experiments were shown in a wide variety of crop plants including wheat, maize, cotton, soybean, tomato, potato, citrus, grape, etc. This system, composed of only two components, the Cas9 protein and gRNA molecule, was found to be both very efficient across a broad range of targets, easy to design and easy to use in a wide range of organisms. Therefore, it has attracted much attention and financial resources from research and industry communities. The bacterial and archaeal CRISPR-Cas systems show high diversity of protein composition and length, locus architecture, Protospacer Adjacent Motifs (PAM) recognition sites (short sequence motifs adjacent to the target), DNA DSB nature (blunt/sticky ends), nucleic acid target (DNA/RNA) etc. All of this diversity of tools may be translated to become a tool kit for genome editing with the ability to induce different types of breaks in many different targets (different PAM sequences) and even to preform few CRISPR mediated breaks at ones. In addition, proteins and RNA molecules can be fused to Cas proteins or gRNA molecules in order to target or manipulate DNA (for review see Adi (2018) and Gao et al. (2020)).

4.3 Non-Homologous End Joining (NHEJ)—Induction of Targeted Mutations

In higher plants somatic cells, NHEJ is the predominant DSB repair mechanism. Commercial use of targeted mutations can speed up breeding processes and create varieties with better traits. One example of such a usage can be found in the breeding of the "waxy corn". In maize, *Waxy* gene (also called *Wx* or *Wx1*) encodes for a protein called Granule-bound starch synthase1, that produces starch with intermediate levels of amylopectin. The starch of Wild-Type (WT) maize seeds is composed of ~25% amylose and ~75% amylopectin, while loss-of-function *waxy* mutants contain 100% amylopectin. Waxy mutants show different properties than normal corn starch such as higher digestibility, easy to gelatinise, etc. Therefore, the starch of *waxy* mutants called "waxy corn" is used mainly as a stabilizer in the food industry, but also in the textile, adhesive and paper industries. A collection of ~ 200 *wx1* mutant alleles that arose spontaneously or created by random mutagenesis, is available for breeders. The commercial "waxy corn" lines are the result of classical breeding process that includes introgression of the *wx1* allele from these mutated lines to elite varieties followed by backcrossing. This process usually takes six to seven generation and the commercial waxy hybrids yield ~5% less than the equivalent non-waxy hybrid. The reason for the reduced yield of "waxy corn" is unknown. It might be due to the direct effect of the *waxy* mutation, or to another gene or genomic segment containing a deleterious allele linked to *waxy* that was introgressed during the backcrossing process. In a recent study, Gao et al. (2020) transformed 12 hybrid maize varieties with *Streptococcus pyogenes* Cas9 protein (spCas9) and two gRNAs flanking the *Wx1* gene and then selected for deletion mutants of 4 Kb which were created by NHEJ repair of the two breaks. The *waxy* phenotype of the CRISPR mutant kernels was similar to the phenotype of commercial used hybrids. After only two generations of backcrosses, 12 lines (one for each variety tested) of transgenic DNA-free plants, homozygote for 4 Kb deletion mutation at *Wx1* were planted at field trials in 25 locations. These field trials showed that the CRISPR-mutants were agronomically superior over the introgressed mutants, with average increase of 5.5 bushels per acre (Gao et al. 2020). This shows that precise mutagenesis has less deleterious hitchhiking effects compared to classical chemical or irradiation shotgun mutagenesis.

Economically important crops such as wheat, oat, peanut, potatoes, bananas, coffee, and sugarcane are polyploids, which means they have more than two paired sets of chromosomes. Classical breeding procedures are much more complex in polyploids since many genes are present in more than two copies. For some genes, mutant plants, homozygous for all gene copies, are not available in the natural or mutagenic reservoirs. Even in the cases where a mutant variety is available, the number of gene copies (or loci) that need to be considered during the breeding procedures is higher and therefore breeding plant populations are bigger, and the whole breeding process becomes more work intensive. Using precision breeding, breeders can induce mutations at all relevant gene copies in one single step and save both time and money. For example, in the hexaploid bread wheat, powdery mildew a fungal disease (*Blumeria graminis f. sp. Tritici*) causes a big yield loss. In bread wheat there are three homeoalleles that encodes for MILDEW-RESISTANCE LOCUS (MLO) proteins. Loss-of-function *mlo* mutants in barley, Arabidopsis and tomato showed resistance to different strains that cause this disease, but there were no bread wheat varieties with mutations in all three homeoalleles available. Using a targeted mutagenesis approach, bread wheat was transformed with TALEN proteins designed to target almost identical sequences at the three *Mlo* homeoalleles (Wang et al. 2014). Mutants with deletions in one, two or three *Mlo* homeoalleles were identified and propagated to create a homozygote mutant. While WT and homozygote TALEN mutants with deletions in one or two homeoalleles did not show resistance to the mildew disease, the homozygote mutant in all three homeoalleles showed high resistance.

Yield, quality, stress response and other beneficial agronomical traits are quantitative. In many cases these traits are controlled by more than one gene located at various loci. Classical breeding process of a few genes located on different chromosomes is complex from the same reasons mentioned above for the polyploids. In rice, triple mutants lines in three different genes that negatively regulate grain weight, GW2, GW5 and TGW6 (for each gene the number represents the chromosome location), were generated using CRISPR and shown to have an increase in grain size and weight (Xu et al. 2016). Another important factor that influences quantitative traits are *cis*-elements which regulate gene expression. Mutations in genes regulating *cis*-elements, hold a great promise for plant breeding, although current knowledge about the exact *cis*-element sequences that regulates each gene is not complete. Using the stochastic nature of NHEJ repair and phenotypes screening approach, Rodriguez-Leal et al. induced multiple DSBs at the promoters of three genes that regulate the tomato fruit size (*CLV3*), the inflorescence architecture (*COMPOUND INFLORESCENCE (S)*) and the plant growth habit (*SELF PRUNING (SP)*) [Rodríguez-Leal et al. 2017]. Due to the Cas9 high DSB efficiency and multiple break induction they obtained allelic series of multiple mutations in the promoters, including big deletions and small indels, single or multiple mutations, etc. After homozygotization, the breeder can screen for desired phenotypes and sequence the mutations. This approach gave rise to diverse phenotypes of quantitative traits due to differences in gene expression levels or timing.

Most of the targeted DNA DSB induced in plant somatic cells, are repaired by the NHEJ machinery. Therefore, applications of precision breeding with desirable end products of NHEJ are simple and easy to produce. Different factors may affect the outcome of the insertion or deletion "scar" at the DSB site: DSB ends (blunt or sticky), Cas protein (for example, Cas9 tends to generate indels of few base-pairs versus Cas12 that tends to leave indels of few tens of base-pairs), the number of gRNAs used to induce single or multiple breaks in the target of interest, the availability of PAM sequences for DSB induction etc. The patterns of NHEJ indels are not controlled and therefore when an exact insertion or deletion is desired, one can use gene targeting methods.

4.4 Gene Targeting

Gene targeting is a process in which exogenous DNA sequence (donor) is introduced to the cell and then replaces the endogenous gene using HR repair. These repair events can lead to insertion, replacement or substitution at specific endogenous DNA fragments, depending on the donor sequence. Therefore, it can serve for breeding when small (up to few Kb) and exact known changes in the genome are required. While gene targeting is very efficient in bacteria and yeast, it was and still is, harder to accomplish in higher plants and animals. Under introduction of DNA segment with homology to the target site, plant cells showed only one targeting event per 10^4 to 10^5 transformed cells. A breakthrough in enhancing gene targeting rates was achieved by mimicking the natural process with induction of a DSB at the endogenous genomic segment in the presence of a donor DNA. Transient expression of the I-SceI mega-endonuclease, enabled increased gene targeting rates at a transgenic locus by 2–3 orders of magnitude (Schmidt et al. 2019). The approach of *in planta* gene targeting was shown to have a similar gene targeting rates with reduced dependencies on the donor transformation rates and on the availability of the T-DNA receptor site for repair (Schmidt et al. 2019). With this method, the transgenic segment contained sequences of the endonuclease I-SceI and donor sequence flanked by two I-SceI targets. Under expression of I-SceI, three DSB events are occurring in parallel: One at the target site and two at the flanking region of the donor sequence. The donor sequence is thus excised from the genome and becomes available to serve as template for HR.

The development of endonucleases and especially of the CRISPR systems improved dramatically the efficiencies of DSB induction in plant cells and the ease of use of these tools. But due to the dominance of NHEJ in plant somatic cells, the levels of gene targeting remained relatively low (one plant per 10^2–10^3 transgenic plants). Mutants in the NHEJ genes (e.g. *ku70* and *lig4*) of Arabidopsis and the *Lig4* gene in rice have shown increased gene targeting efficiencies by 3–16 folds, but showed also genomic instabilities. Another way to take advantage of the prominence of NHEJ in somatic tissues, is to use it for "pseudo gene targeting", through targeted integration of a donor DNA via NHEJ. In rice callus cells, intron targeting was used to create a glyphosate resistant line (Li et al. 2016). Glyphosate is a non-selective herbicide which interferes with the shikimate pathway that produce aromatic amino-acids in plants and microorganisms, but does not exist in mammals. The development of glyphosate-resistant crops can increase yield by decreasing the competition between crops and other weeds in the field. Induction of CRISPR-Cas9 mediated DSBs in two rice 5-enolpyruvylshikimate-3-phosphate synthase (EPSPS) adjacent introns, enabled the deletion of the targeted exon and its replacement by a modified exon containing a substitution of two amino acids that conferred glyphosate resistance. This cut-and-paste process occurred via the NHEJ machinery at an efficiency of up to 2% of the transformed plants. The end joining product still contained the NHEJ indel "scars" but these did not affect the protein sequence, since they were located on the introns flanking the modified exon.

One of the factors that may influence the rate of NHEJ versus gene targeting products is the availability of the donor sequence as a template for repair. Therefore, a simple concept to enhance gene targeting is to increase the number of donor sequences available in the plant cell. The application of geminivirus replicons was found to increase the level of donor sequences and in some of the studies to enhance the gene targeting events in tobacco, tomato, wheat, Arabidopsis and rice (Baltes et al. 2014). The geminivirus replication requires three elements: *cis*-acting large-intergenic region (LIR), short-intergenic region (SIR) and *trans*-acting replication-initiation protein (Rep). In the presence of the Rep protein, the LIR-SIR-LIR sequence generates episomal *trans* replicons that are replicating in the plant cell. The insertion of donor DNA sequence into the replicon sequence leads to an increase in its copy number. When donor amplification is combined with DSB induction in the target, this may lead to levels of gene targeting as high as 25% as was shown at the *CRTISO* locus in tomato (Dahan-Meir et al. 2018).

Using gene targeting methods, breeders can fulfill the dream of precise integration of beneficial genes or mutations to specific sites without any additional changes in the genome. Although gene targeting studies are showing improved efficiencies, there is a wide variation between target loci, species and methods and the technology is not yet mature for wide commercial usage. Most of the studies mentioned above were done with a selectable marker for the gene targeting events. Selection of gene targeting events in the absence of visible phenotypes is still challenging. In addition, delivery of long sequences containing nuclease and donor sequence to the plant is also a limiting factor in some crops due to the lack of transformation protocols or to the low regeneration efficiencies.

4.5 Inter-Homologs Recombination

In classical breeding, the creation of new varieties depends on the meiotic reshuffling of the parental chromosomes. Proper chromosomal segregation during meiosis requires at least one crossover event for each bivalent (pair of homologous chromosomes at meiotic prophase I) (for review on plants meiotic recombination see Mercier et al. (2015)). Meiotic crossover studies in plants showed that both DSBs sites and crossovers are distributed non-uniformly along the chromosome. For example, in wheat chromosome 3B, ~80% of the crossovers are concentrated at the sub-telomeric region which represents ~20% of the chromosome length. Although recombination hot regions are more enriched with genes, recombination cold regions are not gene deserts. Therefore, when breeding program includes segregation of two genes or loci located at the same recombination cold region, higher numbers of F2 plants are needed for selection of the integrated desired allele. One of the optional solution to this problem is to increase the total number of recombination events. Different studies in *Arabidopsis thaliana* have shown increase in crossover rate using mutant lines of anti-crossover genes such as RECQ4, FANCM and FIGL1 or overexpressing pro-crossover genes such as the E3 ligase HEI10. In addition, mutation at the chromatin remodeler *DECREASE IN DNA METHYLATION1 (DDM1)* increases recombination in euchromatic regions, but not in the heterochromatic regions which are "recombination cold". Moreover, crossovers were found to be enriched near cold recombination area of the centromere under loss of H3K9me2 and non-CG methylation (Underwood et al. 2018). All of the above approaches can increase the chances to get a crossover between two linked genes but as the genetic distance between genes get smaller, it becomes harder to obtain a crossover.

Induction of DSB between two linked genes in a hybrid background can potentially direct the IHR machinery to a particular target and lead to the creation of the desired gene conversion or crossover allele. Plant meiosis starts with the induction of hundreds of DSBs by Spo11 (Filler-Hayut et al. 2017) and only one or two breaks per chromosome are resolved into crossover events and the number of gene conversions is in similar order of magnitude to crossovers. Therefore, a DNA DSB that is induced during meiosis competes with hundreds of breaks for IHR repair. In plant somatic cells, NHEJ is considered to be the dominant repair mechanism but there is also evidence for recombination between homologs suggesting that the HR machinery is also active in somatic cells. As opposed to mammalians, in plants the germline differentiation takes place late in plant development and therefore somatic IHR alleles that occurred at early stages of plant development may be inherited. In addition, using regeneration protocols, breeders can regenerate a whole plant from small leaf tissues with IHR allele.

An example for somatic targeted IHR was demonstrated in tomato (Filler-Hayut et al. 2017). In the commercial M82 variety, targeted DNA DSB was induced at the *PHYTOENE SYNTHASE1 (PSY1)* gene using CRISPR-Cas9. A homozygote mutant plant with yellow fruits was crossed with a wild variety, *Solanum pimpinellifolium* (red fruits), and F1 plants were grown. In F1 plants, only the *S. pimpinellifolium* allele was subjected for DSB induction. Without any DSB induction, fruit were red (*psy1* mutation is recessive) but in case of NHEJ or IHR repair, fruits were yellow. Yellow fruits were sequenced and seeds of fruits with homozygote CRISPR mutation at the *psy1* allele were grown to the next generation and sequenced. Using the following method, plants with IHR repair were detected. Molecular analysis uncovered gene conversion events with interrupted conversion tracts which in one case was of >5 Kbp in length and a putative targeted crossover event. In addition, in this system the rate of somatic inter-homologous repair at the tomato *PSY1* allele was estimated to be at least 14% (Filler-Hayut et al. 2017). This approach seems promising for precision breeding.

The feasibility of targeted somatic IHR for precision breeding is still under test and many questions need to be answered. For example, what is the rate of IHR in euchromatic versus heterochromatic targets on a genome? What is the rate of gene conversions versus crossovers? What is the variation between loci in the genome and between crops in targeted IHR frequency? And if and how plant somatic repair mechanisms can be biased toward IHR?

4.6 Conclusion

The technological developments of next-generation sequencing methods and targeted nucleases, especially CRISPR based systems, made the dream of precision breeding achievable. Crop varieties with NHEJ repair products are already available in the market (Fig. 4.1c) and considered as non-genetically modified (non-GMO) in many parts of the world (Globus and Qimron 2018). In addition, the potential of both gene targeting and IHR for precision breeding has been validated in different studies. High resolution identification and characterization of agronomically beneficial genes and genomic elements in different crop plants will help breeders to define the exact targets and exact genomic changes needed for precision breeding. Future research of DNA DSB repair mechanisms in plants will facilitate better control and understanding of how to direct the endogenous repair mechanisms toward the desired repair product.

References

Adli M (2018) The CRISPR tool kit for genome editing and beyond. Nat Commun 9:1–13

Baltes NJ, Gil-Humanes J, Cermak T, Atkins PA, Voytas DF (2014) DNA replicons for plant genome engineering. Plant Cell 1–14

Carroll D (2014) Genome engineering with targetable nucleases. Annu Rev Biochem 83:409–439

Dahan-Meir T, Filler-Hayut S, Melamed-Bessudo C, Bocobza S, Czosnek H, Aharoni A, Levy AA (2018) Efficient in planta gene targeting in tomato using geminiviral replicons and the CRISPR/Cas9 system. Plant J 95:5–16

Filler-Hayut S, Melamed Bessudo C, Levy AA (2017) Targeted recombination between homologous chromosomes for precise breeding in tomato. Nat Commun 8:15605

Gao H, Gadlage MJ, Lafitte HR, Lenderts B, Yang M, Schroder M, Farrell J, Snopek K, Peterson D, Feigenbutz L et al (2020) Superior field performance of waxy corn engineered using CRISPR–Cas9. Nat Biotechnol 38:579–581

Globus R, Qimron U (2018) A technological and regulatory outlook on CRISPR crop editing. J Cell Biochem 119:1291–1298

Li J, Meng X, Zong Y, Chen K, Zhang H, Liu J, Li J, Gao C (2016) Gene replacements and insertions in rice by intron targeting using CRISPR-Cas9. Nat Plants 2:1–6

Mercier R, Mézard C, Jenczewski E, Macaisne N, Grelon M (2015) The molecular biology of meiosis in plants. Annu Rev Plant Biol 66:297–327

Rodríguez-Leal D, Lemmon ZH, Man J, Bartlett ME, Lippman ZB (2017) Engineering quantitative trait variation for crop improvement by genome editing. Cell 171:470-480.e8

Schmidt C, Pacher M, Puchta H (2019) DNA break repair in plants and its application for genome engineering. In: Methods in molecular biology. Humana Press Inc, pp 237–266

Underwood CJ, Choi K, Lambing C, Zhao X, Serra H, Borges F, Simorowski J, Ernst E, Jacob Y, Henderson IR et al (2018) Epigenetic activation of meiotic recombination near Arabidopsis thaliana centromeres via loss of H3K9me2 and non-CG DNA methylation. Genome Res 28:519–531

Wang Y, Cheng X, Shan Q, Zhang Y, Liu J, Gao C, Qiu JL (2014) Simultaneous editing of three homoeoalleles in hexaploid bread wheat confers heritable resistance to powdery mildew. Nat Biotechnol 32:947–951

Xu R, Yang Y, Qin R, Li H, Qiu C, Li L, Wei P, Yang J (2016) Rapid improvement of grain weight via highly efficient CRISPR/Cas9-mediated multiplex genome editing in rice. J Genet Genomics 43:529–532

Shdema Filler-Hayut completed her Ph.D. in the Weizmann Institute of Science studying plants DNA repair mechanisms and genome engineering, under the supervision of Prof. Avraham Levy. She is currently a postdoctoral fellow at MIT (USA).

Cathy Melamed-Bessudo completed her Ph.D. in the Weizmann Institute of Science studying mechanisms of RNA Splicing. She did her postdoctoral research in the department of Plant Genetics on wheat evolution, in the Weizmann Institute, under the supervision of Prof. Moshe Feldman. She joined the laboratory of Avraham Levy where she is currently the Senior Research Scientist.

Avraham A. Levy earned his Ph.D. in plant genetics from the Weizmann Institute of Science. He conducted postdoctoral research at Stanford University and at INRA Versailles, then he joined the Weizmann Institute of Science in Israel where currently serves as Dean of the Faculty of Biochemistry. His research probes the the genetic and epigenetic mechanisms responsible for biodiversity in the plant kingdom and harnesses advanced genetic manipulation techinques, such as genome editing, to engineer plants in a precise manner. He serves on the scientific board of Agro-biotech companies, on the Plant Cell editorial board and was recently elected as associate member to the French Academy of Agriculture.

Chapter 5
CRISPR Technologies for Plant Biotechnology Innovation

Muntazir Mushtaq and Kutubuddin A. Molla

Abstract The invention of CRISPR/Cas technologies and its rapid advancement enables us to modify plant genomes like never before. The Conventional CRISPR/Cas tool has an unparalleled ability to generate targeted mutations in the genome. In contrast, advanced tools like the base editor can perform single DNA letter swap, and prime editing can generate precise insertion or deletion or DNA letter swapping. This chapter intends to provide the readers with the basics of CRISPR/Cas9 genome editing (GE) technologies, a brief introduction to various CRISPR-derived advanced tools, and how they are implemented to generate site-specific DNA modifications for plant biotechnological applications. In addition, we highlight how genome-edited crops are different from genetically modified organisms (GMOs).

Keywords CRISPR/Cas tools · Targeted mutations · CRISPR-mediated plant biotechnology innovation · Crop improvement · Stress tolerant crops · CRISPR in Agriculture

5.1 Introduction

The world has witnessed ample growth in agricultural productivity from the last 50 years. The inception of technologies for crop genetic improvement has especially led to a drastic increase in yield for major staple crops, for instance, wheat and rice. This achievement evolved in the form of the Green Revolution (1966–1985). Later, recombinant DNA-based biotechnology developed in the 1970s gave rise to genetically modified (GM) crops, appreciations to innovators like Marc Van Montagu, Jozef Schell, and Mary-Dell Chilton, who co-developed Agrobacterium-mediated plant transformation technology. While transgenic technology has heralded

M. Mushtaq
School of Biotechnology, Sher-e-Kashmir University of Agricultural Sciences
and Technology of Jammu, Srinagar, Jammu and Kashmir 180009, India

K. A. Molla (✉)
ICAR-National Rice Research Institute, Cuttack 753006, India
e-mail: kutubuddin.molla@icar.gov.in

a new era in crop improvement, GM crops' development is expensive, and they also face societal acceptance issues in many countries. In the meantime, conventional breeding approaches cannot keep pace with biggest challenges such as global population growth and climate change, e.g., the current percentage of annual increase in yield for four major crops (wheat, rice, maize, and soybean) must be doubled to meet the future demand in 2050. All these concerns demand the evolution of new breeding techniques (NBTs) that can likely transform agriculture. GE is one such technology that enables rewriting the code of life, which in most cases depends on the ability to induce DNA double-strand breaks (DSBs) in a sequence-specific manner (Jiang and Doudna 2017). For more details on DSB, see chapter, 'New technologies for precision plant breeding', in this book by S. Filler Hayut et al. Sequence-specific nucleases (SSNs) are molecular clippers that are engineered to create targeted DSBs in DNA. A functional unit of DNA is called a gene. SSNs such as zinc finger nuclease (ZFN), transcription activator-like effector nuclease (TALEN), and CRISPR (clustered regularly interspaced short palindromic repeats)-Cas systems have been successfully used in many plant species to enable efficient genome engineering. Developed in 2012 (Jinek et al. 2012) and applied to eukaryotic cells (mammalian cells) in 2013 (Cong et al. 2013), CRISPR-Cas GE technology has since been transforming plant biology. It improves reverse genetics research in both model and non-model plants and constitutes an efficient breeding tool for crop improvement. In recent years, the number of peer-reviewed papers exploiting CRISPR in plants has skyrocketed. However, it can be challenging and puzzling for new users to select a CRISPR system to achieve a specific GE outcome in a plant of interest. This chapter discusses the development of CRISPR/Cas GE tools with a historical perspective, how these tools precisely alter DNA, and the potential fields of applications in plant biotechnology and crop improvement.

5.2 Genome Editing Technologies: Historical Perspective and the Rise of Genome Editing Tools

Eukaryotic genomes are composed of billions of DNA bases. The ability to change these DNA bases with precision holds tremendous implications for molecular biology, medicine, and agriculture. Introducing desired genomic changes, i.e., "genome editing", has been a long-sought-after goal in molecular biology. An earlier article by Urnov (2018) can be consulted to get an exhaustive historical overview of GE technology. To this end, the breakthrough in the form of restriction enzymes that defend bacteria against invading viruses (bacteriophages) in the late 1970s marked the beginning of an era of recombinant DNA technology. For the first time in history, scientists were successful in manipulating DNA molecules in test tubes. Although such accomplishment drove several discoveries in molecular biology and genetics, the ability to precisely alter DNA in eukaryotic organisms came a few decades later. Recent progress in GE tools brings a new revolution in biological research. The GE toolbox was developed between 1994 and 2010 in a mutual attempt between academia and industry, using meganucleases as a prototype and zinc finger nucleases (ZFNs) to edit native loci. While meganucleases provided valuable information on the efficiency and mechanism of DSB repair, but have not been widely adopted as a gene-editing platform owing to the lack of a clear correspondence between meganuclease protein residues and their target DNA sequence specificity. Even though several techniques have been developed to account for these limitations, assembly of functional zinc finger proteins with the preferred DNA binding specificity remains a major problem that involves an extensive screening process. Zinc finger nucleases (ZFNs) were the first of the genome-editing nucleases to hit the scene. Zinc fingers (ZF) are the most common DNA binding domain in eukaryotes. They usually are comprised of ~30 amino acid units that interact with nucleotide triplets. ZFs have been designed to recognize all of the 64 possible trinucleotide combinations. By stringing different zinc finger moieties, one can create ZFs that specifically recognize any specific DNA triplets' sequence. Each ZF typically recognizes 3–6 nucleotide triplets. ZFN monomer consists of two distinct functional domains: an artificial zinc finger (ZF) domain at the N-terminal portion to bind a target DNA and a FokI DNA cleavage domain (Fok1) at the C-terminal region to create Double-strand break in the DNA. ZFNs have been widely adopted and proved to be the most versatile for GE for more than a decade in living organisms, including animal and plant systems. In 2010–2012, the GE toolkit was rapidly added with a third nuclease class known as transcription activator-like effector nucleases (TALENs). DNA binding characteristic of TALE (transcription activator-like effector) protein is used in constructing TALEN. TALENs are related to ZFNs in that they use DNA binding motifs to direct the same non-specific nuclease (Fok1) to cleave the genome at a specific site. Instead of recognizing DNA triplets as in ZF, each TALE domain recognizes a single nucleotide. In 1987–1989, an "unusual arrangement with repeated sequences" was noticed at a specific locus in the *Escherichia coli* genome—an array now known as a "clustered regularly interspaced short palindromic repeat," or CRISPR. This effort on investigating CRISPR-based bacterial immunity against phages finally led to the discovery, in 2012, that a key enzyme of a specific CRISPR-based system, Cas9, is an RNA-guided endonuclease. CRISPR immunity is RNA-based defensive machinery in bacteria designed to recognize and degrade foreign DNA elements from invading bacteriophage and plasmids. The bacterial genome codes for both Cas endonuclease and the guide RNA by a "CRISPR/array." This system can be co-opted to cut any targeted DNA sequence of choice by modifying it. Thus, in 2012, Martin Jinek, Jennifer Doudna, Emmanuelle Charpentier, and colleagues wrote: "Zinc-finger nucleases and transcription-activator–like effector nucleases have attracted significant attention as artificial enzymes engineered to manipulate genomes. We propose an alternative methodology based on RNA-programmed Cas9 that could offer considerable potential for gene-targeting and genome-editing applications". It was the beginning of a GE revolution that has taken biologists around the world by storm.

5.3 CRISPR/Cas-Based Genome Engineering

GE technologies refer to those enabling gene knockout, chromosomal recombination, and site-directed insertion/substitution at precise gene locations and chromosomal regions. DNA double-strand breaks (DSBs) are one of the powerful forces that figure plant genomes. Briefly, GE technologies create DNA double-strand breaks (DSBs) via sequence-specific nucleases (SSNs), and the DSBs are then repaired either by the error-prone 'nonhomologous end joining' (NHEJ) pathway or by the error-free 'homology-directed repair' (HDR) pathway (Molla and Yang 2020). Error in the coding region during DNA repair may cause codon mutations or frameshift mutations of the gene. Therefore, the gene becomes non-functional. DNA DSB repair systems have been extensively studied in many organisms, including plants. Investigations in plants have characterized the genes associated with DSB repair through Non-Homologous-End-Joining (NHEJ) or Homologous Recombination (HR) and tested the result of DSB repair in both somatic and meiotic tissues. NHEJ has been characterized in a wide variety of species and tissues (mostly somatic), using multiple DSB inducing agents including site specific meganucleases, transposon excision and custom-designed nucleases, such as zinc-finger nucleases, transcription activator-like effector nucleases (TALENs) and Clustered Regulatory Interspaced Short Palindromic Repeat associated protein Cas9 (CRISPR-Cas). In contrast to random mutagenesis by chemical (e.g., EMS, EES, Bisulfan) or physical agents (e.g., gamma-ray, x-ray, UV ray), GE tools can install mutations at specific chromosomal sites; therefore, have loads of advantages in functional genomics and molecular breeding. In fact, the genome fixed-point editing technique not only has a greater chance to cause mutations but also is more specific and more potent than random mutagenesis.

The recently developed CRISPR/Cas9 system has rapidly replaced the earlier ZFNs and TALENs. Owing to its simplicity, high efficiency, low cost, and possibility to target multiple genes at once, the CRISPR/Cas9-based GE platform has become an unprecedented tool in plant science research. The use of the CRISPR/Cas9 system for gene-editing tool was first established in human cells and then utilized in plants. Hitherto, CRISPR/Cas9 vector systems have been applied to generate gene knockouts, deletions, disruption of cis-regulatory elements, as well as gene replacements (knock-ins).

Fundamentally, the CRISPR/Cas9 system is relatively simple and is composed of only two major components viz. the Cas9 protein and a single guide RNA (sgRNA), forming a Cas9/sgRNA complex. The 20-nucleotide sequence at the 5′ end of a sgRNA can precisely hybridize with a homologous DNA sequence. This RNA–DNA hybrid activates Cas9 to generate a DSB in the target sequence. The target sequence must be present immediately upstream of an adjacent protospacer motif (PAM; NGG for SpCas9 from *Streptococcus pyogenes* and TTTV for Cas12a). During DSB repair, NHEJ frequently generates small insertion or deletion (Indel) of nucleobases. Notably, multiple sgRNAs with different target sequences can be designed for the simultaneous editing of more than one gene or DNA region. Once a gene sequence is disrupted due to indel formation, the resultant change in 'the appearance of a plant' (phenotype) is correlated. If the correlation is established, it is assumed that the phenotype is controlled by the gene in the study.

Guide RNAs are artificially designed to explicitly direct Cas9 to the target sequence to be edited. Available bioinformatic programs are used to design candidate guide RNAs while considering the likelihood of off-targets. Plant cell transformation to express guide RNAs and Cas9 follows a procedure analogous to the well-known methods for developing transgenic plants. The expression cassettes contain constitutive or inducible promoters, transcription terminators, and antibiotics and/or herbicide resistance markers used for selection purposes.

A vector harboring the DNA sequences for Cas9 protein, and the guide RNA is then incorporated into *Agrobacterium* bacterial cells. Then, plants are genetically transformed through *the Agrobacterium*-mediated method, and the identification of first-generation transformed plants is made by using either antibiotic or herbicide selection. Cells or calluses carrying sgRNA-Cas9 cassettes can also be distinguished using green fluorescent protein (GFP). In all instances, target gene/DNA region sequencing is necessary to identify Cas9-induced mutations. The plants generated via sgRNA/Cas9 mediated GE is called transgenic as they carry the sgRNA-Cas9 cassette.

On the other hand, the sgRNA-Cas9 transgene cassette can be eliminated through sexual segregation after identifying CRISPR-edited sexually propagated plants. Selection is made for plants that are edited but do not carry the sgRNA/Cas9 cassette. This would lead to transgene removal in the second or subsequent generations, resulting in transgene-free genome-edited plants' production. Hence, they bear a resemblance to those with mutations created by natural means. In some countries, the introduction of the sgRNA-Cas9 DNA cassette as a transgene may be treated as GMOs under existing biosafety regulations.

Protocols have been developed to edit plant genomes using guide RNA-Cas9 ribonucleoprotein (RNA plus protein) complexes or transient expression resulting in DNA-free plants. Preassembled Cas9-guide RNA ribonucleoproteins complexes can be introduced into protoplasts via polyethylene glycol-calcium-mediated transfection.

Notable genetic alterations have been achieved through CRISPR-Cas9 to improve metabolic pathways, biotic (fungal, bacterial, and viral pathogens) and abiotic stress (cold, drought, salt) tolerance, nutritional content, yield, and grain quality, obtain haploid seeds, herbicide resistance, and others. Here, we highlight the technical features of the CRISPR/Cas-based GE systems and their crop improvement applications.

5.4 Various CRISPR/Cas-Based Tools and Their Utilities

Besides the conventional CRISPR/Cas-mediated gene disruption, creative designing and ingenious protein engineering have generated versatile genome disruptors, transcriptional regulators, epigenetic modifiers, base editors, and prime editors (Molla et al. 2020a).

While many Cas proteins have been discovered, Cas9 and Cas12a (also known as Cpf1), are widely used for genome engineering and transcriptional regulation. Cas9 and Cas12a are remarkably easy to program and can be directed to target DNA through Watson–Crick base pairing between the target sequence and gRNA. We refer the readers to an earlier book chapter for more information on orthologous and engineered Cas protein and their utilities in genome targeting (Molla et al. 2020a). In CRISPR systems, gRNAs are composed of a crRNA:tracrRNA complex, that can be fused to form a sgRNA (single gRNA), whereas those for Cas12a consist exclusively of a crRNA. After forming an RNA–protein complex with a gRNA, Cas9 and Cas12a carry out a double-strand break (DSB) adjacent to a protospacer adjacent motif (PAM). Subsequently, DSB repair results in the edited genome.

5.4.1 Multiplex Editing

The Cas9 and gRNA expression cassettes are the two primary components for CRISPR/Cas9-mediated genome editing. In general, Cas9 expression is driven by an RNA polymerase II (Pol II) promoter. The gRNAs are normally expressed by an RNA polymerase III (Pol III) promoter. For editing multiple genomic loci simultaneously, multiple gRNAs need to be expressed. Different gRNA expression systems are used to achieve multiplex genome editing. The Tandem array of single gRNA expression cassettes, where each guide is controlled by a pol-III promoter, is often used. However, a number of techniques have been developed for single promoter driven expression of multiple gRNAs by engineering different RNA processing machineries, including Csy4 RNase from a bacterium, self-cleavable ribozyme from a virus, and the endogenous tRNA processing enzymes (Vicki and Yang 2020). These RNA processing reagents are employed to generate many gRNAs from single primary polycistronic transcript driven by a Pol II or Pol III promoter. Recently, the polycistronic tRNA-gRNA (PTG) system and ribozyme-mediated assembly of multiple gRNAs under the control of a single promoter gained popularity. In the PTG system, post-transcriptional cleavage of tRNAs by cellular RNAse-P and RNAseZ releases individual gRNA. Ribozymes are RNA molecules with nuclease activity that catalyze its own cleavage. Scientists took advantages of ribozyme's self-cleaving activity and designed an array of 'ribozyme-gRNA-ribozyme-gRNA-ribozyme' to produce multiple gRNAs. When transcribed, the primary transcripts contained the designed gRNA flanked with a ribozyme at each end. After self-cleaving of ribozyme, each gRNA becomes free and they guide Cas9 to their respective coded locus. Csy4 based excision system was also developed for multiplexing. On the other hand, the Cas12a system could be used directly with an array of crRNAs without intervening sequences due to its self-processing ability. We recommend the readers to follow a nice review on different strategies for expressing multiple gRNAs (Vicki and Yang 2020).

5.4.2 Transcriptional Activation and Repression

Cas9 and Cas12a are made catalytically inactive by changing critical amino acid residues. They are designated as deactivated Cas (dCas9 and dCas12a) proteins. The fusion of dCas enzymes and effector proteins harnesses efficient transcriptional regulation. Two techniques, CRISPR interference (CRISPRi) and CRISPR-activator (CRISPRa) are widely used for regulating gene expression.

Mechanistically, dCas9 enzymes limit transcription by impeding the binding of RNA polymerase or, if targeted to a coding sequence, by interfering with transcription elongation. In eukaryotes, dCas9 is usually fused to an effector protein to augment repression by recruiting chromatin remodeling proteins. Likewise, effector proteins for CRISPRa work by recruiting endogenous transcriptional activators.

5.4.3 Epigenome Editing

Different epigenetic factors determine distinct phenotypes despite sharing the same DNA. DNA base methylation and histone residue modification are the two well-known epigenetic modifications. CRISPR/dCas system enables transporting a modifier protein to a target epigenetic locus. This could be achieved through the fusion of the modifier protein with the dCas9. The modifier proteins include but not limited to, histone methyltransferase, demethylase, DNA methyltransferase, TET enzyme, etc. The dCas9 tethered with modifier protein, called epigenome editor, can modify epigenetic factors' status and thus alter the phenotypes.

5.4.4 Base Editing

Homology-directed repair (HDR) is utilized to precisely change one nucleotide to another in a target DNA; it needs an adequate supply of donor DNA template harboring the change. However, HDR-mediated editing is extremely inefficient in the plant system. Base editing enables precise alteration of single nucleotides with high efficiency and does not require the donor DNA template. For a comprehensive review of base editing, Molla and Yang (2019) may be consulted. The base editor comprises of two essential components—a deaminase tethered with Cas9 nickase (nCas9). nCas9, a catalytically impaired version of Cas9, generates single-strand nick at the target DNA. Cytosine base editor (CBE) causes C-to-T conversion. Adenine base editor (ABE) performs A-to-G alteration. Three recent studies reported the development of a C-to-G base editor (CGBE) for C-to-G editing in the DNA (Molla et al. 2020b). These CBE, ABE, and CGBE enable us to alter a plethora of functional single nucleotide polymorphism (SNP) to improve desired traits in crops.

5.4.5 Prime Editing

Indeed, the development of base editors reduces our dependency on inefficient HDR-mediated editing. For generating precise indels (as opposed to the random indels caused by Cas9-DSB repair via NHEJ), scientists rely on HDR. A recently developed system, known as prime editing, was demonstrated to efficiently perform all types of base substitutions, 1-44 bp insertions, and 1-80 bp deletions in human cells with far better efficiency than HDR (Anzalone et al. 2019). Several studies also reported the success of prime editing in plant systems, although with lesser efficiency. The prime editing system does not require a supply of donor template. Prime editors contain nCas9 fused with reverse transcriptase (RT). Unlike all other CRISPR-derived techniques that require the same gRNA, prime editing needs a special type of prime editing-guide RNA or pegRNA. Besides specifying the target, PegRNA encodes the reverse transcriptase template (RT), which harbors the desired edit. 'Two extra elements, 10–13 nt primer binding site (PBS) and 10–16 nt long RT template with the desired changes, are added with the traditional gRNA to construct pegRNA' (Molla et al. 2020a). Sequence complementary to the nicked genomic DNA strand acts as a primer binding site (PBS). Hybridization of PBS sequence to the target site serves as the point of initiation for reverse transcription. RT copies the desired changes (coded in the RT template) directly in one strand of DNA. Subsequent nick and flap resolution by cellular repair machinery incorporate and fix the editing into the genome.

5.5 Better Crops with CRISPR/Cas Techniques

The biological world is being modernized through the field of GE technology. CRISPR/Cas9 has been proved to be the best choice for GE with high efficiency, accuracy, and ease of use. World agriculture is witnessing alarming issues, including increasing population growth rate, unpredictable weather, increasing biotic and abiotic stresses, and decreasing arable land availability. To overcome such threatening issues, GE technologies have great potential to be profitable heed to global food security. We recommend readers to consult a recent review article on GE in agriculture (Chen et al. 2019). CRISPR/Cas9 edited non-browning mushroom (*Agaricus bisporus*) by Dr. Yinong Yang from Pennsylvania State University, received a green signal from the United States Department of Agriculture (USDA). The department also approved CRISPR-edited corn, soybeans, tomatoes, pennycress, and Camelina and emphasized that the transgene free-genome edited crops would not be considered transgenic crops. Recently, a CRISPR-breed Petunia plant with pale pinky-purple colored flower has been approved by USDA. CRISPR/Cas9 system of gene editing has been adopted in many crop species such as rice, maize, wheat, soybean, citrus, tomato, potato, cotton, alfalfa, watermelon, grapes, cassava, ipomoea, barley, lettuce, cacao, carrot, banana, flax, rapeseed, *Camelina*, cucumber and many other crops for various traits including yield and nutritional quality improvement, biotic and abiotic stress management (Fig. 5.1).

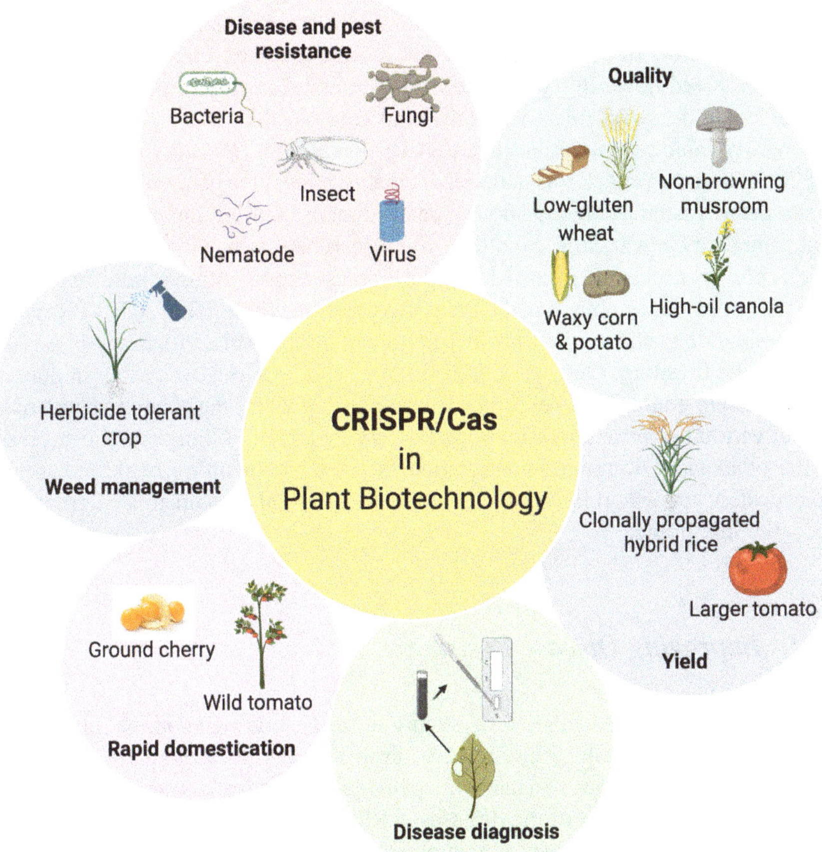

Fig. 5.1 Schematic showing different facets of application of CRISPR-Cas tools for crop improvement. This figure was created using BioRender (https://biorender.com/)

5.5.1 Rapid Domestication of Crops

With the increasing number of crop genomes sequenced, GE offers one the efficient approaches to plant domestication by opening up the vast genetic diversity from wild or semi-domesticated species, thus producing crops with desired traits.

Modern tomato cultivars produced from the long domestication process resulted in the loss of genetic diversity for stress tolerance. Wild tomato plants naturally exhibit a high degree of tolerance to different stresses; thus, they can serve as ideal materials for de novo domestication via targeting of so-called domestication genes using CRISPR. Zsogon et al. (2018) demonstrated accelerated *de-novo* domestication of wild tomato with CRISPR/Cas9-mediated multiplex editing, simultaneously targeting genes involved in flowering time, plant architecture, and fruit size that resulted in loss-of-function mutations. The CRISPR edited plants showed earlier and synchronized flowering, increased fruit size, and determinate plant architecture without losing the stress tolerance of the wild germplasm.

Similarly, rapid *de-novo* domestication of ground cherry (*Physalis pruinosa*) using the CRISPR/Cas approach was undertaken. Inactivating three genes resulted in determinate growth habit, increased flower petal number, and the number of fruits per plant. Another eye-catching candidate for researchers is a winter annual pennycress (*Thlaspi arvense* L.) weed for rapid domestication. Illinois state university researchers have been able to make the pennycress oil edible. Thus, CRISPR mediated *de-novo* domestication events in wild plants offer novel exhilarating possibilities for plant breeding. On one side, exploiting wild crop relatives as an important source of allele mining could expand the germplasm pool for the genetic impoverishment of various crops and resistance against a wide range of biotic/abiotic stresses. On the other side, de-novo domestication facilitates catapulting neglected, semi-domesticated, and wild relatives of crops into the spotlight of mainstream agriculture in a rapid time frame.

5.5.2 Improving Disease Resistance

The goal of providing adequate food supply to feed the growing global population is made even more challenging owing to crop loss due to various diseases. The development of host plant resistance to pathogens is one of the most sustainable ways to reduce the impact of the diseases. CRISPR/Cas9 is being extensively used to enhance the resistance power of host plants. Readers are suggested to consult the review article (Mushtaq et al. 2019). Few notable case studies are discussed below.

5.5.2.1 Bacterial Resistance

Phytopathogenic bacteria are highly diverse, with a high multiplication rate and are difficult to control. For example, in rice, *Xanthomonas oryzae* pv. *oryzae* (Xoo) causes bacterial blight disease resulting in 10–20% yield loss. The virulence of Xoo pathogen strongly depends on secretion of its TALE proteins inside the susceptible host. The secreted TALE proteins bind to the promoter of the SWEET genes of rice and trigger their expression.

Higher expression of SWEET genes promotes a favorable environment for disease development. To interrupt the binding of TALE proteins with SWEET gene promoter, the sequence of the binding region was altered by CRISPR/Cas9. The resultant mutant lines were resistant to bacterial blight. A similar strategy was used to develop resistance against citrus canker bacteria.

5.5.2.2 Fungal Resistance

CRISPR/Cas9 system has shown great potential in mitigating the impacts of fungal diseases as well. Many fungi use host plant genes (known as susceptibility genes) for facilitating their establishment in the host. For example, *mildew resistance locus O (Mlo)* gene, a susceptibility gene, has been mutated through CRISPR/Cas9 system, and the developed mutants were more resistant to powdery disease-causing fungus *Blumeria graminis* f. sp. *tritici* and *Oidiumneo lycopersici* of wheat and tomato, respectively. Targeted mutagenesis of negative regulator *OsERF922* showed disease resistance against rice blast. These examples validate the potential use of the CRISPR/Cas9 system for fungal disease resistance.

5.5.2.3 Virus Resistance

Plant viruses pose a significant threat to modern agriculture. Resistance in *Nicotiana benthamiana* against beet severe curly top virus (BSCTV) was demonstrated by the precise targeting of replication genes of viruses via CRISPR/Cas9. Similarly, rice plants resistant to Tungro virus disease was developed through CRISPR/Cas9 mediated disruption of a gene that is used by the virus for their multiplication. A recent review article may be consulted to get more insight on the application of GE for virus resistance (Mushtaq et al. 2020).

5.5.2.4 Nematode and Parasitic Weed Resistance

Various GE technologies have been adopted to improve crop resistance against nematodes. CRISPR/Cas9-mediated GE technology has been used for targeting *GmSHMT08* to study soybean resistance to soybean cyst nematode (Kang 2016).

CRISPR/Cas9 enabled mutagenesis of the *tomato's CCD8* gene has been used to provide resistance against the weed *Phelipanche aegyptiaca.*

5.5.3 Developing Climate-Smart Crops

Abiotic stresses, including salinity, drought, temperature, and heavy metals, pose a major threat to global food security. Developing cultivars tolerant against these various environmental stresses is the most sustainable and environmentally friendly approach to cope with this challenge. However, little work has been done regarding abiotic stress tolerance development employing CRISPR/Cas tools. Recently, tomato fruit setting under heat stress has been improved by targeting *SlAGAMOUS-LIKE 6* gene using CRISPR-Cas9 mediated genome editing. Researchers have recently developed and tested drought-tolerant maize through precise gene-editing of *AGROS8* (Shi et al. 2017). Nonetheless, the technique is also hugely utilized by researchers to decipher the functions of genes potentially involved in abiotic stress tolerance. Readers may consult a recent review to get more insight (Mushtaq et al. 2018).

5.5.4 Quality Enhancement

5.5.4.1 Color and Texture

The color and texture of fresh tomatoes many times determine consumer preferences. Consumers from diverse areas have distinct color preferences. For example, Americans and Europeans prefer red tomatoes, while Asians like pink-colored tomatoes. Researchers have effectively produced yellow, pink, and purple tomatoes by targeting different pigment biosynthetic genes.

5.5.4.2 Increasing Nutritional Composition and Removing Anti-Nutrient (Allergens) Factors

Consumption of gluten proteins from wheat, rye, and barley causes coeliac disease in genetically predisposed individuals. Patients with a strict gluten-free diet can recover from this disease. As there are a large number of gluten genes and wheat genome being more complex, wheat that is coeliac-safe but retains baking quality cannot be produced by conventional breeding alone. Recently, CRISPR/Cas tool was used to develop low gluten wheat with the baking quality intact (Chen et al. 2019).

Maize (*Zea mays*) is a major cereal crop, and phytic acid-P form more than 70% of the total phosphorus in maize seed. It is supposed to be anti-nutritional because it is not digested by monogastric animals and is also an environmental pollutant. Scientists have used the CRISPR technique for targeted knock out of genes governing phytic acid synthesis in maize.

Anthocyanin, malate, lycopene, and γ-aminobutyric acid (GABA) are the tomato's well-known bioactive compounds. CRISPR-Cas9 technology has been employed to generate tomato fruits with an enhanced level of those bioactive compounds by regulating the key genes responsible for their metabolism.

Potato starch quality is an important trait and a central area of research. Scientists have targeted *the granule-bound starch synthase (GBSS)* gene using CRISPR-mediated gene-editing, which resulted in waxy genotype in potato. Similarly, DuPont/Corteva agriscience developed waxy corn with more than 97% amylopectin by targeted disruption of *the Wx1* gene (https://synbiobeta.com/dupont-pioneer-unveils-first-product-developed-crispr-cas/). These waxy crops have high values in processed food, adhesive, and high gloss paper making industry.

Accumulation of cesium and cadmium in rice plants grown in contaminated soil is a severe health concern. Recently, researchers from two distinct groups used the CRISPR-Cas system to inactivate transporter proteins that carry them to allow their inflow from soil to the plant. Mutant plants exhibited a remarkably lower accumulation of cesium and cadmium.

Recently, Yield10 Bioscience, MA, USA, announced the successful development of CRISPR-edited canola plant lines with increased oil content. They received a non-GMO regulatory response from USDA (https://geneticliteracyproject.org/2020/08/19/crispr-edited-canola-slated-for-2021-field-trials-moving-crop-closer-to-com mercialization/).

Cassava is the primary source of nutrition for ~40% of Africans. It contains toxic cyanogen—excessive consumption of under processed cassava results in an epidemic paralytic disease, Konzo. Researchers from Innovative genomic institutes, California, are working on the removal of cassava cyanogen by disrupting its biosynthetic genes (https://innovativegenomics.org/news/crispr-cyanide-free-cas sava/). Similarly, Tropic bioscience in the UK is working on producing a CRISPR-edited naturally decaffeinated coffee.

5.5.4.3 Enhancing Self-Life

Postharvest browning of mushroom causes decreased consumer acceptance and market value. Yinong Yang and his team at the Pennsylvania state university developed a non-browning mushroom by CRISPR/Cas-induced inactivation of genes responsible for mushroom-browning.

The food industry highly desires prolonged shelf life for fleshy fruits. Tomato lines with extended shelf life were generated by CRISPR/Cas mediated manipulation of ripening pathway genes.

5.5.5 Yield Improvement

Yield is one of the most important traits for crop plants. Traditional breeding has been used for decades to improve yield and develop plants suitable for particular growth environments, which is a time-consuming process. Researchers used CRISPR/Cas9 system to knockout four negative regulators of yield (the genes *Gn1a*, *DEP1*, and *GS3*) in rice and obtained mutants with improved grain number, dense, erect panicles, and larger grain size, respectively. Likewise, another group of scientists used CRISPR/Cas9-mediated multiplex genome-editing system to simultaneously knock out three major rice negative regulators of grain weight (*GW2*, *GW5*, and *TGW6*), resulting in a significant increase in thousand-grain weight. Researchers targeted three homoalleles of *GASR7*, a negative regulator of kernel width and weight in bread wheat, by employing CRISPR/Cas9 that increased the thousand-kernel weight. Researchers at the Cold Spring Harbor Laboratory employed CRISPR-Cas9 gene-editing tool to produce larger tomato fruits by destructing the classical *CLAVATA-WUSCHEL* (*CLV-WUS*) stem cell circuit.

Hybrid varieties provide yield advantages over traditional varieties. However, farmers cannot save hybrid seeds for the next generation because yield advantages are lost in subsequent generations due to genetic segregation. Scientists have long sought for a technology to propagate hybrid seeds clonally. Remarkably, two recent studies demonstrated the clonal seed production of hybrid rice employing CRISPR/Cas9 technology (Chen et al. 2019). For more details on yield improvement using CRISPR/Cas9, see Chen et al. (2019).

5.5.6 Early Disease Detection in Plant

The discovery of orthologous Cas proteins (Cas12a and Cas13) with collateral nucleic acid cleavage activities enabled the development of nucleic acid diagnostic tools. Studies showed they could be used to develop robust, highly sensitive, low-cost, a practical diagnostic tool for disease and pathogen detection. Application of DETECTR (based on Cas12a) and SHERLOCK (based on Cas13) system would enable trait detection, pest surveillance, GMO detection, and pathogen identification (Kocak and Gersbach 2018).

5.5.7 Controlling Invasive Species in Agri Field

Invasive species continue to be one of the greatest challenges to global biodiversity. CRISPR/Cas9-based gene drive is a powerful technology that allows biased inheritance of a gene and spread rapidly through a population. Gene drive is used to insert and spread a desired modification faster than the usual rate of Mendelian inheritance

(50%). Gene drive technology is competent to control pest species to increase agricultural production. In contrast to other pest management strategies, it is cheaper, more precise than pesticide use. Gene drive-mediated pest control can thus be eye-catching for agribusiness, owing to its direct manipulation of pest species. It could also potentially be used against invasive weeds. For instance, pigweed (*Amaranthus*) could be engineered by CRISPR/Cas-based gene drives to become susceptible to the widely used herbicide glyphosate. The principle for CRISPR–Cas9-based suppression of weed species is based on the assumption that gene drives could be used to bring in and spread a fitness load that can limit the establishment, abundance, dispersal, persistence and/or impact of weed populations.

5.5.8 Weed Management

Besides using gene drives to eradicate a target weed population, CRISPR/Cas tools can also be used to generate herbicide-tolerant crop species; so that herbicide could be used to control weeds in the agricultural field effectively. Herbicide usually kills plants by inhibiting one or more crucial plant metabolic enzymes. Herbicide resistance is developed by a single or few point mutations in the enzyme's herbicide binding site. Precise installation of those mutations was achieved either by Base editors or HDR-mediated precise editing. Rice plants resistant to Imazamox, haloxyfop, and bispyribac sodium have been generated by employing those CRISPR tools (Chen et al. 2019).

5.6 Genome Edited Versus Transgenic Crops

Both transgenic and gene-edited crops are developed by forms of genetic engineering that have possible applications in agriculture and plant biotechnology. Still, they both vary in one way or the other. In the process of developing both kind of crops, an initial transformation of foreign DNA constructs is done. For transgenic crops, the construct needs to be integrated and remained present in the genome for expressing the desired trait. On the other hand, in most of the genome-edited crops, the construct is no longer needed after it successfully induces editing. GE is more rapid than conventional breeding, is less controversial than techniques such as transgenesis which are considered as 'GMOs' from a regulatory point of view in many jurisdictions. GMOs involve introducing genes from the same or other species into DNA. Gene editing, in contrast, allows scientists to alter the organism's DNA without inserting genes from a different organism. GE enables modification of an existing gene. Gene-editing technology is advantageous over genetic modification on various grounds. As we discussed earlier in this chapter, editing is done by the influence of Cas9. Once the Cas9 makes a DSB at a target DNA and repaired erroneously, the gRNA/Cas9 transgene is no longer needed in the cell. By sexual inheritance, the transgene cassette can be segregated out. This technology is more specific (precise) than GMO processes, and continue to become more reliable. It is also relatively inexpensive in contrast to other methods, suggesting more scientists could gain access to it. Such advantages represent more potential innovation. How booming gene editing is, however, will also depend in large on how it's perceived. Crops developed by GE could face similar kinds of opposition as perceived by GMOs. Since they don't insert foreign genes into the crop, consumers might find them more natural and consequently more appealing which remains the advantage of GE crops over GMOs. On March 28, 2018, U.S. Secretary of Agriculture declared that the USDA wouldn't regulate crop varieties generated employing GE that would yield plants indistinguishable from those developed through traditional breeding methods.

5.7 Concluding Remarks

Although GE tools existed before CRISPR, it has democratized the field by its efficiency, ease of use, and accessibility. Although it has tremendous potential in human therapeutics, fruits of GE would be visualized quicker in agriculture. It will have many similar applications like GMOs but optimistically with broader public acceptance. While CRISPR could be a major boon for increasing agricultural production, a lack of public acceptance might choke further improvement of CRISPR crops before commercialization can become a certainty. Plant genome editing's societal concerns stem in part due to the unawareness of its principles and applications. Spreading the knowledge to the general public on the GE principles might correct and stop the spread of fallacy. We need to keep in mind that biotechnology familiarity and perceptions of safety, although not sufficient, is a crucial parameter for public acceptance.

References

Anzalone AV, Randolph PB, Davis JR, Sousa AA, Koblan LW, Levy JM, Liu DR (2019) Search-and-replace genome editing without double-strand breaks or donor DNA. Nature 576(7785):149–157

Chen K, Wang Y, Zhang R, Zhang H, Gao C (2019) CRISPR/Cas genome editing and precision plant breeding in agriculture. Annual Rev Plant Biol 70:667–697

Cong L, Ran FA, Cox D et al (2013) Multiplex genome engineering using CRISPR/Cas systems. Science 339:819–823

Jiang F, Doudna JA (2017) CRISPR–Cas9 structures and mechanisms. Annu Rev Biophys 46:505–529

Jinek M, Chylinski K, Fonfara I, Hauer M, Doudna JA, Charpentier E (2012) A programmable dual RNA-guided DNA endonuclease in adaptive bacterial immunity. Science 337:816–821

Kang J (2016) Application of CRISPR/Cas9-mediated genome editing for studying soybean resistance to soybean cyst Nematode. University of Missouri-Columbia: Columbia, MO, USA

Kocak DD, Gersbach CA (2018) Scissors become sensors. Nature 557(7704):168–169

Molla KA, Yang Y (2019) CRISPR/Cas-mediated base editing: technical considerations and practical applications. Trends Biotechnol 37:1121–1142

Molla KA, Yang Y (2020) Predicting CRISPR/Cas9—induced mutations for precise genome editing. Trends Biotechnol. 38:136–141

Molla KA, Qi Y, Karmakar S, Baig MJ (2020a) Base Editing Landscape Extends to Perform Transversion Mutation. Trends Genet. https://doi.org/10.1016/j.tig.2020.09.001

Molla KA, Karmakar S, Islam MT (2020b) Wide horizons of CRISPR-Cas-derived technologies for basic biology, agriculture, and medicine. In: CRISPR-Cas Methods. Humana, New York, NY, pp 1–23

Mushtaq M, Bhat JA, Mir ZA, Sakina A, Ali S, Singh AK, Tyagi A, Salgotra RK, Dar AA, Bhat R (2018) CRISPR/Cas approach: a new way of looking at plant-abiotic interactions. J Plant Physiol 224–225:156–162

Mushtaq M, Sakina A, Wani SH, Shikari AB, Tripathi P, Zaid A, Galla A, Abdelrahman M, Sharma M, Singh AK, Salgotra RK (2019) Harnessing genome editing techniques to engineer disease resistance in plants. Front Plant Sci 10:550

Mushtaq M, Mukhtar S, Sakina A, Dar AA, Bhat R, Deshmukh R, Molla K, Kundoo AA, Dar MS (2020) Tweaking genome-editing approaches for virus interference in crop plants. Plant Physiol Biochem 147:242–250

Shi J, Gao H, Wang H, Lafitte HR, Archibald RL, Yang M, Hakimi SM, Mo H, Habben JE (2017) ARGOS8 variants generated by CRISPR—Cas9 improve maize grain yield under field drought stress conditions. Plant Biotechnol J 15(2):207–216

Urnov FD (2018) GEBC (before CRISPR): lasting lessons from the "old testament." CRISPR J 1(1):34–46

Vicki HF, Yang Y (2020) Efficient expression of multiple guide RNAs for CRISPR/Cas genome editing. aBIOTECH 1:123–134

Zsogon A, Cermak T, Naves ER, Notini MM, Edel KH, Weinl S, Freschi L, Voytas DF, Kudla J, Peres LEP (2018) De novo domestication of wild tomato using genome editing. Nat Biotechnol 36:1211–2121

Muntazir Mushtaq did his M.Sc. and Ph.D. in Plant Biotechnology from Sher-e-Kashmir University of Agricultural Sciences and Technology, Kashmir, India. His research areas of interest are understanding biotic stress and improving plant tissue culture. He has published more than 20 peer reviewed scientific articles on various topic including plant tissue culture, genome editing in plants, genomic selection, disease resistance in plants, and development of double haploids. He is a member of the society of Bioinformatics and biological sciences. He was awarded the Moulana Azad National Fellowship, Mahima Young Scientist Award, and Best PG Thesis Award.

Kutubuddin A. Molla is a scientist at the National Rice Research Institute, Cuttack, India. He has obtained his Ph.D. from the University of Calcutta. Dr. Molla has done his post-doctoral research at the Pennsylvania State University, USA, availing the prestigious Fulbright fellowship. He is interested in genome editing and uses CRISPR-Cas and other advanced tools. Dr. Molla has been awarded the 'INSA Medal for Young Scientist-2020' by the Indian National Science Academy. He is the recipient of the AAAS/Science Program for Excellence in Science from the American Association for the Advancement of Science, Washington D.C. He has >25 peer reviewed publications and one edited book on 'CRISPR-Cas Methods'.

Part II
Contributions to the Society

Chapter 6
Intellectual Property Protection of Plant Innovation

Bernard Le Buanec and Agnès Ricroch

Abstract Plant innovation is expensive. It is therefore crucial to protect the intellectual property of the breeder as all plant varieties are living self-reproducible material. Intellectual property protection of plant innovation varies from country to country. Plant varieties are not patentable in Europe. Protection of biotechnological inventions is obtained by patent if the invention is new, has an inventive step and industrial applicability. The protection conferred by a patent to a biotechnological invention also applies to varieties in which it is inserted, namely transgenic varieties. In Europe the exceptions to the protection of conventional and transgenic varieties are the same, namely, under certain conditions, the possibility of using farm-saved seeds and access to the genetic diversity of the protected variety for breeding. The possibilities of protection depend on national laws. A detailed presentation is made for Europe and the United States of America and short examples of various other countries including countries growing transgenic varieties are also presented.

Keywords Patent · Plant breeders right · Transgenic varieties · Farm-saved seed · Genetic diversity

6.1 Why Protect Intellectual Property?[1]

According to the historian Phylarque the first monopoly was granted to "inventor" around the seventh century BC in the south of Italy, then in Greece, for new cooking recipes. The first comprehensive law on patents, known as *Parte Veneziana* was

---

[1] This chapter has been partially adapted from a text published in French (Le Buanec & Ricroch 2011. Comment protéger les innovations végétales. *In*: Biotechnologies végétales, environnement, alimentation, santé. Publisher Vuibert).

B. Le Buanec (✉)
Academy of Agriculture of France, Paris, France
e-mail: b.lebuanec@orange.fr

A. Ricroch
AgroParisTech and Paris-Saclay University, Paris, France
e-mail: agnes.ricroch@agroparistech.fr

A. Ricroch et al. (eds.), *Plant Biotechnology*,
https://doi.org/10.1007/978-3-030-68345-0_6

promulgated in the Republic of Venice in 1474. It was followed by the Statute of Monopoly passed by the British Parliament in 1623.

Under the influence of the ideas of the Enlightenment, the United States of America voted a law on patents on August 17, 1790, followed closely by France and its decree of January 7, 1791 stating that the inventors are owners of their inventions and providing to their benefit the grant of patents.

The US and French laws referred to the moral approach of the decision to grant intellectual property rights. This is also the approach of Article 27 of the Universal Declaration of Human Rights of 1948 which guarantees everyone "the right to the protection of the moral and material interests resulting from any scientific, literary or artistic production of which he is the author".

The second approach is called "utilitarian": the protection of intellectual property is granted not by moral obligation to reward the inventor but rather because the products he/she creates are useful to society, it is therefore necessary to encourage them to continue their research to foster innovation.

How to encourage innovation? Two approaches are possible: developing public research or encourage private research. These are political choices. Both approaches are not contradictory but complementary. However, in the current States budgetary context, increasing public spending seems difficult. It is therefore necessary to encourage private research and public–private partnerships and ensure a return on investment.

Research in plant biotechnology and plant breeding is expensive, around one million euro for a new variety. It is estimated that seed companies globally spend around € 6.6 billion in research annually, or 10 to 12% of the annual global turnover of the seed industry estimated at € 60 billion.

The table below presents an estimation of the 2019 seed research budgets of large seed companies worldwide (in million of euro):

Bayer crop science (Germany)	Corteva (USA)	Chem China –Syngenta (China)	Limagrain (France)	KWS (Germany)	Florimond desprez (France)
1100	800	650	240	205	42

(*Source* annual reports or personal communication)

In addition it takes capital to produce a new variety and place it on the market. Then success is not necessarily guaranteed, either because the competition has developed equivalent or better products, or the market has changed since the theoretical design of the product. It takes a decade to develop a new variety and as for any applied research plant breeding is a risky business.

If the new variety fits the market, the breeder should be able to exploit it commercially under conditions allowing it to receive the benefits of its investments. Two problems arise: firstly competitors could capture the variety and sell seeds, either identical or with minor improvements, and at a lower price because they do not have to recoup the development costs; on the other hand the "invention" being often capable of self-reproduction (autogamous species or vegetative reproduction) users could reproduce it for their own use without paying the original developer.

Under these conditions, the innovator could not recover costs, it would have no incentive to continue his/her research efforts, and no genetic progress of varieties could be expected from the private sector. To avoid this situation, the breeder must be able to protect his new varieties.

It is therefore necessary to find practical and equitable solutions for the protection of intellectual property to encourage innovation in the context of a private enterprise. One possible solution is secrecy. But the protection of a secret is ineffective in the context of living self-reproducible material.

Moreover, as the secret does not allow the disclosure of the method to get the results, it does not promote the incremental development of knowledge. Therefore, most countries have now put in place systems to protect intellectual property that are true social contracts between the inventor and society.

In particular, a patent is granted only if it has a description of the invention sufficiently clear and complete for a person in the art to reproduce it. Moreover, if the case arises, any description that may prove insufficient or inaccurate results de facto in the cancellation of the protection granted.

It is also important to be noted that, according to the Marrakesh Agreement of April 15, 1994 on Trade-Related Aspects of Intellectual Property Rights (TRIPS Agreement), all member countries of the World Trade Organization (WTO) must grant patents for any inventions of products or processes meeting the conditions of patentability.

However, for the particular case of plants and animals, each country can choose its protection system, if it is effective, as provided for in Article 27.3.b of the TRIPS Agreement. Thus, Member States may exclude from patentability "plants and animals other than micro-organisms, and essentially biological processes for the production of plants or animals [...] However, Members shall provide for the protection of plant varieties by patents or by an effective *sui generis* system or by any combination thereof". This TRIPS article shows the difficulties encountered by legislators to grant intellectual property rights for living organisms. A first attempt was made in the Papal States by an edict of 1833 granting a monopoly of 5 to 15 years to a person who had discovered or introduced a new important type of agricultural plant. That edict is generally considered as the ancestor of the protection of new plant varieties. It was followed almost a century later by the Plant Patent Act in the USA; (see below).

Here we examine how it is possible to protect plant varieties and biotechnological innovations as well as the consequences of the protection on access to genetic resources for breeding and on the conditions under which the varieties can be multiplied by a farmer for its own use (production called "farm-saved seed"). As intellectual property rights are national or regional, and the exercise of the right is territorial and depending on judicial precedents, a global overview is not possible. We will examine in detail the case of Europe and United States of America and some references will be made to the situation in other countries.

6.2 Protection of Plant Innovation in Europe

6.2.1 Protection of Plant Varieties

Europe does not allow the grant of a patent to a plant variety (see *infra* the tomato, broccoli et pepper cases). The only way to protect a plant variety is the Plant Breeders Right (PBR) as defined by the International Union for the Protection of New Varieties of Plants (UPOV) convention. UPOV is an independent intergovernmental organization having legal personality with 76 members on June 2020 (https://www.upov.int). The granting of PBR is regulated by the Council Regulation (EC) N0 2100/94 of 27 July 1994 on Community plant variety rights.

A plant variety protection certificate (PVP) may be obtained if the variety is:

- new, that is to say if the variety constituents or harvested material of the variety have not been sold or otherwise disposed of to others by or with the consent of the breeder. In addition the variety must be,
- distinct of any other variety of common knowledge,
- subject to the variation that may be expected from the particular features of its propagation, sufficiently uniform in the expression of characteristics used for the variety description,
- stable, *i.e.* remaining unchanged after repeated propagation.

Moreover, the variety must be designated by a denomination.

It must be noted that the protection of the harvested material of the protected variety (and possibly the protection of the products obtained directly from the harvested material) applies only if that harvested material was obtained through the unauthorized use of the variety constituents of the protected variety and unless the holder of the PVP certificate has had reasonable opportunity to exercise his right in relation to the said variety constituents. However the mode of implementation of that part of the rights has not yet been adopted.

For most of the species the duration of the protection is 25 years after the grant of the right. The scope of protection is as follows: in respect of varieties constituents or harvested material of the variety the acts listed hereinafter shall require the authorization of the rights 'holder: production or reproduction (multiplication), conditioning for the purpose of propagation, offering of sale, selling or other marketing, exporting from the Community, importing to the Community, stocking for any of the previous purposes, The holder may make his authorization subject to conditions and limitations.

The EC regulation provides exceptions to the breeders' right. Besides the classic exceptions for public order the Community plant variety rights do not extend to acts done privately and for non-commercial purposes, acts done for experimental purposes, acts done for the purpose of breeding or discovering other varieties, those other varieties being free of right expect if they are essentially derived from the initial protected variety (*c.f. infra*). This last exception is known as "the breeder's privilege" allowing access to plant genetic resources for further research and breeding.

In addition, in line with an optional exception of the 1991 act of the UPOV convention the EC Regulation explicitly authorizes, for certain species listed in an annex, the use of "farm saved seed", known as "the farmer privilege". That use is precisely regulated and a farmer using that privilege must pay a certain level of royalties, sensibly lower than the regular royalties on certified seed, to the variety owner. Small farmers, *i.e.* producing less than 92 tons of cereal, are exempted. This provision of the UPOV 1991 act introducing an optional exception for farm saved seed is a compromise between the positions of UPOV members who did not wish to allow it at all, those who wanted that exception but with full royalty rate and those who wanted the authorization without limitations. The collection of royalties on farm saved seeds used by large growers being almost impossible, in several countries a system called "end point royalties" has been put in place. A percentage of the value of the crop delivered to the collecting body is retained and then returned to the breeder. As an example, in France, a part of the money collected for financing private–public research programs.

As indicated above one of the cornerstones of the breeders' right is the breeder privilege. This privilege, without limitation, could allow "plagiarism" of a protected variety in particular by a mere identification and selection of a mutant or of a somaclonal variant within the variety or by the introduction of a specific trait of interest achievable by different ways such as repeated backcrossing or transgenesis. To avoid that risk the EC regulation, also in line with the 1991 act of the UPOV convention, extends the right of the breeder to the varieties that are essentially derived from the protected variety, if that variety is not itself an essentially derived variety. A variety shall be deemed to be essentially derived from another variety called the initial variety when:

(a) it is predominantly derived from the initial variety, or from a variety that is itself predominantly derived from the initial variety;
(b) it is distinct from the initial variety; and
(c) except for the differences which result from the act of derivation, it conforms essentially to the initial variety in the expression of the characteristics that results from the genotype or combination of genotypes of the initial variety.

In fact the main motivation for the introduction of that concept was the development of genetic engineering. Indeed, without that concept the "simple" transfer by a third party of a patented gene in a variety would have allowed the appropriation of the transformed variety, distinct from the initial one, by that third party owner of the gene. This provision allows a balance between the PVP certificate and the patenting of genes of interest (see below protection of biotechnological invention).

6.2.2 *Protection of Biotechnological Inventions*

Protection of innovations in plant breeding, based mostly on the provisions of the UPOV Convention, broadly satisfied the partners involved in the agricultural sector during the second half of the twentieth century. From the 1980s, new techniques used by breeders have emerged, namely genetic engineering and both structural and functional genomics, resulting in particular in the development of transgenic plants, the identification of genes of interest and molecular markers assisted selection. The debate on the protection scheme was launched again.

As we have just seen, the UPOV Convention was amended in 1991 and discussions on a European Directive on the legal protection of biotechnological inventions began in 1988. The debates were long and difficult and it took 10 years of work for a text to be adopted in 1998 (Directive 98/44/EC) (https://eur-lex.europa.eu/).

The first article of the directive stipulates that Member States shall protect biotechnological inventions under national patent law.

The general principles of patentability apply, namely novelty, inventive activity and industrial applicability even if the invention concerns a product consisting of or containing biological material or a process by means of which biological material is produced, processed or used. Any non-confidential publication of research results, both in writing or orally, destroys the novelty and prevents patenting an innovation.

The main provisions relating to plant breeding are as follows:

- Plant varieties are not patentable, but the inventions which concern plants may be patented if the technical feasibility of the invention is not limited to a single variety. This is a complexity that is not always easy to understand. In Europe, transgenic plants carrying a patented event (transgenic trait) fall within the scope of the patent since this element is not limited to a single variety and at the same time a transgenic variety can also be individually protected by a PVP certificate.
- Essentially biological processes for the production of plants and animals which exclusively use natural phenomena such as crossing or selection are not patentable.
- Inventions relating to a product consisting wholly or partly of biological material or to a process by means of which biological material is processed or used. Any material containing genetic information and capable of reproducing itself or being reproduced in a biological system is regarded as biological material. It is for example a DNA fraction, a gene, a cell are patentable.
- The term of protection is 20 years from the date of filing of the application. Discussions took place in Europe for the implementation of supplementary protection certificates (SPC) for biotechnological inventions, due to the length of examinations for obtaining the authorizations of putting a product on the market as for pharmaceutical and plant protection products patents. However these discussions were unsuccessful.
- The protection conferred by a patent on a biological material possessing specific characteristics as the result of the invention shall extend to (apply) to any biological material derived from that biological material through propagation or multiplication and possessing those same properties. It is this provision which allows for real protection of biotechnological invention, noting that in the case of a self-replicating biological material the right exhaustion does not apply at the first sale.

- A patented biotechnological invention incorporated into a variety remains protected in this variety but by no means the variety itself is patented, which would be contrary to EU legislation prohibiting patenting of varieties. Thus the genome of this variety, when it no longer contains the patented biotechnological invention, is completely free of patent rights.

The latter provision was the subject of much debate during the adoption of the Directive, but it was necessary to give a meaning to the protection. Otherwise a patented characteristic introduced in a variety would have lost the benefit of the protection by enabling the creation of a new independent variety containing it. It is important to note that the protection of the patented characteristic is only valid if the genetic information related to the patent performs its function in the variety or the product of the variety.

A judgment on 6 July 2010 by the Grand Chamber of the European Court of Justice clarifies the point well and helps lift the burden of the risk of infringement that were running developing countries exporting to Europe agricultural products from unprotected GM varieties in the country of production. The judgment states that soybean meal imported into Europe, produced in Argentina from GM soybean tolerant to an herbicide and unprotected in this country, does not infringe a patent on soybean in Europe because the gene for tolerance to that herbicide does not exercise its function in the meal.

Authorization of the holder's right is required for making, using, offering for sale, selling or importing for this purposes the product covered by the patent. The scope of the right is substantially the same as that of a PVP certificate though a little less broad because it does not include export.

The rights do not extend to acts done privately and for non-commercial purposes or acts done for experimental purposes relating to the subject matter of the patented invention. In the case of patents of biotechnological invention two exceptions to the rights have been added as follows.

(a) The sale or other form of commercialization of plant propagating material to a farmer by the holder of the patent or with his consent for agricultural use implies authorization for the farmer to use the product of his harvest for propagation or multiplication by him on his own farm, the extent and conditions of this derogation corresponding to those under Article 14 of Regulation (EC) No 2100/94 of 27 July 1994 establishing an EC plant variety right. This is the "farmer's privilege", which applies strictly identical to transgenic and conventional varieties.

(b) The European unitary patent entered formally into force in January 2013, that will apply in all EU countries except Spain and Croatia which acceded to the EU after the unitary patents adoption. It is currently expected to start by the end of 2021, that long delay being due to the lengthy discussions on the creation of the European Unified Patent Court. That European unitary patent has also adopted the exception included in the French and German transpositions of the directive 98/44/EC stipulating that the rights do not extend to acts to breed or discover other varieties, these other varieties being free of rights if they do not express the characteristic of the patented invention. In fact it is the "breeder privilege" that exists in the case of the PVP certificate.

This is in line with the position of the International Seed Federation (www.worlds eed.org/isf/seed_statistics.html) that reads "breeding with a commercialized plant variety comprising a patented gene or trait and non-patented genetic background, should not be considered an infringement of the respective patent on the gene or trait under the following conditions: If a new plant variety, resulting from that breeding, is outside the scope of the patent claims, it should be freely exploitable by its developer provided it is not an essentially derived variety (EDV). However, if the newly developed variety still falls under the scope of patent claims (*i.e.* if the patented gene express itself in the new variety, editor's note), no commercial acts (as defined in article 14(1) of the UPOV 1991 Act) should be undertaken with the new variety without prior consent of the patent holder.

It thus appears that, contrary to what is often said, in the field of plant varieties in most of the EU Member States, the scope of the law in case of patent or PVP certificate is very similar, with even a slightly greater extent in the case of PVP, as it also covers export.

6.2.3 *The Saga of the Broccoli, Tomato and Pepper Cases*

These three cases show the complexity of patenting in Europe and the long time needed to have conclusions.

In 2002 a patent was granted to Plant Bioscience for a method for the production of *Brassica oleacera* (broccoli in this case) with elevated levels of certain glucosinolates comprising crossing wild *Brassica oleacera* species with *Brassica oleacera* breeding lines and selecting hybrids with levels of those glucosinolates elevated above that initially found in *Brassica oleracea* lines and the products thereof. In 2003 a patent was granted to the State of Israel for a similar approach regarding tomatoes with reduced fruit water content permitting the dehydration of the fruit without microbial spoilage. It is interesting to note that the European Patent Office (EPO)'s examiners did not raise the question of article 53(b) of the European Patent convention in particular regarding the exclusion of patentability of essentially biological processes. The two patents were opposed by Syngenta and Limagrain for the broccoli and by Unilever for the tomato on their ground, among others, of exclusion from patentability according to article 53(b). Regarding the broccoli Plant Bioscience modified the claims, in particular in adding that molecular markers where used in the phase of selection of the hybrids. After various steps the cases were considered by the Enlarged Board of Appeal (EBA) of the EPO in 2010. The answers given by the EBA, known as decision G1/O8 are as follows:

1. A non-microbiological process for the production of plants which contains or consists of the steps of sexually crossing the whole genomes of plants and of

subsequently selecting plants is in principle excluded from patentability as being "essentially biological" within the meaning of Article 53(b) EPC.

2. Such a process does not escape the exclusion of Article 53(b) EPC merely because it contains, as a further step or as part of any of the steps of crossing and selection, a step of a technical nature which serves to enable or assist the performance of the steps of sexually crossing the whole genomes of plants or of subsequently selecting plants.

The exclusion from patentability of essentially biological processes was confirmed even if they contain a step of technical nature. However the EBA did not decide on the patentability of the products obtained by the not patentable processes at that stage.

In 2015 the EBA, decisions G 2/12 (tomatoes II) et G 3/13(broccoli II) stated that the patentability exclusion of essentially biological processes for the production of plants in article 53(b) EPC does not have a negative effect on the patentability of products obtained by such processes.

The European Parliament considered that that decision could encourage more patents on natural traits in new plant varieties and asked the European Commission (EC) to review the matter. In 2017 the EC amended the article 53(b) of the European Patent convention in adding *"Under Article 53(b) European patents shall not be granted in respect of plants or animals exclusively obtained by means of essentially biological process"*. After some conflicts with the EPO finally, on May 14 2020, the EBA published its opinion G 3/19(Pepper) regarding a patent granted to Syngenta claiming New Peppers Plants and Fruits with improved nutritional value filed in 2012. That opinion said that Article 53(b) was to be interpreted to exclude from patentability plants, plant material or animals if the claimed product is exclusively obtained by means of an essentially biological process. However this new interpretation of article 53(b) has no retroactive effect on European patents containing such claims granted before 1st July 2017, or on pending application seeking protection for such claims which were filed before that date.

To conclude that broccoli, tomato end pepper saga which shows how complex are those highly debated issues, the claims granted for plants obtained by claims of products filed after 1st July 2017 will not be no longer granted.

6.2.4 The Balance Between the PVP and the Patent

We have seen that in the case of PVP, the introduction of the concept of essentially derived variety (EDV) establishes a balance between the rights of the holder of PVP certificate and the holder of a "gene patent". The developer of a transgenic (GM) crop from a protected variety can obtain a PVP certificate for the new variety but this new variety cannot be exploited without the consent of the holder of the PVP certificate of the initial variety who has the right to subject his authorization to conditions and limitations.

The European Directive on the protection of biotechnological inventions has also, to balance the rights between patents and PVP, introduced two provisions on cross licensing, compliant with article 31.1 of the TRIPS agreement:

a. Where a breeder cannot acquire or exploit a plant variety right without infringing a prior patent, he may apply for a compulsory license for non-exclusive use of the invention protected by the patent inasmuch as the license is necessary for the exploitation of the plant variety to be protected, subject to payment of an appropriate royalty. EU Member States shall provide that, where such a license is granted, the holder of the patent will be entitled to a cross-license on reasonable terms to use the protected variety.

b. Where the holder of a patent concerning a biotechnological invention cannot exploit it without infringing a prior plant variety right, he may apply for a compulsory license for non-exclusive use of the plant variety protected by that right, subject to payment of an appropriate royalty. Member States shall provide that, where such a license is granted, the holder of the variety right will be entitled to a cross-license on reasonable terms to use the protected invention.

c. Applicants for the licenses referred to in paragraphs 1 and 2 must demonstrate that:

   – they have applied unsuccessfully to the holder of the patent or of the plant variety right to obtain a contractual license;
   – the plant variety or the invention constitutes significant technical progress of considerable economic interest compared with the invention claimed in the patent or the protected plant variety.

Although not strictly parallel without that we understand well the reason, these two provisions allow, in principle, a balance between the two rights. Their implementation is however not obvious and will require the courts' decisions that should define what are a *"significant technical progress"* and a *"significant economic interest"*.

6.3 The Protection of Plant Innovation in the USA

Contrary to most of the countries in the world it is possible to patent plant varieties in the US. Three main milestones have gradually developed the right to protect living organisms by patent.

The first milestone is the vote of the "Plant Patent Act" in 1930, allowing the patenting of asexually reproduced plants (except tuber crops).

The second determining milestone that confirmed the patentability of biological matter is the Supreme Court decision in 1980 confirming the patentability of a microorganism, namely a bacterium genetically modified to degrade hydrocarbons. In this decision (Diamond versus Chakrabarty) the Court said that a patent may be obtained on "anything under the sun that is made by man" and that the patentee has produced a new bacterium with markedly different characteristics from any found in nature, and one having the potential for significant utility. His discovery is not nature's handiwork, but his own; accordingly it is patentable subject matter.

This development has paved the way for the patenting of biotechnological inventions as "Utility" patents without need to pass a new legislation as it has been the case in Europe(see *supra*). However, further to the grant of patents to genetic sequences without function indication and the ensuing debate, the USPTO (United States Patent and Trademark Office) published in 2001 new examination guidelines particularly relevant for gene related technology (https://www.uspto.gov/). Under these new Guidelines, the claimed invention must have "specific, substantial, and credible" utility. This is in line with the provisions of the European directive on the Protection of biotechnological inventions and actually is one of the very foundations of the patent. On June 13, 2013, in the case No. 12–398 (Association for Molecular Pathology versus Myriad Genetics) the Supreme Court has invalided patents covering genetic sequences found in nature as 'not made by human'.

The last step regarding patentability of living organisms was made in 1986 with the grant of a patent for a sexually reproduced variety (a high tryptophan corn) based on the decision of the Board of Appeals and Interference of the USPTO. This possibility to patent sexually reproduced variety was confirmed in 2001 by the Supreme Court in the J.E.M. Ag Supply Inc. versus Pioneer Hi-Bred International case.

However a question was still pending: is a patent an efficient protection for living material capable of self reproduction? The problem is well presented by Janis and Kesan (2002). Indeed, according to the patent exhaustion principle also known as the "implied license" the purchaser of a patented product is allowed to use and resell it. When a patented seed grows and produces new seed, is the new seed a new "making" of the patented seed, and hence outside the implied license. Or is it an aspect of the original "using", and hence within the scope of the implied license? In May 2013 the Supreme Court decides on that question in its decision Vernon Hugh Bowman versus Monsanto Company and decided unanimously that the new seed produced by a patented seed was a new making, thus outside the implied license. "Were the matter otherwise [...] patent would provide scant benefit. [...]. The grower could multiply his initial purchase, and multiply that new creation, ad infinitum, each time profiting from the patented seed without compensating the inventor. [...]. The undiluted patent monopoly, it might be said, would extend not for 20 years (as the Parliament act promises), but for only one transaction. And that would result in less incentive for innovation than the congress wanted." In fact that decision is in line with the European Directive which tackled that obvious issue at its very inception.

Parallel to the development of patents, the USA adopted in 1970 the "Plant Variety Protection Act" (PVPA) which is a system of UPOV-type protection for sexually propagated crops. However, the United States become member of UPOV only in 1981 after the 1978 revision of the Convention with the introduction of Article 37 providing for an exemption from the prohibition of plant variety protection in both forms of PVP certificate or patent, waiver made for them. In 1999 the USA ratified the 1991 Act of the UPOV Convention based on its 1997 PVPA act subsequently amended in 2005. The subject matter and the scope of the protection are quite similar to those of the European Regulation detailed above with two significant differences:

- Initially, in terms of subject matter, the US PVPA covered only sexually reproduced or tuber propagated plant varieties (other than fungi or bacteria) and not all genera and species as stipulated in the 1991 of the UPOV Convention and in the EU Regulation. Breeders of asexually reproduced varieties, and in particular horticultural breeders, where not happy with that situation preventing them to benefit from the concept of essentially derived varieties introduced in the 1991 Act of the UPOV Convention. In 2018 the Farm Bill opened the US PVPA to asexually reproduced varieties, solving that problem.
- In terms of scope of protection the US PVP certificate does not extend to farm saved seeds provided that the saved seeds are used on the farm where they have been produced, and this without time limitation. In fact it is a broad farmer privilege, without "reasonable limits and subject to the safeguarding the legitimate interest of the breeder" as required par the UPOV convention. When discussing with the agricultural community in the USA it is rather ironic to hear that farmers would not accept a limitation of farm saved seed for a PVP certificate when there is no possibility of farm saved seed at all, now confirmed by the Supreme Court, in case of patent.

In conclusion, in the USA, the plant breeder has the following options to protect his/her innovations:

(a)   for biotechnological inventions the patent ("Utility Patent"), as in Europe, but with no specific exemptions for plant breeding ("breeder's privilege") or for farm-saved seed.
(b)   for plant varieties different solutions are available:

- for asexually propagated varieties the plant patent act with a scope of protection similar to the one of utility patent.
- for sexually propagated and tuber propagated varieties either the PVPA, with the breeders exception and a broad farm-saved seed exception or the utility patent, with a limited research exception and no possibility for farm-saved seed.

Given the obvious imbalance of the scope of variety protection between PVP and patent, it is not surprising that in the USA breeders are massively applying for patent protection for their varieties. However, as in Europe, the development of transgenic varieties has no particular effect on the use of farm-saved seed and access to genetic diversity, the situation being the same for most the conventional varieties that are patented and transgenic varieties. Indeed, in both cases, farm-saved seed and access to transgenic varieties for further research and breeding are not allowed.

6.4 An Overview of the Situation in Some Other Countries

Utility patent is not allowed for plant varieties in most of the countries. Having said that, it must be noted, that the level of protection of intellectual property depends on the technical, legal and socio-economic conditions of each country. As already indicated, intellectual property rights are national and their implementation is territorial. Protection depends on international treaties, their transposition that is not always entirely consistent in national laws, and on jurisprudence.

In addition, there are in many countries the opportunities to use legal mechanisms other than those provided by the legislation on the protection of intellectual property to protect the breeder's rights, such as the laws on contracts between seller and buyer. For example, the "shrink-wrap agreements", that is to say the tacit agreement on the conditions of use stated on the package when open, are increasingly used in many countries. It is not possible within the scope of this book to review all countries and in particular to analyze the situation in detail for each country. Only a few cases will therefore be presented briefly.

6.4.1 The Least Advanced Countries

In general, the issue of protection of intellectual property arises less acutely in the least advanced countries, which do not have the administrative and legal structures to deal with this issue. In addition, the system of *Humanitarian Use Licenses* should enable the use by farmers producing for home consumption (subsistence) of widely patented technologies globally. Two examples illustrate this possibility further in the context of collaborative public/private sector.

The case of golden rice: it is a transgenic rice enriched with iron and provitamin A which should improve the diet of hundreds of millions of people in Southeast Asia. This rice, originally developed by Professor Ingo Potrykus at the University of Zurich, requires implementation of 70 patents and confidentiality agreements. In order to improve the product, a research project between the University of Zurich, the International Rice Research Institute and Syngenta was initiated.

Companies Bayer, Mogen, Monsanto, Novartis and Zeneca, and a Japanese company wishing to remain anonymous, gave free licenses necessary to launch the project. To enable small farmers in developing countries to benefit from the results in case of success, the company Syngenta has committed itself not to ask for royalties on seeds to farmers with an annual turnover of less than 10,000 US dollars, that is all subsistence farmers. It is interesting to note that the golden rice has been authorized for human consumption in the Philippines in December 2019.

The WEMA, Water Efficient Maize for Africa project, managed by the African Agricultural Technology Foundation (AATF) funded by the Bill & Melinda Gates Foundation, Howard G. Buffet, and by CIMMYT, Monsanto and the agricultural research systems of countries of Eastern and Southern Africa is a second example.

The objective is to develop drought tolerant maize varieties using conventional breeding, marker-assisted breeding and transgenesis. These varieties with patented transgenes will be distributed to African seed companies without request of royalty's payment.

The situation of other countries is highly variable. It is not possible to present here a detailed situation especially as the protection of intellectual property depends on national laws, which vary widely from one country to another. We take a few examples of countries that have great importance in world agricultural production by answering two questions.

6.4.2 Some Other Countries

6.4.2.1 Argentina

Argentina is a member of UPOV since 1994 (Act of 1978) and the WTO since 1995.

(a) Can farmers use farm-saved seed? Yes, without limitation with respect to the non-transgenic varieties under the law of protection of plant varieties; however, to obtain payment of a royalty on farm-saved seed for their new varieties, breeders use the contracts law according to a system known under the name of "Extended Royalties System (ERS)." For transgenic varieties the situation is not clear, but the seed law allows the use of farm seeds of any protected variety, conventional or transgenic as patenting of varieties is not allowed in the country. We need to wait until some jurisprudence to have a definitive answer.

(b) Does the breeder's privilege exist? Yes for non-transgenic varieties. Once again the answer is not clear for transgenic varieties, depending on whether it refers to the seed law or to the patent law. The answer is positive in the first case, negative in the second one.

A new seed law clarifying the situation of farm saved seeds based on the principles of the 1991 UPOV Convention has been under discussion for many years but, until now, this law never passed the parliament.

6.4.2.2 Brazil

Brazil is a member of UPOV since 1999 (Act of 1978) and the WTO since 1995.

(a) Can farmers use farm-saved seed? Yes, at no charge for conventional varieties, with the exception of sugar cane cuttings. For transgenic varieties, farmers can use farm-saved seed but must pay a fee for the patented technology, fee that is retained at the end point delivery.

(b) Does the breeder's privilege exist? Yes for conventional varieties; in addition the concept of "essentially derived variety" applies. It would be possible to use a transgenic variety in a breeding program, but the new variety could be freely marketed only if the transgenic trait is not expressed or if expressed, at the expiration of the patent. Here again we have to wait for law cases to have a definitive answer.

6.4.2.3 China

China is a member of UPOV since 1999 (Act of 1978) and the WTO since 2001. It is difficult to have a clear view of the situation due to the lack of wide dissemination of laws, the importance of Regulations at the provincial level, the rather general lack of enforcement of intellectual property, although situation improves significantly.

(a) Can farmers use farm-saved seed? Yes for conventional varieties without payment of royalties. However, some provincial governments provide subsidies to encourage farmers to buy commercial quality seeds for major species such as cereals and oilseeds. Regarding transgenic varieties, mainly cotton, it seems that the use of farm-saved seed is significantly tolerated.

(b) Does the breeder's privilege exist? Yes for all varieties. Transgenic varieties, after marketing authorization, must obtain a PVP certificate to be protected as plant varieties are not patentable as such.

6.4.2.4 India

India is a member of the WTO since 1995. It is not a member of UPOV despite a membership application, its PVP law having some striking similarities with the UPOV Convention but some others being too far away from some basic principles of UPOV.

(a) Can farmers use farm-saved seed? Yes with no limit for all varieties, whether or not transgenic. They can even exchange, share and sell farm-saved seed as long as the sale does not take place under the brand name of the variety protected by the plant protection law. However, the transgenic varieties are also subject to other laws, especially on the release into the environment. Farmers must respect those other laws, which may limit their ability to use farm-saved seed of such varieties.

(b) Does the breeder's privilege exist? Yes, in principle, regardless of the type of variety. As for farm seeds, in case of transgenic varieties, other laws and restrictions may apply.

6.5 Conclusions

The research in plant biotechnology and the development of a new variety are risky and expensive. The protection of plant breeders is therefore a necessity. Legal instruments for the protection of plant innovations have been implemented through successive conventions of UPOV, the TRIPS Agreement and various regional and national legislations. While maintaining effective protection of biotechnological inventions and new varieties, we must ensure that these instruments allow enriching the genetic diversity available to breeders, ensure access to transgenic varieties for research and creation of new varieties and stimulate public and/or private critical research for future generations.

Reference

Janis MD, Kesan KP (2002) Intellectual property protection for plant innovation: unsolved issues after J.E.M. v Pioneer. Nat Biotech 20:1161–1164

Bernard Le Buanec is a member and honorary secretary of the section "plant productions" of the French Academy of Agriculture, founding member of the French Academy of Technologies. In addition to his professional activities he has been member of: the French High Council for Research and Technology; the French committee for the protection of plant varieties (CPOV); the scientific committees of INRA; the working group of the World Bank on biotechnologies and intellectual property and of the Plant Genetic Resource Committee of the CGIAR Centers. In 2007 he was awarded the UPOV gold medal and, in 2008, the medal of the Agricultural Marketing Service from the United States Department of Agriculture in recognition of his work for the international seed industry.

Agnès Ricroch (Ph.D., HDR) is associate professor in evolutionary genetics and plant breeding at Agro Paris Tech (Paris, France) and adjunct professor at Pennsylvania State University (USA). She was visiting professor at the University of Melbourne (Australia). She has strong experience in benefits-risk assessment of green biotechnology at Paris-Saclay University. She is the Secretary of the Life Sciences section at the Academy of Agriculture of France since 2016. She was awarded the Limagrain Foundation Prize of the Academy of Agriculture of France in 2012. She is a Knight of the National Order of the Legion of Honour of France in 2019.

Chapter 7
Environmental Impacts of Genetically Modified (GM) Crop Use: Impacts on Pesticide Use and Carbon Emissions

Graham Brookes

Abstract This paper estimates some of the key environmental impacts associated with using crop biotechnology (specifically genetically modified crops) in global agriculture. It focuses on the environmental impacts associated with changes in pesticide use and greenhouse gas emissions arising from the use of GM crops since their first widespread commercial use in 1996. The adoption of GM insect resistant and herbicide tolerant technology has reduced pesticide spraying by 775.4 million kg (8.3%) and, as a result, decreased the environmental impact associated with herbicide and insecticide use on these crops [as measured by the indicator, the Environmental Impact Quotient (EIQ)] by 18.5%. The technology has also facilitated important cuts in fuel use and tillage changes, resulting in a significant reduction in the release of greenhouse gas emissions from the GM cropping area. In 2018, this was equivalent to removing 15.27 million cars from the roads.

Keywords GMO · Pesticide · Active ingredient · Environmental impact quotient · Carbon sequestration · Biotech crops · No tillage

7.1 Introduction

GM crop technology has been widely used for more than 20 years in a number of countries and is mainly found in the four crops of canola, maize, cotton and soybean. In 2018, crops containing this type of technology accounted for 48% of the global plantings of these four crops. In addition, small areas of GM sugar beet (adopted in the USA and Canada since 2008), papaya (in the USA since 1999 and China since 2008), alfalfa (in the US initially in 2005–2007 and then from 2011), squash (in the USA since 2004), apples (in the USA since 2016), potatoes (in the USA since 2015) and brinjal (in Bangladesh since 2015) have been planted.

The main traits so far commercialised convey:

G. Brookes (✉)
PG Economics, Dorchester, UK
e-mail: graham.brookes@btinternet.com
URL: https://www.pgeconomics.co.uk

© The Author(s), under exclusive license to Springer Nature Switzerland AG 2021
A. Ricroch et al. (eds.), *Plant Biotechnology*,
https://doi.org/10.1007/978-3-030-68345-0_7

- Tolerance to specific herbicides (notably to glyphosate and to glufosinate and since 2016 tolerance to additional active ingredients like 2,4-D and dicamba) in maize, cotton, canola (spring oilseed rape), soybean, sugar beet and alfalfa. This GM Herbicide Tolerant (GM HT) technology allows for the 'over the top' spraying of GM HT crops with these specific broad-spectrum herbicides, that target both grass and broad-leaved weeds but do not harm the crop itself;
- Protection against/resistance to specific insect pests of maize, cotton, soybeans and brinjal. This GM insect protected/resistance (GM IR), or 'Bt' technology offers farmers protection/resistance in the plants to major pests such as stem and stalk borers, earworms, cutworms and rootworm in maize, bollworm/budworm in cotton, caterpillars in soybeans and the fruit and shoot borer in brinjal. Instead of applying a broad-spectrum insecticide for pest control, an insecticide, far more specific (usually to an insect level) and recognised as safe for humans and other animals is delivered via the plant itself through 'Bt' gene expression.

In addition, the GM papaya and squash referred to above are resistant to important viruses (e.g., ringspot in papaya), the GM apples are non-browning and the GM potatoes (planted in 2016) have low asparagine (low acrylamide which is a potential carcinogen) and reduced bruising.

This paper presents an assessment of some of the key environmental impacts associated with the global adoption of these GM traits. The environmental impact analysis focuses on:

- *Changes in the amount of insecticides and herbicides applied to the GM crops relative to conventionally grown alternatives and*;
- *The contribution of GM crops towards reducing global Greenhouse Gas (GHG) emissions.*

It is widely accepted that increases in atmospheric levels of greenhouse gases such as carbon dioxide, methane and nitrous oxide are detrimental to the global environment (see for example, Intergovernmental Panel on Climate Change 2006). Therefore, if the adoption of crop biotechnology contributes to a reduction in the level of greenhouse gas emissions from agriculture, this represents a positive development for the world.

7.2 Environmental Impacts of Insecticide and Herbicide Use Changes

Assessment of the impact of GM crops on insecticide and herbicide use utilises two measures: the amount of herbicide or insecticide active ingredient used and Cornell University's Environmental Impact Quotient (EIQ) indicator (Kovach et al. 1992). This integrates various environmental impacts of individual pesticides into a single 'field value per hectare' and therefore can be readily used to make comparisons between different production systems across many regions and countries and provides an improved assessment of the impact of GM crops on the environment when compared to only examining changes in volume of active ingredient applied. This is because it draws on some of the key toxicity and environmental exposure data related to individual products, as applicable to impacts on farm workers, consumers and ecology.

7.2.1 GM HT Crops

A key impact of GM HT (largely tolerant to glyphosate) technology use has been a change in the profile of herbicides typically used. In general, a fairly broad range of, mostly selective (grass weed and broad-leaved weed) herbicides has been replaced by one or two broad-spectrum herbicides (mostly glyphosate) used in conjunction with a small number of other (complementary) herbicides (e.g., 2,4-D). This has resulted in:

- Aggregate reductions in both the volume of herbicides used (in terms of weight of active ingredient applied) and the associated field EIQ values when compared to usage on conventional (non-GM) crops in some countries (e.g., herbicide use on soybeans in Canada), indicating net improvements to the environment;
- In other countries (e.g., herbicide use in soybeans in Brazil), the average amount of herbicide active ingredient applied to GM HT crops represents a net increase relative to usage on the conventional crop alternative. However, even though the amount of active ingredient use has increased, in terms of the associated environmental impact, as measured by the EIQ indicator, the environmental profile of the GM HT crop has commonly been better than its conventional equivalent;
- Where GM HT crops (tolerant to glyphosate) have been widely grown, incidences of weed resistance to glyphosate have occurred and have become a major problem in some regions (see www.weedscience.org). This can be attributed to how glyphosate was originally used with GM HT crops, where because of its highly effective, broad-spectrum post-emergence activity, it was often used as the sole method of weed control. This approach to weed control put tremendous selection pressure on weeds and as a result contributed to the evolution of weed populations dominated by resistant individuals. As a result, over the last 15 years, growers of GM HT crops have been (and are increasingly being) advised to use other herbicides (with different and complementary modes of action) in combination with glyphosate and in some cases adopting cultural practices (e.g., revert to ploughing) in more integrated weed management systems (Vencil et al. 2012; Norsworthy et al. 2012). Also, in the last 2–3 years, GM HT crops tolerant to additional herbicides (typically providing multiple tolerances in a crop) such as 2,4-D, dicamba and glufosinate have become available. At the macro level, these changes have influenced the mix, total amount, cost and overall profile of herbicides applied to GM HT crops. This means that compared to the early 2000s, the amount and number of herbicide active ingredient used with GM HT crops in

most regions has increased, and the associated environmental profile, as measured by the EIQ indicator, deteriorated. Nevertheless, the amount of herbicide used on conventional crops has also increased over the same time period and that compared to the conventional alternative, the environmental profile of GM HT crop use has continued to represent an improvement compared to the conventional alternative (as measured by the EIQ indicator, see for example, Brookes and Barfoot 2018). It should also be noted that many of the herbicides used in conventional production systems had significant resistance issues themselves in the mid 1990s and this was one of the reasons why glyphosate tolerant soybean technology was rapidly adopted, as glyphosate provided good control of these weeds.

7.2.1.1 GM HT Soybean

The environmental impact of herbicide use change associated with GM HT soybean adoption between 1996 and 2018 is summarised in Table 7.1. Overall, there has been a small net increase in the amount of herbicide active ingredient used (+0.1%), which equates to 5 million kg more active ingredient applied to these crops than would otherwise have occurred if a conventional crop had been planted. However, the environmental impact, as measured by the EIQ indicator, improved by 12.9% due to the increased usage of more environmentally benign herbicides.

Table 7.1 GM HT soybean: summary of active ingredient usage and associated EIQ changes 1996–2018

Country	Change in active ingredient use (million kg)	% change in amount of active ingredient used	% change in EIQ indicator
Romania (to 2006 only)	−0.02	−2.1	−10.5
Argentina	+9.88	+0.9	−9.2
Brazil	+24.2	+1.7	−7.2
US	−33.3	−2.6	−20.2
Canada	−4.56	−8.8	−24.1
Paraguay	+6.80	+6.5	−8.4
Uruguay	+0.76	+2.0	−8.3
South Africa	−1.00	−9.1	−25.1
Mexico	−0.002	−0.8	−3.7
Bolivia	+2.3	+6.4	7.2
Aggregate impact: all countries	**+5.0**	**+0.1**	**−12.9**

Notes
1. *Negative sign* reduction in usage or EIQ improvement. *Positive sign* increase in usage or worse EIQ value

At the country level, some user countries recorded both a net reduction in the use of herbicide active ingredient and an improvement in the associated environmental impact, as measured by the EIQ indicator. Others, such as Brazil, Bolivia, Paraguayand Uruguayhave seen net increases in the amount of herbicide active ingredient applied, though the overall environmental impact, as measured by the EIQ indicator has improved. The largest environmental gains have tended to be in developed countries where the usage of herbicides has traditionally been highest and where there has been a significant movement away from the use of several selective herbicides to one broad spectrum herbicide initially, and in the last few years, plus complementary herbicides, with different modes of action, targeted at weeds that are difficult to control with glyphosate.

In 2018, the amount of herbicide active ingredient applied to the global GM HT soybean crop increased by 6.8 million kg (+2.4%) relative to the amount reasonably expected if this crop area had been planted to conventional cultivars. This highlights the point above relating to recent increases in herbicide use with GM HT crops to take account of weed resistance issues. However, despite these increases in the volume of active ingredient used, in EIQ terms, the environmental impact of the 2018 GMHT soybean crop continued to represent an improvement relative to the conventional alternative (a 10.6% improvement).

7.2.1.2　GM HT Maize

The adoption of GM HT maize has resulted in a significant reduction in the volume of herbicide active ingredient usage (-242 million kg of active ingredient) and an improvement in the associated environmental impact, as measured by the EIQ indicator, between 1996 and 2018 (Table 7.2).

In 2018, the reduction in herbicide usage relative to the amount reasonably expected if this crop area had been planted to conventional cultivars was 1.8 million kg of active ingredient (-0.9%), with a larger environmental improvement, as measured by the EIQ indicator of 8.4%. As with GM HT soybeans, the greatest environmental gains have been in developed countries (e.g., the US and Canada), where the usage of herbicides has traditionally been highest.

7.2.1.3　GM HT Cotton

The use of GM HT cotton delivered a net reduction in herbicide active ingredient use of 39.5 million kg over the 1996–2018 period (Table 7.3). This represents an 9.6% reduction in usage, and, in terms of the EIQ indicator, a 12.2% net environmental improvement. In 2018, the use of GM HT cotton technology cotton resulted in a 3.8 million kg reduction in herbicide active ingredient use (-14.5%) relative to the amount reasonably expected if this crop area had been planted to conventional cotton. In terms of the EIQ indicator, this represents a 17.7% environmental improvement.

Table 7.2 GM HT maize: summary of active ingredient usage and associated EIQ changes 1996–2018

Country	Change in active ingredient use (million kg)	% change in amount of active ingredient used	% change in EIQ indicator
US	−228.4	−9.5	−13.2
Canada	−6.4	−9.7	−17.8
Argentina	+5.8	+3.0	−4.7
South Africa	−1.9	−1.6	−7.4
Brazil	−8.1	+1.7	−9.1
Uruguay	+0.08	+2.5	−7.2
Vietnam	−0.03	−0.1	−1.3
Philippines	−3.0	−17.7	−36.0
Colombia	−0.3	−13.1	−22.3
Aggregate impact: all countries	**−242.3**	**−7.3**	**−12.1**

Notes
1. *Negative sign* reduction in usage or EIQ improvement. *Positive sign* increase in usage or worse EIQ value
2. Paraguay not included due to lack of data

Table 7.3 GM HT cotton summary of active ingredient usage and associated EIQ changes 1996–2018

Country	Change in active ingredient use (million kg)	% change in amount of active ingredient used	% change in EIQ indicator
US	−28.1	−7.8	−10.0
South Africa	+0.01	+0.6	−9.00
Australia	−5.8	−19.7	−25.8
Argentina	−5.6	−23.7	−28.5
Colombia	−0.04	−5.4	−4.7
Aggregate impact: all countries	**−39.5**	**−9.6**	**−12.2**

Notes
1. *Negative sign* reduction in usage or EIQ improvement. *Positive sign* increase in usage or worse EIQ value
2. Other countries using GM HT cotton—Brazil and Mexico, not included due to lack of data

7.2.1.4   Other HT Crops

GM HT canola (tolerant to glyphosate or glufosinate) has been grown in Canada, the US, and more recently Australia. GM HT sugar beet is grown in the US and Canada. The environmental impacts associated with changes in herbicide usage on these crops in the period 1996–2018 are summarised in Table 7.6. GM HT canola use has resulted in a significant reduction in the amount of herbicide active ingredient used relative to the amount reasonably expected if this crop area had been planted to conventional canola. Its use has also resulted in a net environmental improvement of 31.4%, as measured by the EIQ indicator.

In respect of GM HT sugar beet, the adoption of GM HT technology has resulted in a change in herbicide usage away from several applications of selective herbicides to fewer applications of, typically, a single herbicide (glyphosate). Over the period 2008–2018, the widespread use of GM HT technology in the US and Canadian sugar beet crops has resulted in a net reduction in the total volume of herbicides applied to the sugar beet crop relative to the amount reasonably expected if this crop area had been planted to conventional sugar beet (Table 7.4). The net impact on the environment, as measured by the EIQ indicator has been a 19% reduction in the EIQ value.

In 2018, the use of GM HT canola resulted in a 6.0 million kg reduction in the amount of herbicide active ingredient use (−42%) relative to the amount reasonably expected if this crop area had been planted to conventional canola. More significantly, there was an improvement in associated environmental impact, as measured by the EIQ indicator of 42.5%. The use of GM HT technology resulted in a decrease 65,600 kgof herbicide active ingredient being applied to the sugar beet crops in the US and

Table 7.4 Other GM HT crops summary of active ingredient usage and associated EIQ changes 1996–2018

Country	Change in active ingredient use (million kg)	% change in amount of active ingredient used	% change in EIQ indicator
GM HT canola			
US	−3.3	−28.8	−40.6
Canada	−34.3	−25.2	−35.1
Australia	−1.5	−4.7	−4.2
Aggregate impact: all countries	**−39.1**	**−21.7**	**−31.4**
2GM HT sugar beet			
US and Canada	−1.1	−8.0	−19.0

Notes

1. *Negative sign* reduction in usage or EIQ improvement. *Positive sign* increase in usage or worse EIQ value

2. In Australia, one of the most popular type of production has been canola tolerant to the triazine group of herbicides (tolerance derived from non-GM techniques). It is relative to this form of canola that the main farm income benefits of GM HT (to glyphosate) canola has occurred

3. InVigor' hybrid vigour canola (tolerant to the herbicide glufosinate) is higher yielding than conventional or other GM HT canola and derives this additional vigour from GM techniques

4. GM HT alfalfa is also grown in the US. The changes in herbicide use and associated environmental impacts from use of this technology is not included due to a lack of available data on herbicide use in alfalfa

Canada (-5%) relative to the amount reasonably expected if this crop area had been planted to conventional sugar beet. This also resulted in a net improvement in the associated environmental impact (-5%) as measured by the EIQ indicator.

7.2.2 GM IR Crops

The main way in which these technologies have impacted on the environment has been through reduced insecticide use between 1996 and 2018 (Tables 7.5 and 7.6) with the GM IR technology effectively replacing insecticides used to control important crop pests. This is particularly evident in respect of cotton, which traditionally has been a crop on which intensive treatment regimes of insecticides were commonplace to control bollworm/budworm pests. In maize, the insecticide use savings have been more limited because the pests that the various technology targets tend to be less widespread in maize than budworm/bollworm pests are in cotton. In addition, insecticides were widely considered to have limited effectiveness against some pests in maize crops (e.g., stalk borers) because the pests occur where sprays are not effective (e.g., inside stalks). As a result of these factors, the proportion of the maize crop

Table 7.5 GM IR maize: summary of active ingredient usage and associated EIQ changes 1996–2018

Country	Change in active ingredient use (million kg)	% change in amount of active ingredient used	% change in EIQ indicator
US	−81.6	−53.8	−55.4
Canada	−0.83	−88.7	−62.6
Spain	−0.68	−36.5	−20.7
South Africa	−2.3	−73.3	−73.2
Brazil	−26.6	−92.0	−92.0
Colombia	−0.28	−65.6	−65.2
Vietnam	−0.04	−4.6	−4.6
Aggregate impact: all countries	**−112.4**	**−59.7**	**−63.0**

Notes

1. *Negative sign* reduction in usage or EIQ improvement. *Positive sign* increase in usage or worse EIQ value

2. Other countries using GM IR maize—Argentina, Uruguay, Paraguay, Honduras and the Philippines, not included due to lack of data and/or little or no history of using insecticides to control these pests

3. % change in active ingredient usage and field EIQ values relates to insecticides typically used to target lepidopteran pests (and rootworm in the US and Canada) only. Some of these active ingredients are, however, sometimes used to control to other pests that the GM IR technology does not target

Table 7.6 GM IR cotton: summary of active ingredient usage and associated EIQ changes 1996–2018

Country	Change in active ingredient use (million kg)	% change in amount of active ingredient used	% change in EIQ indicator
US	−28.8	−25.9	−19.6
China	−139.0	−30.9	−30.5
Australia	−19.8	−33.9	−35.3
India	−137.2	−30.4	−38.9
Mexico	−2.7	−13.9	−13.8
Argentina	−1.6	−24.2	−34.0
Brazil	−1.7	−12.7	−17.4
Colombia	−0.2	−24.9	−27.4
Aggregate impact: all countries	**−331.0**	**−32.2**	**−34.2**

Notes
1. *Negative sign* reduction in usage or EIQ improvement. *Positive sign* increase in usage or worse EIQ value
2. Other countries using GM IR cotton—Burkina Faso, Paraguay, Pakistan and Myanmar not included due to lack of data
3. % change in active ingredient usage and field EIQ values relates to all insecticides (as bollworm/budworm pests are the main category of cotton pests worldwide). Some of these active ingredients are, however, sometimes used to control to other pests that that the GM IR technology does not target

in most GM IR user countries that typically received insecticide treatments before the availability of GM IR technology was much lower than the share of the cotton crops receiving insecticide treatments (e.g., in the US, no more than 10% of the maize crop typically received insecticide treatments targeted at stalk boring pests and about 30–40% of the crop annually received treatments for rootworm).

The global insecticide savings from using GM IR maize and cotton in 2018 were 8.3 million kg (−82% of insecticides typically targeted at maize stalk boring and rootworm pests) and 20.9 million kg (−55% of all insecticides used on cotton) respectively of active ingredient use relative to the amounts reasonably expected if these crop areas had been planted to conventional maize and cotton. In EIQ indictor terms, the respective environmental improvements in 2018 were 88% associated with insecticide use targeted at maize stalk boring and rootworm pests and 59% associated with cotton insecticides. Cumulatively since 1996, the gains have been a 112.4 million kg reduction in maize insecticide active ingredient use and a 331 million kg reduction in cotton insecticide active ingredient use (Tables 7.5 and 7.6).

In 2018, IR soybeans were in their sixth year of commercial use in South America (mostly Brazil). During this period (2013–2018), the insecticide use (active ingredient) saving relative to the amount reasonably expected if this crop area had been

Table 7.7 GM IR soybeans: summary of active ingredient usage and associated EIQ changes 2013–2018

Country	Change in active ingredient use (million kg)	% change in amount of active ingredient used	% change in EIQ indicator
Brazil	13.20	13.7	13.8
Argentina	1.04	1.5	0.8
Paraguay	0.54	5.6	2.2
Uruguay	0.14	3.1	1.6
Aggregate impact: all countries	**−14.92**	**−8.2**	**−8.6**

1. *Negative sign* reduction in usage or EIQ improvement. *Positive sign* increase in usage or worse EIQ value
2. % change in active ingredient usage and field EIQ values relates to insecticides typically used to target lepidopteran pests of soybeans. Some of these active ingredients are, however, sometimes used to control to other pests that the GM IR technology does not target

planted to conventional soybeans was 14.9 million kg (8.2% of total soybean insecticide use), with an associated environmental benefit, as measured by the EIQ indicator saving of 8.6% (Table 7.7).

7.2.3 Aggregated (Global Level) Impacts

At the global level, GM technology has contributed to a significant reduction in the negative environmental impact associated with insecticide and herbicide use on the areas devoted to GM crops. Since 1996, the use of pesticides on the GM crop area has fallen by 775.4 million kg of active ingredient (an 8.3% reduction) relative to the amount reasonably expected if this crop area had been planted to conventional crops. The environmental impact associated with herbicide and insecticide use on these crops, as measured by the EIQ indicator, improved by 18.5%. In 2018, the environmental benefit was equal to a reduction of 51.7 million kg of pesticide active ingredient use (−8.6%), with the environmental impact associated with insecticide and herbicide use on these crops, as measured by the EIQ indicator, improving by 19%.

At the country level, US farms have seen the largest environmental benefits, with a 404 million kg reduction in pesticide active ingredient use (52% of the total). This is not surprising given that US farmers were first to make widespread use of GM crop technology, and for several years, the GM adoption levels in all four US crops have been in excess of 80%, and insecticide/herbicide use has, in the past been, the primary method of weed and pest control. Important environmental benefits have also occurred in Chinaand Indiafrom the adoption of GM IR cotton, with a reduction in insecticide active ingredient use of over 276 million kg (1996–2018).

7.3 Greenhouse Gas Emission Savings

Assessment of the impact of GM crop use on greenhouse gas emissions combines reviews on evidence of how GM crop usage has impacted on fuel use and tillage systems. GM crops contribute to a reduction in fuel use from less frequent herbicide or insecticide applications and a reduction in the energy use in soil cultivation. The application of GM HT crops has also facilitated a shift from a plough-based production system to a reduced tillage (RT) or no tillage (NT) production system (CTIC 2002). No-till farming means that the ground is not ploughed at all, while reduced tillage means that the ground is disturbed less than it would be with traditional tillage systems. This shift away from a plough-based, to a RT/NT production system has resulted in a reduction in fuel use. Secondly, the use of RT/NT farming systems increases the amount of organic carbon in the form of crop residue that is stored or sequestered in the soil and therefore reduces carbon dioxide emissions to the environment (Intergovernmental Panel on Climate Change 2006).

7.3.1 Reduced Fuel Use

The fuel savings associated with making fewer spray runs in GM IR crops of maize and cotton (relative to conventional crops) and the switch from conventional tillage (CT) to RT or NT farming systems facilitated by GM HT crops, have resulted in permanent savings in carbon dioxide emissions. In 2018, this amounted to a saving of 2456 million kg of carbon dioxide, arising from reduced fuel use of 920 million litres (Table 7.8). These savings are equivalent to taking 1.63 million cars off the road for one year.

The largest fuel use-related reductions in carbon dioxide emissions have come from the adoption of GM HT technology in soybeans and how it has facilitated a switch to RT/NT production systems with their reduced soil cultivation practices (78% of total savings 1996–2018). These savings have been greatest in South America.

Over the period 1996–2018, the cumulative permanent reduction in fuel use has been about 34,172 million kg of carbon dioxide, arising from reduced fuel use of 12,799 million litres. In terms of car equivalents, this is equal to taking 22.65 million cars off the road for a year.

7.3.2 Additional Soil Carbon Storage/Sequestration

As indicated earlier, the widespread adoption and maintenance of RT/NT production systems in North and South America, facilitated by GM HT crops (especially in soybeans) has improved growers' ability to control competing weeds, reducing the

Table 7.8 Carbon storage/sequestration from reduced fuel use with GM crops 2018

Crop/trait/country	Fuel saving (million litres)	Permanent carbon dioxide savings arising from reduced fuel use (million kg of carbon dioxide)	Permanent fuel savings: as average family car equivalents removed from the road for a year ('000s)
HT soybeans			
Argentina	236	629	417
Brazil	193	516	342
Bolivia, Paraguay,Uruguay	63	169	112
US	39	105	69
Canada	20	55	36
HT maize			
US	144	384	254
Canada	8	21	14
HT canola			
Canada: GM HT canola	81	216	143
IR maize			
Brazil	35	94	62
US/Canada/Spain/SouthAfrica	4	11	7
IR cotton—global	20	52	35
IR soybeans—South America	77	205	136
Total	*920*	*2456*	*1627*

Notes

1. Assumption: an average family car in 2018 produces 123.4 grams of carbon dioxide per km. A car does an average of 12,231 km/year and therefore produces 1509 kg of carbon dioxide/year
2. GM IR cotton. India, Pakistan, Myanmar and China excluded because insecticides assumed to be applied by hand, using back pack sprayers

need to rely on soil cultivation and seed-bed preparation as means to getting good levels of weed control. As a result, as well as tractor fuel use for tillage being reduced, soil quality has been enhanced and levels of soil erosion cut. In turn, more carbon remains in the soil and this leads to lower GHG emissions.

Based on savings arising from the rapid adoption of RT/NT farming systems in North and South America, we estimate that an extra 5606 million kg of soil carbon has been sequestered in 2018 (equivalent to 20,581 million kg of carbon dioxide that has not been released into the global atmosphere). These savings are equivalent to taking 13.6 million cars off the road for one year (Table 7.9).

The additional amount of soil carbon sequestered since 1996 has been equivalent to 302,364 million kg of carbon dioxide that has not been released into the global atmosphere. Cumulatively, the amount of carbon sequestered may be higher than this estimate due to year-on-year benefits to soil quality (e.g., less soil erosion, greater

Table 7.9 Context of carbon sequestration impact 2018: car equivalents

Crop/trait/country	Additional carbon stored in soil (million kg of carbon)	Potential additional soil carbon sequestration savings (million kg of carbon dioxide)	Soil carbon sequestration savings: as average family car equivalents removed from the road for a year ('000s)
HT soybeans			
Argentina	1737	6377	4225
Brazil	1425	5232	3466
Bolivia, Paraguay,Uruguay	468	1718	1138
US	126	463	307
Canada	78	287	190
HT maize			
US	160	5359	3550
Canada	16	59	39
HT canola			
Canada: GM HT canola	296	1088	721
IR maize			
Brazil	0	0	0
US/Canada/Spain/SouthAfrica	0	0	0
IR cotton—global	0	0	0
IR soybeans—South America	0	0	0
Total	*5606*	*20,581*	*13,636*

water retention and reduced levels of nutrient run off). However, it is equally possible that the total cumulative soil sequestration gains have been lower because only a proportion of the crop area will have remained in NT/RT. It is, nevertheless, not possible to confidently estimate cumulative soil sequestration gains that take into account reversions to conventional tillage because of a lack of data. Consequently, the estimate provided of 302,364 million kg of carbon dioxide not released into the atmosphere should be treated with caution.

Aggregating the carbon sequestration benefits from reduced fuel use and additional soil carbon storage, the total carbon dioxide savings in 2018 are equal to about 23,027 million kg, equivalent to taking 15.27 million cars off the road for a year. This is equal to 48% of registered cars in the UK.

7.4 Conclusions

GM crop technology has been used by many farmers around the world for more than twenty years and currently nearly 17 million farmers a year plant seeds containing this technology. This seed technology has helped farmers be more efficient with their application of crop protection products, which not only reduces their environmental impact, but saves time and money. The technology is also changing agriculture's carbon footprint, helping farmers adopt more sustainable practices such as reduced tillage, which has decreased the burning of fossil fuels and allowed more carbon to be retained in the soil. This has led to a decrease in carbon emissions. In relation to GM HT crops, however, over reliance on the use of glyphosate by farmers, in some regions, has contributed to the development of weed resistance. As a result, farmers have, over the last 15 years, adopted imore integrated weed management strategies incorporating a mix of herbicides and non-herbicide-based weed control practices. This means that the magnitude of the original environmental gains associated with changes in herbicide use with GM HT crops have diminished. Despite this, the adoption of GM HT crop technology in 2018 continues to deliver a net environmental gain relative to the conventional alternative and, together with GM IR technology, continues to provide substantial net environmental benefits. These findings are also consistent with analysis by other authors (Klumper and Qaim 2014; Fernando-Cornejo et al. 2014).

References

American Soybean Association Conservation Tillage Study (2001). Available on the Worldwide Web at https://soygrowers.com/asa-study-confirms-environmental-benefits-of-biotech-soy beans/

Brookes G, Barfoot P (2018) Environmental impacts of genetically modified (GM) crop use 1996–2016: impacts on pesticide use and carbon emissions. GM Crops & Food 9(3):109–139. https://doi.org/10.1080/21645698.2018.1476792

Conservation Tillage and Plant Biotechnology (CTIC) (2002) How new technologies can improve the environment by reducing the need to plough. Available on the World Wide Web: https://www.ctic.purdue.edu/CTIC/Biotech.html

Fernandez-Cornejo J, Wechsler S, Livingston M, Mitchell L (2014) Genetically engineered crops in the United States. USDA Economic Research Service report ERR 162. www.ers.usda.gov

Intergovernmental Panel on Climate Change (2006) Chapter 2: Generic methodologies applicable to multiple land-use categories. guidelines for national greenhouse gas inventories, vol 4. Agriculture, Forestry and Other Land Use. Available on the World Wide Web: https://www.ipccnggip.iges.or.jp/public/2006gl/pdf/4_Volume4/V4_02_Ch2_Generic.pdf

Klumper W, Qaim M (2014) A meta analysis of the impacts of genetically modified crops. PLOS One. https://doi.org/10.1371/journal.pone.0111629

Kovach J, Petzoldt C, Degni J, Tette J (1992) A method to measure the environmental impact of pesticides. New York's Food Life Sci Bull. NYS Agriculture Experiment Station Cornell University, Geneva, NY, 139:8 pp and annually updated. Available on the Worldwide Web: https://www.nysipm.cornell.edu/publications/EIQ.html

Norsworthy J, Ward S, Shaw D, Llewellyn R, Nichols R, Webster T et al (2012) Reducing the risks of herbicide resistance: best management practices and recommendations. Weed Sci 60(SP1):31–62. https://doi.org/10.1614/WS-D-11-00155.1
Vencill W, Nichols R, Webster T, Soteres J, Mallory-Smith C, Burgos N et al (2012) Herbicide resistance: toward an understanding of resistance development and the impact of herbicide-resistant crops. Weed Sci 60(SP1):2–30. https://doi.org/10.1614/WS-D-11-00206.1

Graham Brookes is an agricultural economist and consultant with more than 30 years' experience of examining economic issues relating to the agricultural and food sectors. He is a specialist in analyzing the impact of technology, policy changes and regulatory impact. He has, since the late 1990s, undertaken a number of research projects relating to the impact of agricultural biotechnology and written widely on this subject in peer reviewed journals. This work includes frequent updates of a global economic and environmental impact of GM crops report plus individual GM crop impact studies in Spain Romania and Vietnam. He has examined the impact of GMO-related regulations on trade and economies in the EU, the UK and Turkey and undertaken studies of the potential impact of using crop biotechnology in the Ukraine Russia Thailand and Indonesia.

Chapter 8
Is It Possible to Overcome the GMO Controversy? Some Elements for a Philosophical Perspective

Marcel Kuntz

Abstract The main belief systems that express themselves over GMOs are summarized. The existence of these different modes of thought (termed modernism, postmodernism, environmentalism and religious views) partially explains the reason why it has not been possible to overcome the public controversy despite the accumulation of scientific data. In addition, the divergent views on GMOs often reflect more general value judgments on the free market economy and on the integration of agriculture and food production in a globalized economy. In this context, it has proven difficult for most people to distinguish genuine scientific controversies from political ones.

Keywords Modernism · Postmodernism · Environmentalism · Political dispute · Scientific controversy

8.1 A Dispute that is Not Just a Scientific Controversy

As indicated by the rapid adoption rate of GMOs in countries where their cultivation is permitted, the current transgenic varieties appear to match farmers' needs. In addition, agricultural biotechnology holds numerous promises but there are still gaps between farmers need and biotechnological research (Ricroch et al. 2015). A significant number of scientific publications are now available related to the debated potential risks linked to the commercial use of transgenic varieties (see for example Petrick et al. 2019). Nevertheless, the use of GMOs is still fiercely opposed by certain organizations. Obviously, views on agricultural biotechnology do not converge towards a consensus in the media, or internet etc., despite the accumulation of scientific knowledge. This suggests that this controversy is not primarily a scientific one. Controversies in biology do not normally last more than 15 years. For example, the Monarch butterfly controversy ignited by a 1999 scientific paper suggesting that Bt corn pollen can harm this butterfly was largely extinguished by a series of six papers published in 2001 (see Minorsky 2001). An alarmist article on the consumption of

M. Kuntz (✉)
CNRS, Grenoble, France
e-mail: marcel.kuntz@cea.fr

A. Ricroch et al. (eds.), *Plant Biotechnology*,
https://doi.org/10.1007/978-3-030-68345-0_8

a transgenic maize published in 2012 triggered a worldwide controversy, but was retracted from the journal, and finally refuted by scientific studies financed by public subsidies published in 2018 and 2019 (see Kuntz 2019).

In contrast to the normal evolution of scientific research (new data will often open new questions), the public controversy appears highly entrenched. To understand the real nature of the dispute, it is important to examine the various belief systems that express themselves over GMOs.

8.2   An Overview of Various Modes of Thought

8.2.1   The 'Modern' Thought

This term is used here for the school of thought inherited from the Enlightment. This rational view of the world has actually been built over millennia (starting with Ancient Greek philosophers) to access objective reality. It is the traditional rational basis of scientific activities. Often, scientists will support an evidence-based judgment and a case-by-case assessment of GMOs (and other technologies) and consider that risk will be reduced, and even appropriately managed, by increasing scientific knowledge. It should be emphasized that the 'modernist' attitude has profoundly changed since the philosophical views termed 'scientism' or 'positivism' during the end of nineteenth and early twentieth centuries; nowadays very few 'modernists' still believe that science and technology will necessarily lead to social improvement. Rather, they generally consider that without science and technology, social (and environmental) progress would be impossible in the context of a growing human population and climate change.

8.2.2   The 'Environmentalist' Thought

The dominant 'environmentalist' views (this term is used here in a philosophical sense) are that human technologies are now so powerful that they can cause not only local damage but potentially the destruction of the planet Earth. 'Environmentalism' has growing support since the 1970s because of the awareness of the impact of human activities on the environment, the arising of a different attitude towards nature and a distrust of artificial processes and products by many consumers. The view that GMOs are 'unnatural' has had a profoundly negative impact on their acceptance and ignores the fact that no conventional crop variety is actually natural, but rather has been subject to an artificial (human) selection process (many crop species, notably corn, rice, wheat, would not have existed without human intervention). Despite the fact that the 'unnatural' argument against GMOs is not unanimously shared amongst environmentalists, it remains important for many and for consumers (we tend to consider as 'natural' something we are familiar with, although, strictly speaking, it is often artificial).

To 'save the planet', environmentalism has adopted a strategy which is often denounced as 'fear mongering' by their detractors. The philosopher Jonas (1984) provided the theoretical background for this strategy in his 'Heuristics of Fear' (where fear is considered to be a better motivator than positive incitement).

8.2.3   The 'Postmodern' Thought

'Postmodern' philosophy attempts to deconstruct the foundation of the Western (modern) philosophy and its tendency to promote universal values. Within this postmodern movement, the 'science studies' school of thought (see Barnes et al. 1996) claims that scientific truth is merely a 'cultural construction' of truth by a scientific community bound together by allegiance to a shared paradigm. This social science movement also criticizes the scientific method and its universality. It exerts a strong influence on academic thought in the Western world, despite being often criticized as representing a form of relativism.

Postmodern sociologists consider that public distrust of some technologies is not due to a lack of knowledge (the 'deficit model') but to the fact that the public was not involved in discussion on the technology and decision-making (they advocate an 'upstream public engagement model'). Thus, to deal with 'controversies' related to technologies (GMOs, nanotechnology, synthetic biology), postmodernists recommend 'citizen participation' in science and 'coproduction' of scientific programs (often with opponents). They are also critical of the scientific risk assessment of these technologies and its separation from the socio-political world. A criticism of this approach is that there is currently no convincing evidence that controversies over technologies, especially over GMOs (Kuntz 2012a), have actually been appeased by following these recommendations. Nevertheless, public policies are often embedded in a postmodern ('participative') doctrine. The ideological shift towards postmodernism in scientific institutions is illustrated by a report on 'gene drive', published by the National Academies of Sciences, Engineering, and Medicine of the USA, and which recommended *"Aligning Research with Public Values"*, in contrast with the 'modern' view which considers that it is rather the public who should rely on the judgments of experts (see Kuntz 2016).

8.2.4   Religious Views on GMOs

There is no clear consensus view on GMOs among Christian, Jewish or Islamic religious leaders.

A 2009 study on GMOs sponsored by the Pontifical Academy of Sciences concluded favorably on the technology, viewing it to be praiseworthy for improving the living conditions of the poor (see Coghlan 2010). However, the Popes' positions were more ambivalent. In 2000, Pope John Paul II stated that "application of biotechnology [...] cannot be evaluated solely on the basis of immediate economic interests". Pope Benedict XVI's position has been diversely interpreted, and the position of Pope Francis (2015) does not seem fundamentally different when saying that "*it is difficult to make a general judgement about genetic modification (GM), whether vegetable or animal, medical or agricultural, since these vary greatly among themselves and call for specific considerations. The risks involved are not always due to the techniques used, but rather to their improper or excessive application*". However, Pope Francis also wrote in the same encyclical that "*the scientific and experimental method in itself is already a technique of possession, mastery, and transformation*", which was criticized as "*postmodern*" and "*a pessimistic post-humanist Western sentiment rather than the older, confident humanism*" (Reno 2015).

8.3   Why is There No Consensus on GMOs?

At first glance, not all the views summarized above seem to be incompatible. It is difficult to envisage a GMO that is more "socially equitable" than Golden Rice. However, this humanitarian rice is targeted by radical opponents in the same way as Monsanto seeds are. Academic and governmental research projects on GMOs have been subject to around 80 acts of destruction in Europe alone (Kuntz 2012b). Most of these experiments were designed to assess the safety of GMOs. It is evident that even these trials are not acceptable to those who argue that GMOs have not been tested enough for their potential effects. Therefore, it can be suggested that further improvement in the safety of GMOs and their social benefits, or addressing sensible questions such as diversification in agriculture, is unlikely to lead to a change of mind for the most determined opponents, simply because their primary motivation may well not lie there.

It is relatively easy to find statements by opponents which are of clear political nature. For example, Bruno Rebelle, a spokesman for Greenpeace in France, explained on 2nd February 2002 during an official audition by a State Council: "we are not afraid of GMOs. We are just convinced it is the wrong solution [...] GMOs may be a wonderful solution for a certain type of society project. But it is precisely this type of society we do not want" (translated from French). Thus, since the divergent views on GMOs appear to reflect more general value judgments on the free market economy and on the integration of agriculture and food production in a globalized economy, it is difficult to envisage how a consensus could actually be reached.

Therefore, it can be predicted that plant biotechnology will remain a battlefield for the divergent visions of good and evil. Industrialized countries will remain divided on the topic with poor countries having to choose one side. However, the possibility that some may change position (in either direction) cannot be ruled out if the political balance of power changes.

One may also wonder why the accumulation of sound scientific data on GMOs does not overcome political views. In fact, scientists, scientific risk assessment and even science itself are dragged into this political battle. The widespread postmodern view contributes to the idea that science may be considered as an opinion among many other opinions, which needs to be debated by 'stakeholders' with divergent agendas. Thus, it has proven difficult for most people to distinguish genuine scientific controversies from political ones.

An example of a genuine scientific question is gene flow and its consequences in terms of agronomy or biodiversity, while views on 'purity' of corn landraces for example tend to be a matter of cultural 'identity' for some Mexican farmers. In an interesting article, Bellon and Berthaud (2004) distinguished scientific questions from value judgments on this topic. However, such considerations will rarely be used in the 'debate'.

Others topics such as for example lifestyle choices among farmers or economical protection of domestic markets by some governments are quite distinct from environmental or food safety issues. However, the latter are often used to justify restrictions on GMO marketing.

Finally, one can wonder why Europe is often at the origin of restrictions regarding biotechnology, including gene editing. It can be proposed that this phenomenon has deeply-embedded roots, i.e. in the broader historical background of this continent (Kuntz 2020). More precisely, that its tragic history gave rise to a new ideology (postmodern in a general sense) that aims to avoid repetition of these tragedies (e.g. World Wars, totalitarian states, etc.). This new ideology has been transposed to science and technology, with the aim of avoiding at all costs other tragedies, namely those potentially arising from the use of technology. It is difficult to change this ideology since Europe is convinced to be of great virtue. In such a context, scientific data are of little value. Scientists should show that adopting biotechnologies is of greater virtue than restricting this technology.

References

Barnes B, Bloor D, Henry J (1996) Scientific knowledge: a sociological analysis. University of Chicago Press, Chicago

Bellon MR, Berthaud J (2004) Transgenic maize and the evolution of landrace diversity in Mexico. The importance of farmers' behavior. Plant Physiol 134:883–888

Coghlan A (2010) Vatican scientists urge support for engineered crops. New Sci https://www.new scientist.com/article/dn19787-vatican-scientists-urge-support-for-engineered-crops/

Jonas H (1984) The imperative of responsibility: in search of ethics for the technological age (Das Prinzip Verantwortung). University of Chicago Press, Chicago

Kuntz M (2012a) The postmodern assault on science. If all truths are equal, who cares what science has to say? EMBO Rep 13:885–889

Kuntz M (2012b) Destruction of public and governmental experiments of GMO in Europe. GM Crops Food 3(1–7):166

Kuntz M (2016) Scientists should oppose the drive of postmodern ideology. Trends Biotechnol 34(12):943–945

Kuntz M (2019) The Séralini affair—the dead−end of an activist science. Study for Fondapol, Sept 2019. https://www.fondapol.org/en/etudes-en/the-seralini-affair-the-dead-end-of-an-activist-science/

Kuntz M (2020) Technological risks (GMOs, gene editing), what is the problem with Europe? A broader historical perspective. Front Bioeng Biotechnols 8:557115

Minorsky PV (2001) The monarch butterfly controversy. Plant Physiol 127:709–710

Petrick JS, Bell E, Koch MS (2019) Weight of the evidence: independent research projects confirm industry conclusions on the safety of insect-protected maize MON 810. GM Crops Food 11(1):30–46

Pope Francis (2015) Encyclical letter Laudato si'. On care for our common home. https://w2.vatican.va/content/francesco/en/encyclicals/documents/papa-francesco_20150524_enciclica-laudato-si.html

Reno RR (2015) The return of catholic anti-modernism. First Things. https://www.firstthings.com/web-exclusives/2015/06/the-return-of-catholic-anti-modernism

Ricroch A, Harwood W, Svobodová Z, Sági L, Hundleby P, Badea EM, Rosca I, Cruz G, Salema Fevereiro MP, Marfà Riera V, Jansson S, Morandini P, Bojinov B, Cetiner S, Custers R, Schrader U, Jacobsen HJ, Martin-Laffon J, Boisron A, Kuntz M (2015) Challenges facing European agriculture and possible biotechnological solutions. Crit Rev Biotechnol Jul 1:1–9

Marcel Kuntz is a Director of research at Centre National de la Recherche Scientifique (CNRS) and also teaching at Université Grenoble-Alpes (France). His research as a plant biologist are conducted in the Laboratory of Plant & Cellular Physiology, a joint laboratory of CNRS, University Grenoble-Alpes, Institut National de Recherche pour l'Agriculture, l'alimentation et l'Environnement (INRAE) and Commissariat à l'Energie Atomique et aux Energies Alternatives (CEA).

Marcel Kuntz is also publishing articles in scientific journals on the political controversy over GMOs and on aspects relevant to the philosophy of science (foundations, methods, implications of science) with examples taken from this controversy.

Part III
Sustainable Management

Chapter 9
Sustainable Management of Insect-Resistant Crops

Shelby J. Fleischer, William D. Hutchison, and Steven E. Naranjo

Abstract Sustainability is a goal-oriented process that advances with new knowledge. We discuss factors relevant to insect-resistant crops and sustainability: adoption patterns, insecticide use patterns and their influence on humans, biological control, areawide effects, and evolution of populations resistant to genetically engineered (GE) crops. GE insect-resistant crops were introduced at a time when insecticide options and use patterns were changing. Management of lepidopteran and coleopteran pests has been achieved through constitutive expression of proteins derived from the crystalline spore and the vegetative stage of various strains of *Bacillus thuringiensis*. Management of aphid-transmitted viruses has been achieved through expression of viral coat proteins. Adoption patterns have been rapid where use is allowed. Areawide reductions in pest populations have occurred in cotton and maize in multiple parts of the world, enabled eradication programs, and conferred significant economic benefits to crops that are not GE. Insecticide use has decreased dramatically in cotton, leading to improved biological control, reductions in pesticide poisonings, and changes in species composition that achieve pest status. Pro-active resistance management programs, the first to be deployed in all of agriculture, has slowed but not stopped the evolution of resistant populations. Nine pest species have evolved resistance to one or more *Bt* proteins. Future constructs may provide induced or tissue-specific expression or use RNAi to deliver protection from insect pests. Constructs that alter plant metabolism, to achieve drought tolerance, nitrogen-utilization, or biomass conversion efficiency, may also affect insect populations and communities. Sustainable management of insect-resistant GE crops requires consideration of regional effects of both the genetics and densities of mobile target insect populations. The underlying assumption of IPM, that multiple and diverse management tactics are more sustainable, continues to be highly relevant, and necessary, to

S. J. Fleischer (✉)
Penn State University, State College, USA

W. D. Hutchison
University of Minnesota, Minneapolis and Saint Pauls, USA

S. E. Naranjo
USDA-ARS, Arid-Land Agricultural Research Center in Arizona, Maricopa, USA

A. Ricroch et al. (eds.), *Plant Biotechnology*,
https://doi.org/10.1007/978-3-030-68345-0_9

maintain the utility of GE crops, to manage the wider community of species relevant to agroecosystems, and to enable agriculture to adapt to change.

Keywords IPM · Areawide · *Bacillus thuringiensis* · Insecticide · Resistance

9.1 Introduction

Genetically engineered (GE) crops with resistance to insects or insect-vectored viruses have been used worldwide since 1996 (Tabashnik et al. 2013). Commercial plantings include cotton, maize, potato, soybean, eggplant, rice, papaya, cowpea, and squash; potential commercial lines also exist for broccoli and plum (e.g., ISAAA 2018; Romeis et al. 2019; Naranjo et al. 2020). Globally, the vast majority of insect-resistant GE crops, now grown in 24 countries include maize, soybean and cotton (ISAAA 2018). To date, genes have targeted above- and below-ground herbivores from two taxonomic orders of insects (Lepidoptera and Coleoptera), and successfully managed aphid-transmitted viruses. We anticipate genes targeting two more insect orders (Heteroptera and Thysanoptera) to be commercialized soon. All these examples—indeed, any change in plant phenotype—affect both the cropping system and the insect populations and communities that utilize those crops. The practice of Integrated Pest Management (IPM), rooted in the science of applied ecology and entomology, provides our context for describing effects on insect populations and communities, as well as an essential framework for maintaining sustainable agro-ecosystems. Here, we briefly summarize IPM and applied entomology concepts and existing GE crops, then discuss opportunities and challenges for their sustainable management at field, landscape, and regional scales.

9.2 Insect Resistance Traits

Insect resistance has been categorized as conferring antibiosis, antixenosis, or tolerance. Antibiosis traits directly reduce fitness of the insect, such as decreasing survivorship, prolonging development, or reducing fecundity. Plant expression may be continuous or induced (expressed in response to specific stimuli). Current commercially deployed GE crops that express proteins from *Bacillus thuringiensis* (*Bt*) express constitutive antibiosis. The concentration of these proteins, however, varies within the plant, through time as the plant develops and senesces, and across the landscape depending on adoption patterns (e.g., Hutchison et al. 2010). The interaction of the protein concentration with the degree to which it affects insect fitness is critical to both effectiveness and sustainability of insect resistant crops. Engineering crops with induced antibiosis may be deployed in the future. Induced proteins would affect the spatio-temporal dynamics of insect exposure, and thus the selective pressure for resistance, possibly lowering selective pressure if expressed only when needed.

The additional categories—antixenosis and tolerance—also affect insect populations. Antixenosis refers to phenotypic traits that affect insect behavior, and tolerance refers to traits that affect the way in which the plant allocates resources to compensate for pest attack: for example, compared to older cultivars, modern cultivars of maize may produce higher grain yields in the presence of low to moderate amounts of stem-boring by lepidopterans (=caterpillars) due to a wide range of structural and biochemical traits that compensate for damage. Both the transgene that, for example, reduces survivorship of an herbivore, and other phenotypic traits that influence insect behavior and plant resource allocation, are integrated into elite hybrids during modern plant breeding. In addition, when considering insect-resistant crops in the future, it is important to realize that traits that may not be directly targeting insects, such as drought-tolerance or nutritional content, may also affect insect populations and communities through their effects on insect behavior and fitness.

9.3 Insecticides and Their Integration into IPM

GE plants with insect resistant traits were commercialized while the types and availability of commercial insecticides were changing rapidly. Advances in insect physiology, toxicology, and formulation technology led to improved targeting and delivery of insecticidal molecules. Increased ecological and human safety is achieved, in part, through development of selective insecticides. Today's insecticides are classified into 32 chemical classes, and multiple subclasses, based on their modes-of-action, defined globally.[1]

Insecticides made from the bacterium *Bacillus thuringiensis* (*Bt*) achieve high levels of selectivity. This microbe produces biodegradable protein crystals (termed Cry proteins), with typically three components (termed domains) during sporulation; some strains also produce additional insecticidal proteins during vegetative growth (termed vegetative insecticidal proteins, or Vips). The Cry proteins separate into domain subunits in the micro-environmental conditions of the insect gut, and the subunits bind directly to protein receptors on the microvilli of the insect midgut lining. Effective binding results in pore formation and osmotic shock, which is followed by septicemia of the insect, probably involving microbes beyond the *Bt* species. Selectivity is achieved through specificity of micro-environmental conditions, and binding properties of specific Cry proteins with specific receptor proteins, all associated with the insect gut. While the degree of selectivity varies, and thus some non-target species can be affected, high degrees of selectivity are common, often at the species level. Thus, a given Cry protein may be effective on one species of caterpillar but not a related species of caterpillar. Furthermore, effectiveness often varies with the life stage of an insect. Many *Bt* materials need to be acquired by immature (larval) life stages and are viewed as larvicides. Selectivity is further achieved through the requirements needed to deliver the protein to the target site: acquisition

---

[1] www.irac-online.org.

must be through ingestion, in contrast to modes-of-action of many insecticides that can be delivered through contact.

Many insecticidal proteins derived from bacteria, including the Cry and Vip proteins, are databased and defined by their structural similarities.[2] As of 2020, there are over 700 Cry and 100 Vip proteins in the pesticidal protein database. For example, Cry1Ab, commonly used in agriculture, refers to category 1, subgroup A, and subgroup b within A. Sprayable formulations of a few *Bt* groups have been used for over 70 years in agricultural production, protection of stored grain, and mosquito control. When used as sprayable formulations, typically produced in fermentation culture, the *Bt* insecticides require precise targeting because microbes can be sensitive to solar irradiation and they require ingestion by early insect life stages. By 1987 GE plants had been created that produced Cry proteins. This enabled efficient targeting of insects through ingestion by immature insect life stages. Commercial lines were first available in 1995, and current lines are summarized in Naranjo et al. (2020).

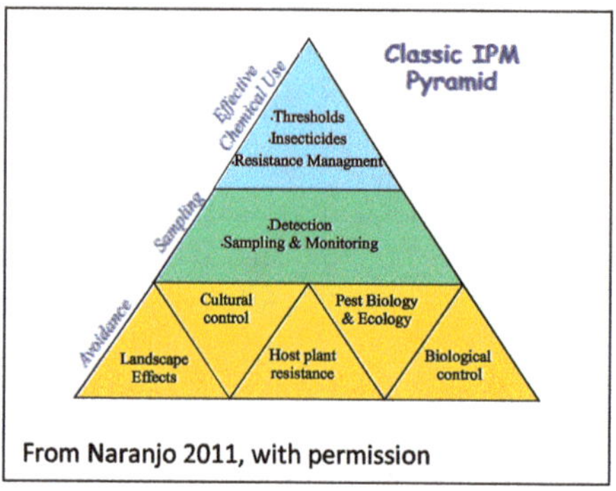

From Naranjo 2011, with permission

The integration of insecticides with other tactics for insect management, notably biological control, driven by problems of resistance and additional species achieving pest status, was a primary basis for the emergence of Integrated Pest Management (IPM) programs during the last half of the twentieth century. An underlying assumption is that multiple and diverse management tactics are more sustainable when applied as a package than any one tactic would be when relied on in isolation. A classic IPM pyramid (Fig. 1, Naranjo 2011) shows a base, designed to minimize the effect of a pest upon the crop, built from knowledge of pest biology and ecology, biological control, host plant resistance, cultural control, and landscape factors. Monitoring, decision-making, and the use of insecticides in response to economically threatening population densities that have developed despite avoidance tactics are used in IPM. Insect resistance management (IRM)—efforts to delay the evolution of resistance—is also now formally integrated into IPM programs. Management may be at the field scale, or at larger geographic scales. Areawide management programs strive to remove, reduce, or slow the geographic expansion of pest populations at

---

[2] https://www.bpprc.org/.

wide geographic scales (e.g., Hutchison 2015). One way to focus the debate about the use of *Bt*-transgenes is to ask if they represent host-plant resistance, or pro-active deployment of an insecticide. Both are true.

9.4 Emergence of Insect-Resistant Crops, Pyramids, Stacks, and Coupled Technologies

During the breeding process that leads to GE insecticidal plants, genes are isolated, connected to markers, and inserted into plants. The final construct after successful insertion is called an event, and government registrations are issued for specific events. Early constructs, and those still in use in some crops today, include a single event which codes for a single protein, such as Cry1Ab or Cry1Ac. Constructs can also be pyramided with multiple genes targeting the same pest (or a slightly overlapping group of related pest species), to broaden its activity and reduce the likelihood of resistance or stacked with other traits such as herbicide tolerance. Pyramided constructs generally have different modes-of-action targeting the same species and are replacing single gene plants. For example, the MON89034 event is a pyramided stack which codes for two Cry proteins, Cry1A.105 and Cry2Ab, targeting a group of lepidopterans. Almost always, the insecticidal trait, conferred with either a single gene or through a pyramided event, is being stacked with events that confer herbicide tolerance.

Sweet corn provides a simple example. In the U.S., cultivars with the *Bt*11 event that codes for Cry1Ab have been in use since 1996. By 2012, cultivars became available that also code for the Cry3Bb protein which confers resistance to rootworm larvae, plus CP4 which provides tolerance to glyphosate herbicide. Also, cultivars became available in 2012 that produce Cry1A.105 and Cry2Ab, which provide resistance to several additional caterpillar species through different modes-of action, and cultivars that include Vips are also commercially available (e.g., Shelton et al. 2013).

Field maize presents a much wider array of GE cultivars. In 2010, sources from 5 commercial enterprises provided 22 trait groups, some of which involve licensing agreements among several companies. All but one of these cultivars stacked insect resistance with herbicide tolerance. Insect resistance was conferred with nine different proteins in varying combinations, aimed at insect pests from two different taxonomic orders. The range of Cry proteins expressed includes several from the Cry1A group, at least one Cry1F and one Cry2, several from the Cry3 group, and cultivars that express Vips. Early cultivars provided resistance to moths in the family Crambidae. Newer cultivars add resistance to several moth species in the family Noctuidae, and/or larval stages of beetles in the family Chrysomelidae with Cry3 proteins. Cry3 proteins had been introduced earlier, in potato in the mid-1990s, to control another Chrysomelidae species (Colorado potato beetle), and later potato cultivars included traits that conferred resistance to several aphid transmitted viruses.

Stacked and pyramided constructs are found in cotton. The initially introduced events expressed a single Cry protein (Cry1Ac). Cry1Ac was then pyramided with either Cry1F or Cry2Ab2 to provide for better resistance management and to enhance the spectrum of efficacy within the lepidopteran group, and some recent cultivars express Vips. China and India have cultivated a few unique events including a Cry1Ab + Cry1Ac pyramid and a pyramid involving a fusion protein (Cry1A) combined with a cowpea trypsin inhibitor. Notable among insect-resistant crops beyond maize and cotton, China is developing insect-resistant rice with stacked and pyramided constructs (Chen et al. 2011).

In addition to constructs modeled from proteins derived from *Bt*, insect management is influenced by constructs derived from viral coat proteins (see the chapter by Rosa in this book). Expression of those coat proteins result in activation of a plant immune response, mediated by small RNA molecules, providing protection against infection by the virus of origin. This strategy has been used to achieve control of papaya ringspot virus in papaya since 1997, and one or more strains of four viruses in squash or zucchini since 1994. Aphids transmit these viruses by first acquiring them from an infected host. The virions adhere to receptor proteins in the needle-like mouthparts of the aphid. In subsequent feeding probes by the aphid, the virions are injected into a new plant. Where aphids pose a threat of pathogen transmission, tolerance of aphid populations by farmers is very low, resulting in a higher incidence of insecticide use. In contrast, when aphids do not pose a threat of pathogen transmission, tolerance of aphid feeding itself can be very high, and aphid management tends to rely primarily on biological control through natural enemies and entomopathogenic fungi. There are multiple other examples in which GE can contribute to the control of plant pathogens that are transmitted by insects.

RNA-mediated processes are also being developed to target insects that are direct pests of the plant, and mites that are pests of honeybees. These involve a different modes-of-action than achieved with proteins. They can result in high degrees of specificity and can interfere with expression of specific genes in the insect (they are termed RNAi, for RNA-interference), although their stability of expression can be a limiting factor. As of 2020, however, none of the RNAi events for insect resistance have been commercialized.

9.5 Sustainable Management of Insect-Resistant Crops

The agroecosystems in which GE crops are introduced are dynamic, and components do not operate independently. GE crops often involve coupled technologies, including stacks or pyramids of insect-resistant genes and stacks of herbicide-tolerant genes. GE as well as some non-GE cultivars are increasingly (currently almost always in the U.S.) being coupled with systemic insecticidal seed treatments, and may include seed coatings to help with mechanized planting, protect against soil-borne pathogens, or as biostimulants that aim to induce up-regulation of resistance genes. Sustainability, which we recognize as a process with inherent goals and values, is affected by all these technologies and their interactions with socioeconomic factors. Here, we illustrate factors particularly relevant to insect-resistant crops and the broad definition of sustainability: adoption patterns, insecticide use patterns and their influence on human welfare and biological control, areawide effects, and evolution of insect populations that are resistant to GE crops.

9.6 Adoption Patterns

Adoption patterns are defined overwhelmingly by social, political and economic factors (Naranjo et al. 2020). By 2018, adoption rates for soybeans, maize, and canola approached saturation in the 5 countries that have the largest acreage (ISAAA 2018). Adoption of GE papaya reached about 80% in Hawaii, and 12% of the squash in the U.S. utilized transgenes in 2005 (NAS 2016). In contrast, commercial sales of *Bt*-potato in the U.S. were halted after about six years after their initial introduction. These cultivars had resistance to Colorado potato beetle and several viruses, but processors declined to accept market risk, and growers tended to adopt systemic neonicotinoid insecticides that were introduced at the same time (NAS 2016). Neonicotinoid insecticides controlled a much broader array of insects and thus were easier to use with a much wider array of potato cultivars. More recently, GE potatoes with other traits have been moving forward. *Bt*-eggplant to control stem borers has been deployed in Bangladesh and is projected to dramatically reduce insecticide use (Shelton et al. 2018). Multiple *Bt*-crucifer crops have advanced to commercially relevant lines, within the context of international public–private partnerships, although to date none are being produced commercially (see Shelton et al. 2008 for a review of GE vegetables and fruit relevant to insect management). Market forces, political and business decisions, prohibitions, and labeling requirements are among the primary factors slowing or stopping the commercialization of insect-resistant transgenes in vegetable and fruit crops. Insect-resistance achieved with genetic engineering techniques is totally prohibited in certified organic production for any crop, although *Bt* proteins can be sprayed onto the plant. In cotton and maize, adoption rates are influenced by the interest growers place on stacked traits such as herbicide-tolerance or traits aimed at multiple insect species. Adoption is also being influenced by the availability of seed: seed without GE traits, or without stacked traits, may be hard to obtain. In the future, adoption rates may decline in response to decreasing pest populations resulting from areawide effects, discussed below.

9.7 Insecticide Use

Where insecticide inputs have been low on a per-acre basis prior to the adoption of GE cultivars, as in maize, changes in insecticide inputs are less clear, and may increase, in part due to the coupling of neonicotinoid seed treatments with GE crops.

This coupling is a common, but not an inherent property of GE technology; seed treatments are found in non-GE seeds as well. Changes in use patterns are often driven by market factors interacting with factors driving the intensification of agriculture.

GE crops have influenced insecticide use patterns both directly and indirectly through their coupling with neonicotinoid seed treatments. Insecticide quantity dramatically declined in cropping systems that were heavily dependent on insecticides prior to the introduction of GE crops, such as cotton. Reductions due to *Bt*-cotton have been profound (Naranjo 2011; Brookes and Barfoot 2020a, b). For 1996–2018, insect resistant maize, cotton and soybean have reduced global insecticide (active ingredient) use by 112, 331 and 14.9 million kilograms, respectively, with associated reductions of 63, 34 and 8.6% in the environmental impact quotient (a measure of the pesticide's impact on the environment and human health) (Brookes and Barfoot 2020a, b). Metrics based on quantity, however, do not consider changes in potency. Also, the addition of neonicotinoids as seed treatments on GE maize and soy increased the acreage receiving neonicotinoids. When both potency and acreage are considered, the potential honeybee toxic load may increase due to the increase in neonicotinoid treated GE seed (Douglas et al. 2020), resulting in a 'potency paradox', where the dose-hazard to honeybees increased while insecticide quantity, in term of active ingredient, decreased. Insecticide risk is a function of both the hazard (dose potency) and exposure, and future research is needed to include the exposure aspect of the risk equation.

Because most insecticides are petroleum-based, savings on the "carbon footprint" or greenhouse gas (GHG) emissions have recently been calculated as well. For 2018 alone, fuel savings via reduced transport and application of insecticides in maize, cotton and soybean were 105, 52, 205 million kg CO_2, respectively, which equates to the removal of ~240,000 automobiles from roadways (Brookes and Barfoot 2000b). Reductions in insecticide use have clearly led to improved biological control of several pest species (Romeis et al. 2019). In contrast, however, neonicotinoid seed treatments in maize have led to increased slug damage due to reductions in slug biocontrol. Gains to human health can be dramatic when adoption of insect-resistant genotypes reduces insecticide use. Examples are well-documented in small-holder production systems. In India, where pesticide applications were reduced by 50% in *Bt*-cotton, with larger reductions of the more toxic materials, the technology is decreasing the incidence of pesticide poisonings by several million cases per year (Kouser and Qaim 2011). Studies also document fewer pesticide poisoning events in China and South Africa. Reductions of insecticide exposure to farmworkers and insecticide poisoning are consistent with values embedded in the process of sustainability. Socio-economic studies, controlled for other factors, have also documented improved dietary quality and caloric value, along with reduced food insecurity, among smallholder households that adopted *Bt*-cotton (Qaim and Kouser 2013). In contrast to field crops, the per-acre insecticide load is highest in vegetable and fruit crops, where manual labor is much more prevalent and insecticide problems related to human safety tend to be the most dramatic. Ironically, this is where market and regulatory forces are slowing development or adoption of insect-resistant GE cultivars. However, in a recent study in five US states across multiple years, *Bt* sweet corn performed better and required fewer sprays than conventional sweet corn to meet market standards, thus reducing hazards to farm workers and the environment (Shelton et al. 2013). Unfortunately, debates about sustainability or desirability of GE crops rarely elaborate on effects on farmworkers. More recently, Shelton et al. (2020) documented the adoption patterns and farmer perceptions of *Bt* brinjal (eggplant) in Bangladesh. The technology is being adopted rapidly due to high efficacy as well the opportunity to reduce a historically high insecticide use rate for this crop; it is common in this region for farmers to apply 35–45 insecticide sprays against the primary pest, the fruit and shoot borer (*Leucinodes orbonalis*).

9.8 Areawide Effects

Females deposit eggs equally among cultivars, and if the GE cultivar reduces survivorship, then the GE cultivar acts as a population sink. The degree to which it drives down populations depends on rates of insect dispersal and adoption of the GE cultivar. For pink bollworm, a specialist herbivore, Carrière et al. (2003) showed adoption rates of ~65% *Bt* cotton would drive down regional populations of the pest. Regional reductions have also occurred with polyphagous species, including *Heliothis virescens*, and to a lesser extent *Helicoverpa zea*, in cotton in the eastern U.S. In China, GE cotton dramatically reduced *Helicoverpa armigera* populations both in the cotton crop, and in the surrounding matrix of vegetable, corn, peanut, and soybean (Wu et al. 2008). Even in the presence of complex cyclical dynamics, Hutchison et al. (2010) documented how *Bt*-maize reduced population growth rates of the highly damaging European corn borer, driving populations to historically low levels in large and multiple areas of the Midwestern U.S. Given the polyphagous nature of the corn borer, a 40-year study (data collected before and after commercialization of *Bt* corn) in 5 U.S. Atlantic coast states, confirmed a "spill-over" benefit of areawide suppression of the pest in three major vegetable crops (i.e., non-*Bt* crops: peppers, snap beans, sweet corn, Dively et al. 2018). The areawide suppression also resulted in a significant reduction in estimated insecticide use, and therefore an additional environmental benefit on these high-value food crops.

The areawide effects of *Bt* plants are influencing IPM in ways relevant to values associated with sustainability. For pink bollworm, GE cultivars enabled an organized eradication program that integrated GE cultivars with mating disruption via pheromone technologies, sterile insect release, cultural controls, and insecticides if needed. Consequently, pink bollworm was declared eradicated from the U.S. and northern Mexico in October of 2018. In China, where reductions in insecticides due to *Bt*-cotton enabled other species (mirid bugs) to emerge into pest status, the technology also significantly increased populations of beneficial arthropod predators, which reduced herbivorous (aphid) prey populations, both in the cotton crop and surrounding maize, peanut and soybean crops (Lu et al. 2012). In combination with other IPM tactics, *Bt* cotton in the western U.S. has dramatically enabled biological control of non-lepidopteran pests such as whiteflies and mirids and driven overall insecticide use down by nearly 90% (Naranjo 2011). An extensive, updated meta-analysis for 5 major *Bt* crops globally, confirmed overall positive results of the technology on beneficial predators and parasitic wasps of crop pest species, confirming a high level of compatibility between conservation biological control and GE crops (Romeis et al. 2019, also see the chapter by Naranjo in this book).

In the Midwestern U.S., economic analyses considered effects to both land planted to *Bt*-maize, and to the land planted to non-*Bt* cultivars. Cumulative benefits were $9.6 billion across 5 states, with a surprisingly high percentage (66%) accruing to non-*Bt*-maize growers, or $6.3 billion (Hutchison et al. 2010). The benefits to non-*Bt* maize growers were attributed to the substantial areawide pest suppression effect (European corn borer), yet the non-*Bt* acreage did not carry the additional expense of the *Bt* technology fee. Moreover, Brookes and Barfoot (2020a) recently updated their global analysis of cumulative economic benefits of GE crops (1996–2018), and found that the insect-resistant crops contributed ~$123.3 billion additional net revenues to farmers ($225.1 billion for all GE crops); much of this is now benefitting farmers in developing countries (16–17 million), as well as growers in industrial countries (ISAAA 2018).

Numerous environmental benefits of GE crops have been estimated, in large part due to reduced insecticide use; globally, reductions in foliar and soil-applied insecticide use in maize and cotton have been estimated at 8.3 million kg (−82%), and 20.9 million (−55%), respectively (Brookes and Barfoot 2020b). This outcome has resulted in less environmental impacts such as reduced reliance on petroleum-based insecticides, less fuel consumption necessary to transport and apply pesticides, and subsequent reductions in CO_2 emissions, alleviating impacts on climate change. Clearly, areawide effects—including eradication programs and reductions in insecticide use, increases in biocontrol, and economic savings—extend well beyond the boundaries of the planted crop.

Adoption patterns in the future could also be influenced by areawide effects. Theoretically, as populations decline, growers could shift to non-GE cultivars if they are available as elite hybrids (e.g., inherent genetic potential for high yield), thus saving the appreciable cost of *Bt*-seed; however, some question if the non-*Bt* hybrids will be available at a wide scale. Theoretically, both resistance management (discussed below) and maintenance of low populations could be achieved through spatio-temporal dynamics in adoption patterns at landscape and regional scales.

9.9　Evolution of Populations Resistant to GE Crops

Deployment of insecticides or insect-resistant germplasm has never been static. For example, to manage Hessian fly, over 60 wheat cultivars have been released with antibiosis resistance. The pest, in turn, has evolved over 16 biotypes that can overcome antibiosis, and management programs include variable spatial deployments of resistant germplasm. Insects are incredibly adaptable, and 550 species include populations with resistance to one or more insecticides. Sole reliance on antibiosis traits, regardless of the plant-breeding technology or insecticide mode-of-action, often creates a "treadmill": a race between evolution of resistance and new trait development and deployment. Models to help manage this evolutionary process were established prior to the deployment of GE crops. These models estimate time to acquire resistance, defined as an increase in the frequency of a resistant allele, as a function of life history, fitness, and population genetics. Simulations and experiments considered varying deployment options, assumptions regarding initial gene frequencies for *Bt* resistance, and how they affected the time to acquire resistance.

Insect-resistant GE crops were deployed in the U.S. only after resistance management plans were defined and accepted by the U.S. Environmental Protection Agency. Although heavily critiqued, and often lacking enforcement, to our knowledge this is the first, and only, regulatory-mandated use of resistance management plans prior to deployment of any technology in agriculture. These plans typically rely on refuges of non-*Bt* hosts, and assume that alleles conferring resistance are rare, so that very few individuals survive on the *Bt*-crop. The non-*Bt* hosts provide a population of susceptible individuals, and the plans assume the rare survivor on the *Bt*-crop will have a much higher probability of mating with a susceptible individual, resulting in individuals that are heterozygous for the resistant allele. Expression of *Bt* is typically targeted sufficiently high to kill the heterozygote offspring. This is termed the "high-dose refuge" strategy (e.g., Tabashnik et al. 2013; Tabashnik and Carrière 2017). Additional assumptions inherent to the high-dose refuge strategy include random mating and single alleles conferring resistance. Additional factors that can contribute to delayed resistance include lower fitness or competitive abilities of individuals that manage to develop on the *Bt*-crop. In practice, there have been many variations of refuge design, in terms of the percent of the crop ("structured refuge"), or non-crop alternative host ("unstructured refuge"), which serves as a source of susceptible individuals, and their spatial placement. The area required for planting to non-*Bt* maize has varied from 5 to 50%. For cotton, structured refuge has varied from 5 to 20%. Spatial placements of structured refuges have varied from nearby blocks to seed mixes termed "refuge-in-a-bag". Refuge requirements for pink bollworm were suspended as the eradication program was deployed, with the assumption that sterile male releases were providing susceptible phenotypes. In one case for *H. zea*, carbon-isotope studies documented that non-crop plants were providing susceptible individuals, leading to inclusion of "non-structured refuges" in resistant management plans under certain circumstances. Work with *Bt*-crucifer crops as a model system demonstrated that deployment of pyramided constructs prior to the deployment of single constructs delays resistance, and pyramided deployments are becoming more common. For certain cotton cultivars planted east of west Texas, where unstructured refuges contributed susceptible phenotypes and the cultivars included pyramided resistant genes, the structured refuge requirement has dropped to 0%. Stacked constructs aimed at multiple insect species require refuge designs appropriate to each of the targeted species, which can be difficult due to their differing behaviors (e.g., dispersal patterns and how that influences mating probabilities). Migratory species may experience selection pressure in a southern geographic area and bring resistant phenotypes to northern areas. Similarly, polyphagous pests may experience selection pressure in one host crop and carry resistant phenotypes to another host crop. Different life stages of the insect may have different susceptibilities to the resistant trait, which may also be expressed at variable levels within the plant or during the plant's development, all of which affect the ability to consistently achieve a dose that kills heterozygous individuals. Thus, refuge designs change as new GE cultivars become commercialized, often with considerable debate among parties with conflicting interests.

Tabashnik et al. (2013) suggested that field-evolved resistance has been delayed when the allele conferring resistance has a low initial frequency, refuges are abundant, and pyramided toxins are used. They define resistance as the "…genetically based decrease in susceptibility of a population…caused by exposure to the toxin in the field", regardless of whether there are reductions in expected levels of control, or whether the insect was a pest that was expected to be controlled. By this definition, and additional research since 2013 (Smith et al. 2017, 2019), at least 9 insect species now include populations in specific locations that are resistant to one or more *Bt* proteins in GE crops. The degree to which this has affected pest control varies among populations and species. In maize, seven species now exhibit sufficiently high resistance to a single protein in some populations to affect control. In the case of *Bt* corn targeting corn rootworm larvae, the dose is not sufficiently high to meet criteria typically assumed to be necessary to achieve the "high-dose refuge" strategy, mating may not be random, resistance appears to be caused by more than a single allele and some of these may not be rare, and frequencies of fields with unexpected damage has recently been increasing. In cotton, two species have evolved a level of resistance to result in significant field damage in specific locations of the world and to specific events. With the exception of western corn rootworm (*Diabrotica virgifera virgifera*), plants with pyramided *Bt* proteins are currently effective, although the reduced selection process that should be conferred by separate modes-of-action may be compromised when the efficacy of one of the proteins is compromised; i.e., cross-resistance among *Bt* toxins is common among several *Bt* resistant pest species. There are additional cases where the frequency of resistant alleles has increased, but not at a level that has affected pest control as of 2020. Various tactics have been implemented to manage *Bt* resistant fall armyworm (*S. frugiperda*) populations when resistance resulted in significantly reduced field control. In the first clear case of resistance resulting in field failure, one case of resistance on a Caribbean island, the GE cultivar was removed from the market in Puerto Rico. In another, for western corn rootworm, there has been increased emphasis on crop rotation, and rotation among cultivars that express different *cry* genes. There has also been increased emphasis on development and deployment of pyramided constructs and adherence to established refuge requirements.

9.10 Summary

GE crops affect population densities of pest and beneficial insect species, biological control services, insecticide use patterns, pesticide poisoning of humans, and economics. Values relevant to discussions about sustainability exist for deployment of insect-resistant genetically engineered crops, many examples document environmental and human health benefits in the first 25 years of adoption, which has been remarkably rapid where the technology has been allowed. The need for resistance management makes it clear that we are also dealing with effects on population genetics. The effects often occur at scales that transcend the land planted to the GE crop. Missing from many discussions, however, are the impacts on farmworkers, particularly for vegetable and fruit crops when less insecticide, or insecticides with lower mammalian toxicity, is used. In the future, sustainable management of insect-resistant GE crops will continue to require consideration and management of regional effects of both densities and genetics of mobile insect populations, as well as the broader socio-economic impacts.

The underlying assumption of IPM, that multiple and diverse management tactics are necessary to be more sustainable, continues to be highly relevant. Widescale adoption and over-reliance on only host plant resistance, especially when conferred via a single protein, creates exceptionally strong selection pressure, and insects have and will adapt with heritable changes in their genotypes and phenotypes. Insect resistance management (IRM), a component of IPM, is an integral part of the deployment strategy for GE cultivars. Sole reliance on a treadmill strategy with GE traits is not sustainable, and widespread coupling with insecticidal seed treatments affects insecticide use patterns. Integration of insect-resistant traits with diverse pest management methods, through IPM, enables agriculture to also adapt and evolve, for management of the species targeted by the transgene(s), but also for the wider community of pest and beneficial species in agroecosystems, and in the wider realm of changing markets, policies, and social and economic structures in which farmers operate.

Acknowledgements The authors thank T. Sappington (USDA-ARS, IA, USA) and A. Shelton (Cornell University, NY, USA) for valuable reviews of earlier versions of this manuscript.

References

Brookes G, Barfoot P (2020a) GM crop technology use 1996–2018: farm income and production impacts. GM Crops & Food 11(4):242–261. https://doi.org/10.1080/21645698.2020.1779574

Brookes G, Barfoot P (2020b) Environmental impacts of genetically modified (GM) crop use 1996–2018: impacts on pesticide use and carbon emissions. GM Crops & Food 11(4):215–241. https://doi.org/10.1080/21645698.2020.1773198

Carriere Y, Ellers-Kirk C, Sisteron M, Antilla L, Whitlow M, Dennehy TJ, Tabashnik B (2003) Long-term regional suppression of pink bollworm by *Bacillus thuringiensis* cotton. Proc Natl Acad Sci 100:1519–1523

Chen M, Shelton A, Gong-yin Y (2011) Insect-resistant genetically modified rice in China: from research to commercialization. Annu Rev Entomol 56:81–101

Dively GP, Venugopal PD, Bean D, Whalen J, Holmstrom K, Kuhar TP, Doughty HB, Patton T, Cissel B, Hutchison WD (2018) Regional pest suppression associated with widespread *Bt* maize adoption benefits vegetable growers. Proc Natl Acad Sci 115:3320–3325

Douglas MR, Sponsler DB, Lonsdorf EV, Grozinger CM (2020) County-level analysis reveals a rapidly shifting landscape of insecticide hazard to honeybees (*Apis mellifera*) on US farmland. Sci Rep 10:797. https://doi.org/10.1038/s41598-019-57225-w

Hutchison WD (2015) Integrated pest management and insect resistance management: prospects for an area-wide view. In: Soberon M, Gao A, Bravo A (eds) *Bt* resistance: characterization and Strategies for GM crops providing *Bacillus thuringiensis* toxins. CABI, Wallingford, UK, pp 186–201

Hutchison WD, Burkness EC, Mitchell PD, Moon RD, Leslie TW, Fleischer SJ, Abrahamson M, Hamilton KL, Steffey KL, Gray ME, Hellmich RL, Kaster LV, Hunt TE, Wright RJ, Pecinovsky K, Rabaey TL, Flood BR, Raun ES (2010) Areawide suppression of European corn borer with *Bt* maize reaps savings to non-*Bt* growers. Science 330:222–225

ISAAA (2018) Global Status of Commercialized Biotech/GM Crops in 2018: biotech crops continue to help meet the challenges of increased population and climate change. ISAAA Brief No. 54. Intern. Service for Acquisition of Agri-biotech Applications: Ithaca, NY. (https://www.isaaa.org)

Kouser S, Qaim M (2011) Impact of *Bt* cotton on pesticide poisoning in smallholder agriculture: a panel data analysis. Ecol Econ 70:2105–2113

Lu Y, Wu K, Jiang Y, Guo Y, Desneux N (2012) Widespread adoption of *Bt* cotton and insecticide decreases promote biocontrol services. Nature 487:362–365

Naranjo SE (2011) Impacts of *Bt* transgenic cotton on Integrated Pest Management. J Agric Food Chem 59:5842–5851

Naranjo SE, Hellmich RL, Romeis J, Shelton AM, Vélez AM (2020) The role and use of genetically engineered insect-resistant crops in integrated pest management systems. In: Kogan M, Heinrichs EA (eds) Integrated management of insect pests. Burleigh Dodds Publishing, UK.

National Academies of Sciences (2016) Genetically engineered crops: experiences and prospects. The National Academies Press, Washington, DC., 584 pp. https://doi.org/10.17226/23395.

Qaim M, Kouser S (2013) Genetically modified crops and food security. PLoS ONE 8(6):e64879. https://doi.org/10.1371/journal.pone.0064879

Romeis J, Naranjo SE, Meissle M, Shelton AM (2019) Genetically engineered crops help support conservation biological control. Biol Control 130:136–154

Shelton AM, Fuchs M, Shotkoski FA (2008) Transgenic vegetables and fruits for control of insects and insect-vectored pathogens. In: Romeis J, Shelton AM, Kennedy GG (eds) Integration of insect-resistant genetically modified crops within IPM programs. Springer Science+Business Media B. V.

Shelton AM, Olmstead DL, Burkness EC, Hutchison WD, Dively G, Welty C, Sparks AN (2013) Multi-state trials of *Bt* sweet corn varieties for control of the corn earworm. J Econ Entomol 106:2151–2159

Shelton AM, Hossain MJ, Paranjape V, Azad AK, Rahman ML, Khan ASMMR, Prodhan MZH, Rashid MA, Majumder R, Hossain MA, Hussain SS, Huesing JE, McCandless L (2018) *Bt* eggplant project in Bangladesh: history, present status, and future direction. Front Bioeng Biotechnol 6:106. https://doi.org/10.3389/fbioe.2018.00106

Shelton AM, Sarwer SH, Hossain MJ, Brookes G, Paranjape V (2020) Impact of *Bt* brinjal cultivation in the market value chain in five districts of Bangladesh. Front Bioeng Biotechnol 8:498. https://doi.org/10.3389/fbioe.2020.00498

Smith JL, Rule DM, Lepping MD, Farhan Y, Schaafsma AW (2017) Evidence for field-evolved resistance of *Striacosta albicosta* (Lepidoptera: Noctuidae) to Cry1F *Bacillus thuringiensis* protein and transgenic corn hybrids in Ontario, Canada. J Econ Entomol 110:2217–2228

Smith JL, Farhan Y, Schaafsma AW (2019) Practical resistance of *Ostrinia nubilalis* (Lepidoptera: Crambidae) to Cry1F *Bacillus thuringiensis* maize discovered in Nova Scotia, Canada. Sci Rep 9:18247. https://doi.org/10.1038/s41598-019-54263-2

Tabashnik BE, Carrière Y (2017) Surge in insect resistance to transgenic crops and prospects for sustainability. Nat Biotechnol 35:926–935

Tabashnik BE, Brevault T, Carriere Y (2013) Insect resistance to *Bt* crops: lessons from the first billion acres. Nat Biotechnol 31:510–521

Wu KM, Lu YH, Feng HQ, Jiang YY, Zhao JZ (2008) Suppression of cotton bollworm in multiple crops in China in areas with *Bt*-toxin-containing cotton. Science 321:1676–1678

Shelby J. Fleischer is a Professor of Entomology at Penn State University. His research deals with the structure, dynamics, and management of insects based on biological and ecological processes. Priority is placed on economically feasible management that improves worker and environmental safety. He has authored or co-authored over 100 publications, been an invited speaker at over 30 venues, and served as a subject editor for *Environmental Entomology*. Dr. Fleischer conducts educational programs with farmers and agricultural industry personnel, where he integrates scientific principles with problem-solving and natural history of arthropods.

William D. Hutchison is Professor and Extension Entomologist in Entomology at the University of Minnesota. He led long-term studies documenting areawide suppression of *Bt*-corn on European corn borer and resulting economic benefits in the U.S. Corn Belt. Current work includes assessment of transgenic crops on insect populations and insect resistance management for *Bt* corn. He has published over 160 journal articles, several book chapters and extension publications on sampling; integrating reduced-risk insecticides with biocontrol; invasive species; and early detection of *Bt* resistance in lepidopteran pests. He is co-editor for Radcliffe's World IPM textbook, an interactive teaching resource used worldwide, and *Integrated Pest Management: Concepts, Tactics, Strategies and Case Studies*. He served as an Editor for the *Journal of Economic Entomology* and *Crop Protection* and is currently a Chief Editor for the *Invasive Insects* section of the new journal, *Frontiers in Insect Science*. He is on the Editorial Boards of *Crop Protection, Insects*, and *GM Crops and Food*.

Steven E. Naranjo is Director of the USDA-ARS, Arid-Land Agricultural Research Center in Arizona, and adjunct faculty with University of Arizona. His research interests include biological control, IPM, sampling, population ecology, and environmental risk assessment of genetically engineered crops. He has conducted long-term studies on non-target effects of Bt cotton with emphasis on natural enemy abundance and function and led meta-analyses to quantify non-target effects of Bt crops worldwide. Dr. Naranjo served as Co-Editor-in-Chief for *Crop Protection*, and currently serves as Subject Editor for *Environmental Entomology* covering Transgenic Plants and Insects.

Chapter 10
Effects of GE Crops on Non-target Organisms

Steven E. Naranjo

Abstract Genetically engineered (GE) crops have now been part of the agricultural landscape for 25 years and are important tools in crop production and Integrated Pest Management (IPM) in 26 countries. Considerable research has addressed many associated issues including environmental and food safety, as well as economic and social impacts. Non-target effects have been a particularly intensive area of study, and extensive laboratory and field research has been conducted for transgenic Bt crops that produce the insecticidal proteins of a ubiquitous bacterium, *Bacillus thuringiensis*. This body of evidence and the quantitative and qualitative syntheses of the data through meta-analysis and other compilations generally indicate a lack of direct impacts of Bt crops on non-target macro-invertebrates. The data also clearly show that Bt crops are much safer to non-target organisms than the alternative use of traditional insecticides for control of the pests targeted by the Bt proteins. Some indirect effects on arthropod natural enemies associated with reduced abundance or quality of Bt target herbivores have been shown, but the ramifications of these effects remain unclear and would be shared by other pest control technologies. As one tactic in the IPM toolbox, Bt crops have contributed to large reductions in insecticide use. While reduced insecticide use and reduced herbivory may be involved in precipitating new pest problems in Bt crops, it also has broadened opportunities for deployment of another key IPM tactic, biological control.

Keywords Transgenic Bt crops · Risk assessment · Meta-analysis · Ecological guilds · Biological control · Integrated pest management

10.1 Introduction

Genetically engineered (GE) crops have now been part of the agricultural landscape for 25 years and their geographic scope and breadth of traits continues to advance. By 2018, nearly 192 million hectares of GE crops were cultivated in 26 countries, with 21

S. E. Naranjo (✉)
USDA-ARS, Arid-Land Agricultural Research Center, Maricopa, AZ, USA
e-mail: Steve.naranjo@ars.usda.gov

   127
A. Ricroch et al. (eds.), *Plant Biotechnology*,
https://doi.org/10.1007/978-3-030-68345-0_10

of these developing nations. The USA leads the world in adoption of GE crops with Brazil, Argentina, Canada and India among the top five. Spain and Portugal are the only European Union countries growing GE crops, and this is limited to relatively small areas (< 121 K hectares) of insect-resistant maize. Another 40 countries or so allow for the importation of GE crop products for food, animal feed and other processing uses.

The primary GE crops currently under cultivation involve those that have been engineered to either display tolerance to several broad-spectrum herbicides or selective resistance to specific insect pest groups, primarily those belonging to the Orders Lepidoptera and Coleoptera. GE cotton, maize and soybean often include varieties that offer both traits. Major GE crops include soybean, maize, cotton, and canola (oilseed rape), grown in many adopting countries, with much smaller plantings of herbicide-tolerant alfalfa and sugar beets, virus-resistant papaya and squash, insect-resistant eggplant (brinjal), sugarcane and cowpea, and quality enhanced traits in apple, potato and pineapple in a smaller number of countries including the USA, Canada, China, Bangladesh, Costa Rica, Indonesia and Nigeria.

Consistent with this 25-year adoption, there has been considerable research addressing many associated issues including environmental and food safety, economic and social impacts, and effects on crop production and protection. The potential negative effects of GE crop technology have been perhaps most visible and controversial in the area of environmental and food safety. The GE crops that have been scrutinized most in this regard are those with insect-resistance, and that will be the major focus of this chapter.

10.2   Insect-Resistant Crops

At present, all insect resistant crops are based on the production of one or more of the crystal (Cry) and vegetative (VIP) proteins of a ubiquitous gram-positive bacterium, *Bacillus thuringiensis* (Bt). These so called Bt crops comprise about 54% of all GE crops produced globally and are grown in 22 countries. The insecticidal properties of this bacterium have been known for more than 100 years and commercial products based on this organism have been available since the 1940s. Bt spray products occupy >90% of the bio-pesticide market and are an important tool for pest control in organic farming and stored grain, and for control of larval mosquitos. Presently, Bt cotton and Bt maize are the dominant forms of transgenic, insect-pest control technology globally. Bt soybean has been grown in several South American countries since 2012, Bt sugarcane was approved for production in Brazil in 2017, Bt cowpea was approved for Nigeria in 2019 and several countries are evaluating Bt rice for potential production in the future. Bt eggplant (brinjal) was initially granted approval for cultivation in India in 2009, but a governmental moratorium was imposed shortly thereafter citing the need for more testing and evaluation. In 2014, Bangladesh began to allow cultivation of Bt brinjal and adoption rates have grown quickly, with > 20,000 farmers growing about 1200 ha (2.5% of total crop) in the 2018–19 growing season.

A recent summary of all insect-resistant GE crops, year of approval and adoption rates are provided in Naranjo et al. (2020).

Several other crops producing Bt Cry proteins are under research and development in different parts of the world. One relatively unique crop is a Bt cotton with resistance to plant bugs (Heteroptera) and thrips (Thysanoptera) that is currently under regulatory review and expected to be available to USA farmers in 2021 (Naranjo et al. 2020). Several non Bt approaches are currently under investigation and development. One of these is RNA interference (RNAi), a conserved immune response in eukaryotes whereby double-stranded RNA (dsRNA) produced in the organism itself directs the repression of a specific gene sequence. In crop biotechnology, the approach identifies a gene that controls a vital biological function in the target organism and then produces the associated dsRNA in the plant where it is taken up by the target insect through feeding. This process has the potential to be very selective to the target organism because of the gene specificity. The first commercial event with this technology was approved in 2017 as an additional approach for control of corn rootworm but is not yet being commercially grown due to some lingering trade issues. This RNAi trait will be added to maize already producing multiple Bt and herbicide tolerance traits. Finally, CRISPR-based approaches that allow genome editing have the potential to further revolutionize pest control, but the technology is in the early stages of development for this purpose and there remain many biological and regulatory challenges (Naranjo et al. 2020).

10.3 The IPM Context

The breadth and scope of GE crop technology is undeniably large on the world stage as are the potential solutions they contribute to agriculture in the face of a rapidly growing human population. However, it is important to keep their role in focus when thinking about crop productivity and crop protection, especially with Bt crops. Whether one considers Bt crops to be a form of host plant resistance or alternatively a convenient method for the delivery of a selective insecticide, they represent only a single tactic within the integrated pest management (IPM) toolbox. Effective and sustainable crop protection must include multiple tactics that are carefully integrated to manage multiple pests within agricultural landscapes. Nonetheless, some global compilations based on the adoption of Bt cotton and maize suggest that they have contributed significantly to economic and environmental gains. For example, Brookes and Barfoot (2020b) estimate that Bt cotton and Bt maize have increased global farm level incomes by \$63.6B and \$59.6B, respectively, from the period 1996-2018 with 55% of the benefits derived by farmers in developing countries. The associated environmental gains over this same time period in terms of insecticide use reductions also is large. Brookes and Barfoot (2020a) further estimate that foliar and soil insecticide use in Bt crops has declined by a total of 331 and 112 M kilograms (active ingredient) in Bt cotton and Bt maize, respectively. The associated decline in insecticide use for Bt soybeans over the period 2013–2018 in South America is

14.9 M kg. These insecticide reductions have paid dividends, particularly in cotton by facilitating improved biological control of non-target pests (Naranjo et al. 2020; Romeis et al. 2019).

Though not unique to Bt crops, seeds treated with neonicotinoid insecticides for seedling and early plant stage pest control have become nearly ubiquitous in field crop production, particularly for maize and cotton in the USA. While the non-target impacts of this technology remain unclear, this trend has the potential to partially erase some of the positive gains in reduced foliar and soil insecticides use with adoption of Bt crops. Growing target pest resistance to several Bt Cry proteins also has the potential to erode some of these gains in insecticide reductions in both maize and cotton (see Chapter 9 by Fleischer et al.).

10.4   What Is a Non-target Organism?

The focus of this chapter is to consider what we know about the effects of GE crops on non-target organisms. What is a non-target organism? Very simply, a non-target organism is broadly defined as any organism that the transgenic technology was not intended to control. Given that the intended targets of Bt crops are quite narrow, for example, several species of corn rootworm beetles (*Diabrotica* spp.) for Cry3 Bt maize, and several dozen species of caterpillars (various bollworms, defoliators and stalk borers) for Cry1, Cry2 and VIP Bt maize and cotton (Naranjo et al. 2020), the list of non-targets is potentially quite extensive. In Bt crops, non-targets include other arthropod crop pests that are not susceptible to Bt proteins and a wide range of organisms, many of which provide important ecosystem services such as biological control, pollination and decomposition. Much of the research focus has been placed on arthropods and other invertebrates, but some attention has been placed on vertebrates and it is common for regulatory agencies to require testing on a wide range of organisms including birds, mammals, fish and multiple invertebrate groups as part of the registration process for Bt crops. For instance, the US-Environmental Protection Agency considers Bt engineered into crops to be plant-incorporated-protectants (so-called PIPs), and regulatory oversight includes a process similar to that required for pesticides. This process often involves the use of surrogate species in a tiered testing system (see below) starting with laboratory experiments under extreme-dose, worst-case exposure conditions, but is increasingly emphasizing more extensive evaluations on non-target organisms in crop fields.

This chapter will focus primarily on the effects of Bt crops on non-target arthropods. These organisms are often among the most abundant and important residents of agricultural fields where they serve a wide variety of ecosystem functions and represent a significant portion of agroecosystem biodiversity. The focus here on insect resistant Bt crops stems from the fact that much of the non-target research conducted has focused on these crops. It is recognized that Bt proteins engineered into crops represent a different risk to non-target organisms and overall biodiversity than Bt proteins applied as foliar spray treatments. For example, Bt proteins are continually produced in Bt crop plants and these proteins are protected from the environmental degradation (e.g. rain, UV exposure) common in sprayable products applied to the plant surface. To date, over 700 scientific studies have been completed to assess effects of Bt crops on non-target invertebrates in both the laboratory and in the field. These collective data have been the subject of dozens of review articles. The data also have been used in more quantitative, synthetic studies called meta-analyses, which is simply a way to enhance the rigor and power of testing for non-target effects by statistically combining the results of multiple studies. The most recent synthesis that looked at multiple Bt crops and examined both field and laboratory studies was Naranjo (2009). The study including two dozen individual and pyramided (two proteins) Bt Cry proteins, eight Bt crops in 20 countries, and over 300 species in three Phyla (Arthropoda, Annelida, Mollusca). Other meta-analyses have since been published but with a narrower focus on single Bt crops or restricted regions of the world. A summary of these meta-analyses will be presented and discussed. For coverage of the other environmental risk issues associated with GE crops, including gene flow, invasiveness and soil ecosystem effects, the reader is directed to several recent reviews (Guan et al. 2016; Naranjo et al. 2020; Romeis et al. 2019).

10.5   Effects on Non-target Organisms

10.5.1   How to Characterize Risk

Globally, an environmental risk assessment (ERA) is generally conducted before any GE crop can be approved and released for production in the field. While each country has their own set of processes, there are many similarities among them (Schiemann et al. 2019). The approach commonly used involves a problem formulation process to establish protection goals and to assess current knowledge and identify areas of concern or uncertainty. Biodiversity and its associated ecosystem services is a typical protection goal evaluated in the ERA. Through problem formulation, risk hypotheses are developed and subsequently tested. Most regulatory bodies use conventional tier testing that starts with worse-case exposure in the laboratory and escalates through more complex and realistic tiers only if the null hypothesis of no risk is rejected or other uncertainties exist. Several considerations are important in identifying non-target species to assess, including their potential sensitivity to the insecticidal compounds in the GE crop. This is often predicated on phylogenetic considerations, especially for Bt where there is a long history of known effects on specific taxonomic groups (e.g. Lepidoptera). Another consideration is the relevance of the non-target group and this is established from knowing what taxa can be found in and around crop fields and if they are at risk of exposure. Relevance might also be based on consideration of the important ecosystem services provided, for example arthropods involved in biological control, pollination or decomposition. A further, practical consideration is the availability of select taxa and the ease with which high quality organisms can be reared and maintained. This might necessitate the use of representative surrogate species. Finally, it is critical that studies conducted in support of risk assessment are rigorous and can meet minimum quality standards (Romeis et al. 2019; Schiemann et al. 2019). In the end, a tiered approach is a balance between ecological reality and practicality. Regardless of the process it is ultimately up to decision-making bodies of each jurisdiction to determine the balance of risks and benefits to society as a whole.

The approach for assessing the non-target risk of the newer generation of GE crops such as those based on RNAi is still a developing field. Many of the same considerations employed for Bt crops would likely apply but nuances in how the technology works to effect pest control require careful consideration (Schiemann et al. 2019). For the one crop utilizing RNAi technology so far approved in the USA (corn rootworm control), the ERA was similar to the typical tiered system used for other crops with PIPs such as Bt. One additional supporting evaluation that can be added to identify species most likely to be at risk is to use bioinformatics to determine if the non-target organism shares a sufficient genomic match in the affected genomic sequence to that in the target.

10.5.2 General Non-target Effects

Although the topic area of non-target effects of Bt crops has enjoyed its share of controversy and debate, the extant body of research supports the conclusion that these crops have minimal negative effects on non-target organisms, and certainly less impact that the alternative use of insecticides to control the same target pests. Three broad and several more specific meta-analyses have been published in the past 15 years, beginning with Marvier et al. (2007). Based on funding from the US-EPA, this group developed the first database that attempted to compile the global English-language published research on the effects of Bt crops and Bt cry proteins on non-target organisms (primarily Arthropods but also including Annelida and Mollusca) in 2005. The database included studies conducted in both the laboratory and the field, although the Marvier et al. (2007) study examined only field studies. Their analyses showed that the abundance of all non-target invertebrates combined was slightly lower in Bt maize and cotton compared with non-Bt crops, but that abundances were much higher in Bt crops compared with non-Bt crops that had been treated with insecticides to suppress Bt targeted pests. They further concluded that taxonomic affiliation did not alter these general findings and that it was unclear if the observed reductions of abundance in Bt crops were due to direct toxicity or indirect effects causes by lowered target prey/host availability in the case of natural enemies.

Two subsequent and more detailed meta-analyses followed, including Naranjo (2009), who updated the Marvier database and examined both laboratory (discussed below) and field studies from an ecological rather than a taxonomic context. In general, analyses of field studies showed little difference in the abundance of various ecological guilds when insecticides were not applied to either the Bt crop or its non-Bt counterpart (Fig. 10.1a). This comparison tests the hypothesis that the plant itself, either directly or indirectly, affects non-target organism abundance. The one notable

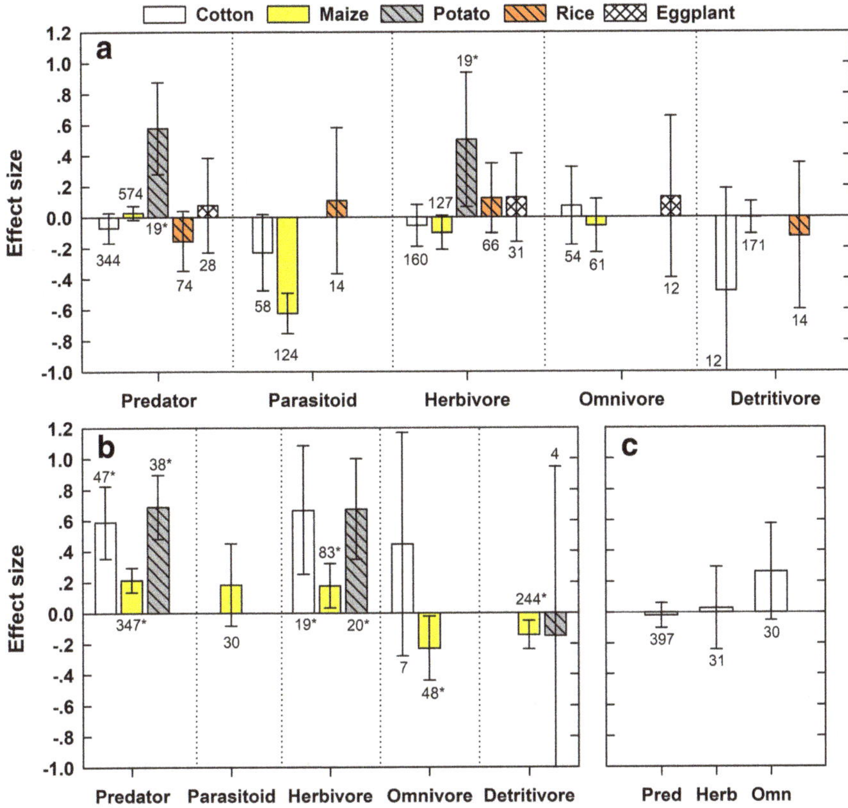

Fig. 10.1 Meta-analyses of field studies that examined the comparative abundance of non-target invertebrates in Bt and non-Bt crops. Meta-analysis quantitatively combines the results of multiple studies using a metric called the effect size that takes into account the variability, sample sizes and the magnitude of differences in individual comparative studies. The data are plotted such that a negative effect size denotes a lower abundance in the Bt crop compared with the non-Bt crop; a positive effect size denotes the opposite. Here the data are parsed into ecological guilds that represent different ecosystem functions. **a** Neither the Bt nor non-Bt crop received any insecticide treatments. These analyses test the hypothesis that the Bt protein or any other differences in the Bt plant affected non-target abundance either directly or indirectly. **b** Here the non-Bt crop was sprayed with insecticide and these analyses test the hypothesis that the method used to control the Bt targeted pest affects non-target abundance. **c** Finally, both the Bt and non-Bt crop are treated with insecticides to control both target and non-target pests and these analyses test the realistic hypothesis that management of pests in both Bt and non-Bt crops affect non-target abundance. The numbers above or below the bars denote sample size and the asterisks denote statistical significance of the effect size, i.e. significantly lower or higher than zero *Data modified* from Naranjo (2009) to include additional studies up to 2013

exception was a large reduction in the abundance of insect parasitoids in Bt maize. This pattern was found to be entirely due to a large number of USA based studies that examined densities of a specialist exotic parasitoid that attacks the European corn borer, a primary target of Bt maize. With effective control of the parasitoid's host there was an expected reduction in abundance in Bt maize fields. This is an example of an indirect ecological effect as parasitoids require their hosts to survive but are not necessarily directly affected by Bt proteins. Such indirect ecological effects would be expected of any tactic that lowers the target pest (the goal of pest management) and is not something unique to the deployment of a Bt crop. Another result of note is the effect of Bt potato on predators and herbivores (Fig. 10.1a). Here, the abundance of both groups was higher in the Bt crop. This is another example of an indirect ecological effect in which higher herbivore populations in Bt potato, primarily sucking insects (insects that feed by inserting their straw-like mouthparts into plant parts) led to a corresponding increase in predators responding to higher prey availability. The reason for increased sucking insect populations has not been studied, but it has been suggested that it is related to the lack of induced plant defenses in Bt potato when its primary targeted defoliator is controlled, and/or the lack of collateral control previously provided by insecticides (see Non-target Pests below). Other functional guilds (herbivores, omnivores and detritivores) were unaffected in Bt maize, cotton or potato in comparison with untreated non-Bt controls. At the time of this early meta-analysis, fewer studies had been conducted in Bt rice and eggplant, but results indicated no effects of Bt crops on any ecological guild (Fig. 10.1a). Most ecological guilds were more abundant in Bt maize, cotton and potato when the comparative non-Bt crop was treated with a variety of insecticides for control of target pests (Fig. 10.1b). The results for detritivores in Bt maize provides another example of indirect ecological effects wherein springtails, the primary detritivores in the system, were released from control by soil dwelling predatory beetles when insecticides were applied. Why omnivores were less abundant in unsprayed Bt maize compared with sprayed non-Bt maize is not completely understood. Heterogeneity analysis indicated that this pattern was due to omnivorous ants. Ants were not affected by Bt maize when either the Bt or non-Bt crop were insecticide-free suggested it is not a direct effect of the Bt proteins. When insecticides are used in both the Bt and non-Bt crops, a common situation in cotton, which harbors multiple pests, the abundance of the ecological guilds available for analyses were the same in both crops. While different pest complexes would have been targeted in Bt and non-Bt cotton, both systems rely on relatively broad-spectrum insecticides for non-target pest control, albeit generally fewer applications are needed for the Bt crop (Fig. 10.1c).

Several other crop and region targeted meta-analyses have subsequently been conducted. Comas et al. (2014) focused on arthropod herbivores, predators and parasitoids common to maize systems in Spain and generally found no effects of Bt maize. Although Bt rice has yet to be commercially approved for widespread cultivation in China, there have been numerous laboratory and field studies to examine potential non-target effects. Using a similar ecological guild approach to Naranjo (2009), Dang et al. (2017) conducted meta-analyses on laboratory and field studies of Bt rice in China. This extended on previous meta-analyses involving rice, adding new studies published in both English and Chinese. Their synthesis found patterns of lower herbivore and parasitoid density, and higher abundance of detritivores in Bt rice in the field but no differences for predators. Aside from parasitoids, these results differed from previous meta-analyses. Unfortunately, this study lacks methodological details; for example, it is unclear if they included studies involving insecticide use and if those were parsed in their analyses. Furthermore, the study authors do not provide any clear discussion on the underlying reasons for such differences. A global meta-analysis for Bt maize generally supported prior analyses in finding a general lack of differences in most non-target groups with the exception of expected declines in parasitoids associated with European corn borers Pellegrino et al. (2018). Krogh et al. (2020) published a systematic review and global meta-analyses focused on soil invertebrates in Bt maize. As in prior studies on maize, they generally found no effects of Bt maize on the non-target fauna and their inclusion of soil invertebrates extends and complements existing syntheses.

Meta-analyses also have been conducted to examine the relationship between laboratory and field studies. As noted above, many agencies that regulate GE crops used a tiered system to test for safety and to assess risk. Very often, field studies are conducted regardless of the outcome of early tier testing, specifically by academic and other public research organizations. Field studies also are increasingly being requested of industry by regulatory authorities as part of the registration process. Thus, there are robust datasets from both the laboratory and the field that allows a way to test the validity of the tier system. One such study found that "laboratory studies of transgenic insecticidal crops show effects that are either consistent with, or more conservative than, those found in field studies" (Duan et al. 2010). These findings suggest that the tier system can function to identify harm or the lack thereof in the environment.

10.5.3 *Non-target Pests*

The unique physiological effect of Bt proteins currently found in GE crops, a characteristic governed by the specific receptors and conditions in a caterpillar's or beetle's gut allowing activation of the Bt proteins, limits their activity to relatively few arthropod pests of crops. Thus, there are often a wide range of other insect and mite pests not affected by Bt crops, particularly in long season crops like cotton and soybean grown in lower latitudes. Many of these pest species are managed much as they were before the advent of Bt crops and represent an equal threat to Bt as well as conventional non-Bt crops. It is these pests that force a greater focus on the principles of IPM, which calls for a suite of integrated tactics to provide effective overall crop protection.

In general, meta-analyses of field abundance studies in Bt cotton, maize, rice and eggplant have shown that non-target herbivores, that would include non-target pests, are no more abundant in Bt crops compared with non-Bt crops when no insecticides are used (see Fig. 10.1a). That is, there is nothing specific about Bt crops themselves that would alter herbivore communities. This conclusion is further supported by meta-analyses of laboratory studies, which clearly show a lack of toxicity of Bt crops to non-target pests (Naranjo 2009). Alternatively, when insecticides are used in non-Bt crops and arthropod abundances are compared to untreated Bt crops, then herbivores, again including non-target pests, are considerably more abundant in Bt crops (Fig. 10.1b) (Naranjo 2009). This does not mean that all these pests are necessarily more problematic in Bt crops but that additional management tactics may be required to suppress their numbers as noted above.

However, some non-target pests (in this case secondary or induced pests) have become more problematic in Bt crops in some production systems. Some of the most visible examples have arisen within sucking insects. This includes, for example, plant bugs in China, Australia, and the USA in cotton. The best documented case comes from China, where multiple species of plant bugs have become more pestiferous in cotton but also in a number of other crops cultivated in the same region (see Chapter 9 by Fleischer et al.). Both plant bugs and stink bugs also have risen in importance as pests of cotton in parts of the mid-southern and southeastern USA. The causes for these increases are not completely understood, but in some areas like China and the USA, the problem appears to be ironically associated with the general reduction in broad-spectrum insecticides that were once used to manage caterpillar pests now effectively controlled by Bt cotton. These insecticides would often provide collateral control of these non-target pests. Similarly, in Australia it is thought that reduced insecticide use for bollworms has allowed plant bugs, stink bugs, leafhoppers and thrips to become more prominent. Sprays now applied to these pests have in turn disrupted a complex of natural enemies and lead to secondary outbreaks of pests such as spider mites, aphids and whiteflies. Growers in India are facing similar issues with mealy bugs, thrips and leafhoppers.

The increased emergence of non-target pests in Bt maize has been relatively minor in comparison to the situation with Bt cotton. Western bean cutworm (a caterpillar, but sensitive to only certain Cry proteins in Bt maize) has been less problematic with the introduction of pyramided cultivars producing at least two Bt proteins but it has been used negatively as an example of the side-effects of Bt crop adoption. While the use of Bt maize and the associated reduction in insecticide use has contributed to the range expansion of this pest, there are many other factors to consider. Among those are basic insect biology, pest and maize phenology, increasing use of conservation tillage afforded by herbicide tolerant maize, soil properties in the expanded range, insect genetics, insect pathogens, pest replacement and climate change (Hutchison et al. 2011). Issues with minor pests like wireworms and grubs (beetles) have been addressed mainly through insecticide-treated seed, a now common practice in the USA as noted before.

The rising importance of some non-target pests in Bt crops is likely associated with the large reductions in insecticides previously applied to control Bt targeted pests. The fact that reduced insecticides and associated conservation of natural enemies did not enhance control of these non-target pests suggests that biological control does not strongly operate for these pests. The induction of natural plant defenses is another factor that may play a role. Both maize and cotton are known to produce defensive compounds in response to certain types of herbivory. For instance, it is well known for maize and cotton that caterpillar feeding leads to the release of volatile compounds that act as attractants for natural enemies, thus facilitating biological control. In addition to such volatile signaling, herbivory, particularly by chewing herbivores, also induces plants to produce defensive compounds that can have negative effects on other herbivores feeding on the plant. In Bt crops this induction by chewing herbivores (caterpillars) is lessened significantly. Studies in cotton showed that a group of chemicals called terpenoids have lower levels of induction in Bt cotton compared with non-Bt cotton in the presence of caterpillar feeding. This difference allows better survival and growth in other pests such as aphids and plant bugs in Bt cotton. Reduced competition from target pests also may play a role and allow non-target pests to perform better in Bt crops.

10.5.4   Valued Non-target Organisms

While all non-target organisms could be considered valuable for multiple reasons, there are several groups that hold special significance because of the way they are valued by both agriculture and the public. Such groups include pollinators (e.g. honeybees), charismatic butterflies (e.g. Monarchs) and moths of special economic value (e.g. silk moths). Natural enemies that provide biological control services also would fall into this group, but they will be discussed separately below due to the key roles they play in crop protection. One charismatic insect came to represent the debate about the safety of GE crops more than all others, the Monarch butterfly, a well-known resident of North America. In 1999, a laboratory study presented in the prominent science journal Nature suggested that pollen from a Bt11 event of Bt maize in the USA could cause larval mortality when applied in large quantities to the surface of the butterfly's milkweed host plant. Interestingly, the pollen of Bt11 contains very low levels of Bt proteins but the anthers contain high levels. It is speculated by some that there was anther contamination of the pollen during the study. In addition to lots of negative popular press coverage, this study also precipitated a large research effort by multiple scientific groups in the mid-western USA to examine many aspects of this issue in both laboratory and field studies. Data from these studies and others was then used to construct a robust risk assessment that took into account many variables including factors relative to hazard (toxicity) and exposure. Ultimately, hazard was found to be low, especially with the primary Bt maize events under production that contained very little Bt in their pollen. This coupled with the very low potential for exposure (timing and extent of pollen dispersal from corn, proportion of Bt maize in the butterfly's breeding habitat, etc.) to the toxin in the butterfly's habitat ultimately led to a conclusion of negligible risk in the field (Sears et al. 2001). This was the same conclusion reached by the US-EPA for valued non-target butterflies during the registration process prior to 1996. The susceptibility of the Monarch to Bt proteins was never doubted given its taxonomic affinity with target caterpillars and a meta-analysis of laboratory studies on Monarch and other valued Lepidoptera showed this to be true (Naranjo 2009). Some recent work suggests that drift from commonly used insecticides for soybean pest management may be more toxic to monarch larvae that pollen drift from Bt maize. The larger focus currently is on the widespread use of glyphosate and other herbicides on herbicide-tolerant maize leading to a significant reduction in the abundance of the butterfly's host plant (milkweed) within and bordering maize fields.

Pollinators are another important non-target group, and awareness has been heightened even more with the current issues surrounding declining honeybee health and colony collapse. A meta-analysis based on 25 laboratory studies showed that survival of neither adult or larval stage honeybees was affected by Bt proteins targeting either caterpillar or beetle pests (Duan et al. 2008). An independent meta-analysis that included honeybees as well as bumble bees reached the same conclusion based on both survival and development in the laboratory (Naranjo 2009). Relatively few field studies have examined pollinators in general, but laboratory studies on a few species of bees indicate a lack of hazard from Bt proteins.

10.5.5 Non-target Effects on Arthropod Natural Enemies

Arthropod natural enemies represent another valuable group of organisms that require consideration in assessing risks from GE crops. They can potentially provide biological control services critical to controlling target and non-target pests, may help to ameliorate the evolution of resistance to Bt crops, and represent important members of communities in natural and managed habitats overall. Due to their importance there has been considerable research in assessing the impact of Bt crops on biological traits (e.g. survival, development, reproduction), abundance, and to a more limited degree, biological control function. There are multiple pathways by which natural enemies can be potentially exposed to Bt proteins (Fig. 10.2). First, most predators and parasitoids directly feed on vegetative and reproductive plant tissue or plant products such as nectar and pollen; some species such as hover flies and certain lacewings feed exclusively on nectar and pollen as adults. Such an exposure route is frequently called bi-trophic—plant to natural enemy. Secondly, predators and parasitoids can be exposed to Bt proteins through their prey or host, which have fed directly on the plant. This route is referred to as tri-trophic—plant to prey to natural enemy. A third pathway is related to tri-trophic exposure, but involves natural enemies feeding on honeydew produced by certain plant-sucking insects. Soil dwelling natural enemies can potentially be exposed to Bt proteins entering the soil through root exudates, decaying plant material or dead arthropods—a combination of bi- and tri-trophic exposure. Natural enemies on the borders of a crop or in adjacent habitats may be exposed to Bt proteins that have left the field through various mechanisms in the soil or air such as transport by ground water or dispersion of pollen and plant debris— again, bi-trophic exposure. Finally, predators, parasitoids and herbivores from the

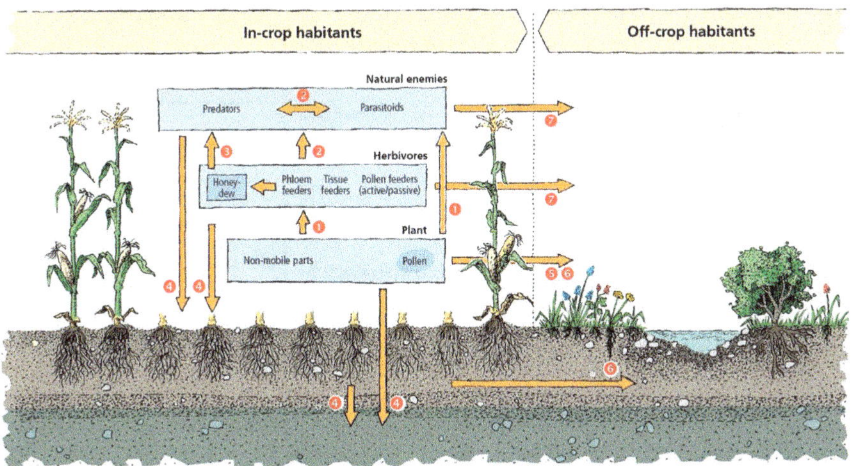

Fig. 10.2 Conceptual diagram showing the potential exposure pathways for natural enemies in Bt crop fields: (1) herbivores and natural enemies can feed directly on pollen and other plant parts; (2) predators and parasitoids can be exposed to Bt proteins in plants by consuming herbivores that have fed on the plant; (3) natural enemies can feed on honeydew secreted by various heteropteran insects that have feed on Bt plants; (4) soil dwelling natural enemies may be exposed to Bt proteins entering the soil through root exudates, decaying plant material or dead arthropods; (5,6) natural enemies outside the crop may be exposed to Bt proteins that have left the field through various mechanisms in the soil or air; (7) trophic interactions may occur outside the crop field via organisms originating in the crop. From Romeis et al. (2019), drawing by Ursus Kaufmann, Agroscope, Switzerland

crop can engage in trophic interactions outside the crop field as they move within the agroecosystem.

Many studies have examined both bi- and tri-trophic exposure pathways in a number of species in the laboratory. Meta-analyses and other data reviews have shown that bi-trophic exposure, direct feeding on either the plant or artificial diets containing Bt, has no effect on important biological parameters such as development/growth, survival or reproduction (Naranjo 2009; Romeis et al. 2019).

Interpreting the results from exposure studies examining tri-trophic interactions, or feeding on prey that have ingested Bt proteins, has been more problematic. An issue that has not consistently been factored into the interpretation of study results is that prey that are susceptible to Bt proteins (e.g. target caterpillars or beetles) are frequently affected by this feeding even if they do not die from the exposure. Those that survive are typically smaller and grow slower, a sign of sublethal effects from the Bt protein. Natural enemies that in turn use these compromised prey often suffer as well. The question of whether this is a direct or an indirect effect of the Bt protein is important but sometimes muddled. In order to establish that effects are direct, i.e. toxicological, it is necessary to control for prey quality effects. Two approaches have been employed, including the use of prey that are not susceptible to Bt proteins because they are unrelated taxonomically to the target insects, or the use of target insects that have been selected to be resistant to Bt proteins. Both of these strategies

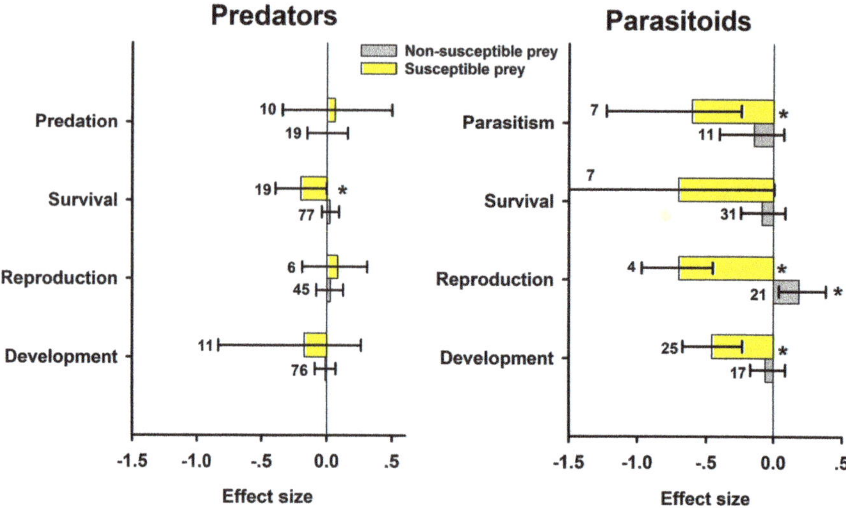

Fig. 10.3 Meta-analyses of laboratory studies that examined the non-target effects of Bt crops on arthropod natural enemies exposed to Bt proteins via their prey or host that had fed on Bt plants (tri-trophic exposure). Bt susceptible prey often suffer sub-lethal effects that degrade their quality as food for natural enemies while non-susceptible or resistant prey are normal. The data are plotted such that a negative effect size would denote a negative impact on performance in the Bt crop compared with the non-Bt crop; a positive effect size denotes the opposite. The numbers next to the bars denote sample size and the asterisks denote statistical significance of the effect size, i.e. significantly lower or higher than zero. Figure modified from Naranjo (2009) and Romeis et al. (2019)

have been used effectively to eliminate prey quality effects and enable a clear testing of direct effects of Bt proteins. Meta-analyses have compared studies where prey quality effects were apparent or were eliminated using non-susceptible or resistant prey (Fig. 10.3) (Naranjo 2009; Romeis et al. 2019). Analyses show that the use of susceptible prey results in slower development and reduced reproduction and parasitism in parasitoids and lower survival in predators. If the effects of prey quality are removed by using non-susceptible or Bt resistant prey then these parameters were either not affected or were affected positively in the case of parasitoid reproduction (i.e. better performance on prey containing Bt proteins). These results demonstrate that Bt proteins do not by themselves have any toxicological effects on the arthropod natural enemies. However, much like the field-based results discussed above there are indirect effects because biological attributes can be negatively affected if natural enemies use compromised prey. The impacts of these indirect effects in the environment are not clear and they are not limited to cases in which Bt crops are being deployed. Any tactic that affects the target prey (previous parasitism, insecticides, other host plant resistance factors, etc.) would likely yield the same indirect effect on the associated natural enemy. It also is unclear if such effects would have any ramification for the biological control services provided by natural enemy populations (discussed below). A key question is if there are enough of these compromised prey or hosts after the action of a control tactic (Bt or otherwise) to materially affect natural enemy dynamics in the field.

Issues with differing interpretations of data from tri-trophic studies have created debates in the scientific community. One of the most widely known cases involves the green lacewing, a common and important predator found in many cropping systems. In the late 1990's a group showed that certain biological attributes of green lacewing larvae were negatively affected when feeding on caterpillar prey that have been exposed to certain Bt proteins. They also showed that bi-trophic exposure routes resulted in negative biological effects. Numerous issues with experimental design were identified in these studies, but work conducted in the same laboratory and many others since this initial report have failed to duplicate any of these direct negative findings for several Bt proteins (Romeis et al. 2014). Another debate involved a laboratory-based meta-analysis that reported direct negative effects on arthropod predators and parasitoid by various Bt proteins. This result was surprising and not consistent with many other reviews and meta-analyses, including those discussed here. A rebuttal identified a number of statistical and logical issues with the study but one of the overriding factors was that the study authors failed to account for prey quality issues when examining tri-trophic studies (Romeis et al. 2019). The data presented in Fig. 10.3 shows how different the results can be when prey-mediated effects are not taken into account. Most laboratory studies being done today are cognizant of prey/host quality issue and use proper controls to eliminate their spurious effects.

10.5.6 Effects on Biological Control Function

The impacts of Bt crops on arthropod natural enemies have already been discussed (see Figs. 10.1 and 10.3). Cases where abundance was reduced were associated with indirect ecological effects such as prey scarcity, or possibly with the indirect effects resulting from preying on compromised, Bt susceptible prey. While measures of abundance and general biodiversity are a simple means to gauge non-target effects in the field, the more critical question for natural enemies is whether or not the biological control services they provide have been compromised. Compared with abundance studies relatively few studies have examined some measure of function. Such studies have used a variety of techniques including simple measures of parasitism from field samples, and measurement of predation or parasitism rates on prey artificially placed in the field, to more comprehensive life tables quantifying predation and parasitism rates on natural prey populations. Except for a few cases in which parasitism by specialist parasitoids attacking target pests have been reduced, there is no evidence that biological control capacity differs between Bt and non-Bt crop fields. Even in cases were natural enemies might be less abundant in the Bt crop there is no evidence that biological control services are reduced. For example, a long-term study in Bt cotton showed that a group of five common predators were reduced by about 20% in the Bt crop, but rates of natural enemy induced mortality on two key pests

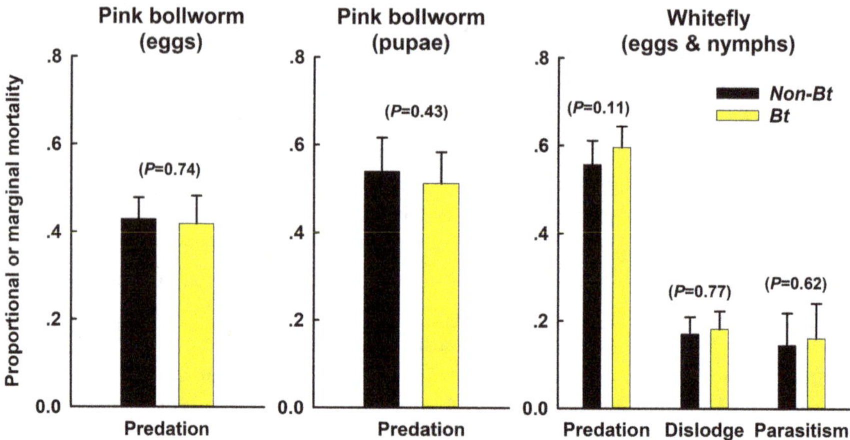

Fig. 10.4 Comparative measurement of biological control function in cotton fields using sentinel pink bollworm eggs and pupae or natural infestations of whitefly nymphs. Both the Bt cotton and the non-Bt cotton were unsprayed. Results for pink bollworm are summarized over four trials in each of two years; results for whitefly are based on two trials in each of three years. Statistical p-values are provided, and error bars denote 95% CIs. Mortality of pink bollworm represented both disappearance and chewing predation; no parasitism was observed for either life stage. For whitefly, dislodged nymphs disappeared from the leaf surface as a result of weather or chewing predation, predation indicates mortality by predators with sucking mouthparts, and parasitism is by several native and exotic aphelinid parasitoids. Data compiled from Naranjo (2005)

of cotton, pink bollworm and whitefly, remained unchanged compared to the non-Bt crop (Fig. 10.4). Overall, opportunities for enhanced biological control in Bt crops have been demonstrated in several systems (see Chapter 9 by Fleischer et al.). These benefits are derived not from anything special about Bt crops, but through the selective control of key pests afforded by the GE technology that allows natural enemy populations to flourish and provide critical ecosystem services within an overall IPM strategy.

10.6   Conclusion

GE crops have become important tools in crop production and protection in many countries and contribute significantly to overall IPM programs. Extensive laboratory and field data have been generated relative to the assessment of ecological risk in these crops, particularly for non-target organisms in Bt crops. This body of evidence and the quantitative and qualitative syntheses of the data through meta-analysis and other compilations generally indicate a lack of direct impacts of Bt crops and the insecticidal proteins they produce on non-target invertebrates. The data also clearly show that Bt crops are much better than the alternative use of traditional insecticides for control of the pests targeted by Bt crops. Some indirect effects on natural enemies associated with reduced abundance or quality of Bt target herbivores have been shown, but the ramifications of these effects are unclear. Crops developed with new technologies based on RNAi gene silencing and CRISPR gene editing have the potential to further revolutionize pest control, but the technologies, particularly CRISPR, are in the early stages of development for crop improvement and there remain many biological and regulatory challenges. As one tactic in the IPM toolbox, Bt crops have had a profound effect on insecticide use patterns. While reduced insecticide use may be involved in precipitating new pest problems in Bt crops it also has broadened opportunities for deployment of biological control.

Acknowledgements   I thank Joerg Romeis, Richard Hellmich and Bill Hutchison for reviewing an earlier draft of this manuscript.

References

Brookes G, Barfoot P (2020a) Environmental impacts of genetically modified (GM) crop use 1996–2018: impacts on pesticide use and carbon emissions. GM Crops Food 11:215–241

Brookes G, Barfoot P (2020b) GM crop technology use 1996–2018: farm income and production impacts. GM Crops Food 11:242–261

Comas C, Lumbierres B, Pons X, Albajes R (2014) No effects of *Bacillus thuringiensis* maize on nontarget organisms in the field in southern Europe: a meta-analysis of 26 arthropod taxa. Transgenic Res 23:135–143

Dang C, Lu ZB, Wang L, Chang XF, Wang F, Yao HW, Peng YF, Stanley D, Ye GY (2017) Does Bt rice pose risks to non-target arthropods? Results of a meta-analysis in China. Plant Biotechnol J 15:1047–1053

Duan JJ, Lundgren JG, Naranjo SE, Marvier M (2010) Extrapolating non-target risk of *Bt* crops from laboratory to field. Biol Let 6:74–77

Duan JJ, Marvier M, Huesing J, Dively G, Huang ZY (2008) A meta-analysis of effects of Bt crops on honey bees (Hymenoptera: Apidae). PLoS ONE, p e1415

Guan Z-J, Lu S-B, Huo Y-L, Guan Z-P, Liu B, Wei W (2016) Do genetically modified plants affect adversely on soil microbial communities? Agric Ecosyst Environ 235:289–305

Hutchison WD, Hunt TE, Hein GL, Steffey KL, Pilcher CD, Rice ME (2011) Genetically engineered Bt corn and range expansion of the western bean cutworm (Lepidoptera: Noctuidae) in the United States: a response to Greenpeace Germany. J Integr Pest Manage 2:B1–B8

Krogh PH, Kostov K, Damgaard CF (2020) The effect of Bt crops on soil invertebrates: a systematic review and quantitative meta-analysis. Transgenic Res, https://doi.org/10.1007/s11248-11020-00213-y

Marvier M, McCreedy C, Regetz J, Kareiva P (2007) A meta-analysis of effects of Bt cotton and maize on nontarget invertebrates. Science 316:1475–1477

Naranjo SE (2005) Long-term assessment of the effects of transgenic Bt cotton on the function of the natural enemy community. Environmen Entomol 34:1211–1223

Naranjo SE (2009) Impacts of *Bt* crops on non-target organisms and insecticide use patterns. CAB Reviews: perspectives in agriculture, veterinary science, nutrition and natural resources 4, No. 011 (DOI:010.1079/PAVSNNR20094011)

Naranjo SE, Hellmich RL, Romeis J, Shelton AM, Vélez AM (2020) The role and use of genetically engineered insect-resistant crops in integrated pest management systems. In: Kogan M, Heinrichs EA (eds) Integrated management of insect pests: current and future developments. Burleigh Dodds Science Publishing Limited, Cambridge, UK, pp 283–340

Pellegrino E, Bedini S, Nuti M, Ercoli L (2018) Impact of genetically engineered maize on agronomic, environmental and toxicological traits: a meta-analysis of 21 years of field data. Sci Rep 8:3113

Romeis J, Meissle M, Naranjo SE, Li Y, Bigler F (2014) The end of a myth-Bt (Cry1Ab) maize does not harm green lacewings. Front Plant Sci 5:391

Romeis J, Naranjo SE, Meissle M, Shelton AM (2019) Genetically engineered crops help support conservation biological control. Biol Control 130:136–154

Schiemann J, Dietz-Pfeilstetter A, Hartung F, Kohl C, Romeis J, Sprink T (2019) Risk assessment and regulation of plants modified by modern biotechniques: current status and future challenges. Ann Rev Plant Biol 70:699–726

Sears MK, Hellmich RL, Stanley-Horn DE, Oberhauser KS, Pleasants JM, Mattila HR, Siegfried BD, Dively GP (2001) Impact of Bt corn pollen on monarch butterfly populations: a risk assessment. Proc Natl Acad Sci 98:11937–11942

Steven E. Naranjo is Director of the USDA-ARS, Arid-Land Agricultural Research Center in Arizona, and adjunct faculty with University of Arizona. His research interests include biological control, IPM, sampling, population ecology, and environmental risk assessment of genetically engineered crops. He has conducted long-term studies on non-target effects of Bt cotton with emphasis on natural enemy abundance and function and led meta-analyses to quantify non-target effects of Bt crops worldwide. Dr. Naranjo served as Co-Editor-in-Chief for *Crop Protection*, and currently serves as Subject Editor for *Environmental Entomology* covering Transgenic Plants and Insects.

Chapter 11
Virus-Resistant Crops and Trees

Cristina Rosa

Abstract Plant viral and other plant microbial diseases cause significant economic losses every year and limit food supplies worldwide. To control viral diseases, we can use a variety of strategies, including various forms of genetic resistance. Genetic resistance can be manipulated to control viruses by exploiting a natural eukaryotic defence system called RNA interference or gene silencing. This system can be additionally exploited to control insect vectors of viruses, broadening the impact of transgenic technologies. An overview on plant defense, plant viruses and integration of transgenic technologies in virus resistant crops is given in this chapter.

Keywords RNA interference · Gene silencing · Plant viruses · Plant defenses · Viral induced gene silencing

11.1 Introduction

In the natural environment, plants are colonized by a microbiota (a microbial community) composed of various organisms including viruses, bacteria and fungi. The majority are not pathogenic and do not harm their host. On the contrary, they are necessary for plant wellness and participate in a mutualistic interaction beneficial to themselves and their plant host. For instance, some plant viruses confer tolerance to stress, such as heat or drought stresses (Marquez et al. 2007; Xu et al. 2008), allowing the infected plants to colonize extreme environments (such as the extremely hot soil in the Yellowstone Park, US) and to survive sudden changes in their habitat (such as water fluctuations due to tide surges).

C. Rosa (✉)
The Pennsylvania State University, University Park, PA, United States
e-mail: czr2@psu.edu

© The Author(s), under exclusive license to Springer Nature Switzerland AG 2021   145
A. Ricroch et al. (eds.), *Plant Biotechnology*,
https://doi.org/10.1007/978-3-030-68345-0_11

Viruses are ancient microbes and shape the evolution of all living organisms by their host:virus interactions and by transferring genetic material among species. Today many viruses are used for beneficial applications including in medicine to fight bacterial infections, in forest ecosystems to kill defoliating caterpillars, and in nano/biotechnology as carriers for genetic information, delivering drugs into the right type of cells or expressing proteins in plants, insect cells or bacteria. In contrast, and this is true especially in the agricultural landscape, some plant viruses, as well as bacteria and fungi, can cause diseases that become eventually lethal to their hosts, and most importantly can cause a substantial yield reduction in crops that are important for food and fiber. Deleterious viruses seem to have evolved as consequence of the advent of agriculture, and are relatively evolutionarily new, compared to viruses found in natural landscapes. **The negative impact of plant viral diseases, even to the point of limiting food supply, is the reason we are concerned about growing healthy crops.** It is estimated that plant diseases in general cause up to a 14% loss in total crop production every year, a percentage that equals hundreds of billions of US dollars in lost revenues. Ten to 15% of this loss can be attributed to viruses, but for specific crops and in specific locations, like in the Asian and African continents where food supplies are already limited, the losses can have severe direct effects on human health. To control viral diseases, we routinely use a variety of strategies, ranging from the application of pesticides aimed to reduce the number of insect vectors that spread viruses, the adoption of Integrated Pest Management (a system designed to use multiple pest management practices in an environmentally sound manner), the establishment of plant and animal quarantine areas, the use of certified germplasm material (a collection of genetic resources, for instance seeds and tubers) and various forms of genetic resistance.

11.2 What are Plant Viruses, and do Plant Viruses Differ from Animal Viruses?

Plant viruses are generally smaller and less complex than animal and bacterial viruses. The sizes of the viral particles range from ~ 20–200 nm, and their chemical composition is generally simple. Virus particles typically have an outer shell made of proteins arranged in a geometrical form, either in a rod-shaped or in an isometric "coat".

Representative plant virus particles purified from infected leaves. **a** Tobacco streak virus (isometric).
b Pea seed-borne mosaic virus (rod shaped)

The viral shell is composed of repetitive small protein subunits and for some plant viruses the shell is surrounded by a lipid (fatty) membrane derived from the host cell. The viral genome is protected inside the shell, and can be made of either RNA or DNA.

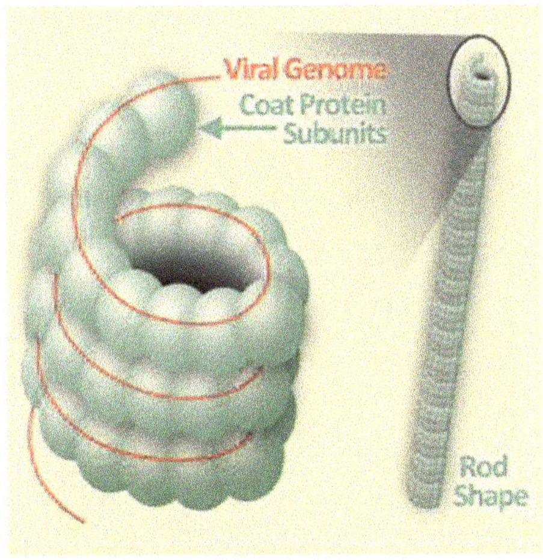

1. The **capsid** is the protein shell that encloses the nucleic acid. It is built of structural subunits.
2. **Coat protein subunits** are the smallest functional equivalent building units of the capsid.
3. The capsid with its nucleic acid is called the **nucleocapsid**.
4. The nucleocapsid may be protected by an **envelope**.
5. The **virion** is the infective virus particle

The viral genome can be circular or linear, can be composed of one or more "chromosomes" and can be contained in single or multiple viral particles or virions. The plant viral genome size is generally small, usually ranging from 3000 to 30,000 nucleotides (blocks that build RNA and DNA), while animal virus genomes can consist of 800 thousand nucleotides and encode for up to a few hundreds proteins. The virions of a few viruses may contain viral proteins necessary to initiate viral replication or multiplication, but viruses are unable to replicate without the host's cellular machinery; they are intracellular molecular obligate parasites. They exploit and highjack the host cell to multiply and they cannot perform this outside living cells. Since plant viruses are so small, they can express only a few proteins, and are thus amazing in their ability to replicate and respond to plant defenses, considering their limited genomic arsenal. Viruses rely on multifunctional proteins, protein modifications and on timely regulated protein expression and genomic replication to successfully complete their life cycle.

11.3   Can Plants Defend Themselves Against Viruses?

Let's have a first look at how plant defences function, and how they differ from animal defences.

Plants are multicellular organisms and their cells are interconnected by channels called plasmodesmata. Plasmodesmata serve as freeways that allow small molecules to travel between cells, or even throughout the entire plant. Plant viruses take advantage of the plasmodesmata to spread within the infected plant.

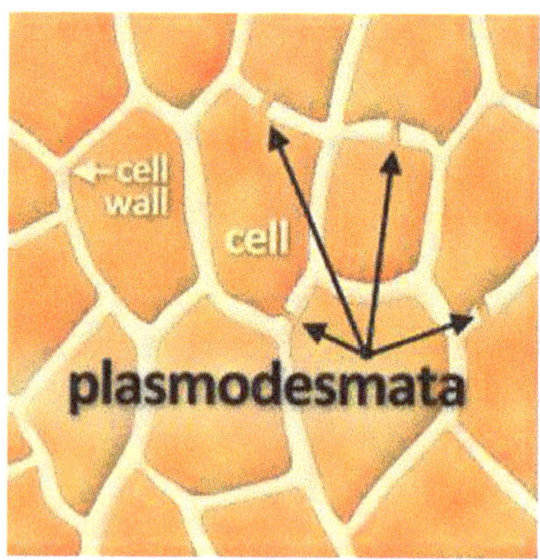

Plant cells are protected by cell walls, and connected by plasmodesmata

Plants possess constitutive and inducible defences against insects, microbes and other stressors (Dangl and Jones 2001; Howe and Jander 2008). Constitutive defences are always present; they include physical barriers such as modified cell wall composition or the presence of leaf trichomes (hairs or appendages), and chemical barriers such as production of chemical deterrents. Inducible defenses are those that are 'turned on' only as consequence of pathogen attack, when the plant cell's innate immune system recognizes pathogen-encoded effectors that are signature molecules specific of a class of organisms. Hypersensitive Response, Reactive Oxygen Species release and Programmed Cell Death are the names given to plant reactions linked to Resistance (R)-gene mediated resistance. Here upon recognizing the pathogen, plant cells release chemical defenses and "commit suicide", in the attempt to contain infections from spreading into healthy tissues. The leaves of plants expressing these reactions show necrotic spots, corresponding to the points in which the attack started.

Pathogens are perceived by plants and activate local responses as well as systemic responses in the attacked plant

Signals originated from the site of infection then stimulate the Jasmonic Acid, Ethylene and or Salicylic Acid dependent or independent defense pathways in distal parts of the plants. These pathways are sequential events that turn on and off other genes and their products to help the plant in its fight against pathogens. Viruses and other plant pathogens have evolved means to counteract plant defeneses, thus plant resistance and pathogen counter defense represent an ongoing evolutionary "arms race".

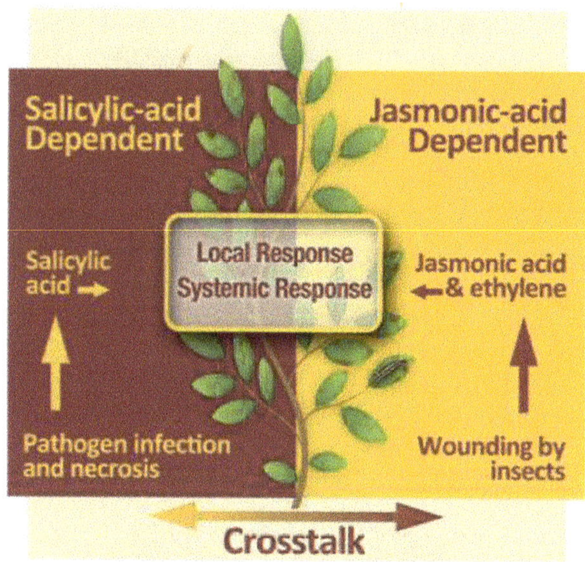

Pathogen infection, cell necrosis, or insect wounding activate the salicylic acid or the jasmonic acid and ethylene pathways, leading to local and systemic plant responses. These defense pathways are interconnected

Finally, plants lack the somatic adaptive immune system (antibodies) typical of animals and do not possess lymphocytes (type of white blood cells).

11.4 Are Cultivated Plants More Susceptible to Viruses Than Their Wild Relatives?

Individual plants in a natural population are not genetically identical and usually differ in their resistance to pathogens. By contrast agricultural species grown in monocultures (where all the plants belong to the same species and are essentially identical) are equally susceptible or resistant to specific pathogens. Thus, if wild relatives that show degrees of resistance or tolerance to the same pathogen can be identified, these can serve as sources of genetic resistance for cultivated crop plants. Resistance can be conferred by expression of a single host gene (R-gene), or by multiple genes, and plant breeders try routinely to introgress (move) genes for resistance found in wild plants into cultivated varieties via traditional breeding, or via transgenesis.

11.4.1 Examples of Natural Resistance

Many plants show R-gene mediated natural resistance to viruses. For example, in many *Arabidopsis thaliana* accessions (a collection of plants from the same location), some proteins called the Restricted Tobacco etch virus Movement RTM1, RTM2 and RTM3 restrict long distance movement of plant viruses called potyviruses. This resistance is not present in *A. thaliana* accessions showing amino acid changes in the RTM proteins. Since the viral coat protein is the effector recognized by the RTM proteins, potyviruses that naturally show changes in their outer shell structure are not recognized by *A. thaliana* and can infect plants with *Rtm* genes.

Many tomato *(Solanum lycopersicum)* species show natural resistance to viruses called tospoviruses, as well as to other plant pathogens. *S. peruvianum*, *S. chilense*, *S. habrochaites* and *S. pimpinellifolium* are used as source of resistance genes to be introgressed into cultivated tomatoes. A single gene called *Sw-5b* introgressed from *S. peruvianum* into *S. esculentum* cultivar Stevens shows broad-spectrum resistance to tospoviruses. The *Sw-5b* gene belongs to a particular class of plant resistance genes and is similar to other resistance genes such as the tomato nematode and aphid resistance gene *Mi*. The resistance conferred by *Sw-5b* seems can elicit a hypersensitive response in virus inoculated tissues, where it blocks the spread of the virus, but the resistance is absence in tomato fruits.

Some plant proteins called Endosomal Sorting Complex Required for Transport (ESCRT) are involved in endosome maturation. Endosomes are cellular vesicles that help in transporting proteins to different destinations in the cells and to or from the cell surface. Most viruses exploit the ESCRT system during their replication and movement, and impairment in ESCRT interferes directly with the ability of the viruses to replicate and move. In fact, *Arabidopsis* plants modified to lack ESCRT show inhibited viral replication and infection for a group of viruses called tombusviruses.

The resistance to a virus called *Tobacco mosaic virus* in the plant *Nicotiana glutinosa* is due to the 'Necrotic-type response to infection with TMV' *N* gene. This gene product also interferes with viral replication and as the name suggests induces plant cells to necrotize (commit suicide) in order to stop viral infection.

11.4.2 Examples of Transgenic Resistance

In absence of natural resistance, today we can sometimes use transgenesis to incorporate viral resistance traits into crop plants. In some instances, we can move natural R-genes from one crop to another crop, but this is not always effective, probably due in part to the genetic background of the recipient crop plant. In the late 1980s, scientists started exploring the idea that inserting viral genes into plants could trigger the transgenic plants to become 'immune' to viruses, in a kind of self-perpetuating plant vaccination. In 1986 Powell-Abel et al. successfully generated the first transgenic tobacco plants expressing the TMV coat protein. In 1988 Nelson et al. engineered whole transgenic tomatoes to express the TMV coat protein. Some of the transgenic lines were partially resistant to TMV infection. Since then, many plants (barley, canola, corn, oat, rice, wheat, chrysanthemum, dendrobium, gladiolus, grapefruit, grapevine, lime, melon, papaya, pineapple, plum, raspberry, strawberry, tamarillo, walnut, watermelon, alfalfa, sugarcane, bean, clover, groundnut, pea, peanut, soybean, lettuce, pepper, potato, squash, sugar beet, sweet potato, and tomato) have been transformed to be virus resistant. However, of all the plants that have been generated and tested in laboratory or greenhouse settings for their viral resistance and here are listed the few plants that reached the market.

The transgenic summer squash line ZW-20 resistant to *Watermelon mosaic virus* and *Zucchini mosaic virus*, and the transgenic squash line CZW-3 resistant to *Cucumber mosaic virus*, *Watermelon mosaic virus* and *Zucchini mosaic virus* were commercially released in the US in 1994 and 1996, respectively (Tricoli et al. 1995). A transgenic papaya (Gonsalves et al. 1997) resistant to *Papaya ringspot virus* (PRV) was released in 1998, while transgenic green pepper varieties and tomato resistant to *Cucumber mosaic virus* are today released in the People's Republic of China. Two potato lines resistant to *Potato virus Y* were deregulated in Canada (1998) and the US (1999) but were later abandoned because of the extremely negative public opinion (Kaniweski and Thomas 2004). Of the transgenic plants released to the market, the most successful story comes from the use of PRV transgenic papaya in Hawaii. This event saved the papaya industry from complete destruction due to PRV, allowed the cultivation of non-transgenic papaya cultivars in between transgenic fields, and increased the cultivar diversity in the islands.

The virus resistant transgenic plant that has been deregulated (2011) in the US is the *Plum pox virus* resistant 'HoneySweet' plum. Research to establish the safety and characteristics of this plum variety took more than 20 years, and it has been particularly important since there is no high level of PPV resistance known in *Prunus domestica, P. spinosa and P. insitita*. Only *P. cerasifera* offers a cultivar that is hypersensitive to PPV inoculation, and young plants of this cultivar naturally die when exposed to the virus. 'HoneySweet' plums score high in fruit quality and yield, and are today crossed with other plum varieties since they can transmit the dominant resistant trait as a single locus. Scorza et al. in 2013 wrote a highly remarkable review on the process of deregulation on HoneySweet.

The mechanism that allows these transgenic plants to be resistant to viruses is not always known, but in most of the cases the resistance is due to a host natural defense system called gene silencing or RNA interference.

11.4.3 A New, Nucleic Acid Sequence-Based Inducible Defense Mechanism

The eukaryotic inducible defense mechanism that evolved specifically against viruses, called RNA interference (RNAi), or gene silencing (Voinnet 2001; Waterhouse et al. 2001), works against specific nucleotide sequences, and it is thus atypical when compared to the classic effector-mediated plant defense system described above.

11.5 How Does RNAi Work?

When viruses replicate in eukaryotic cells, a double-stranded RNA (dsRNA) form of the viral genome is produced by the viral enzyme RNA dependent RNA polymerase that uses the RNA genome as template and copies it in an RNA molecule of opposite polarity. Since the two RNA molecules are complementary, they can anneal to each other and form a double stranded RNA helix. DsRNA can also be generated from the pairing of complementary stretches of RNA on the same molecule, and is not always linked to viral replication. Large dsRNA molecules, such as those generated during virus infections, are not found in healthy cells and their presence is recognized as foreign and serves as the trigger of the eukaryotes' (e.g. plants) RNAi pathway. A plant enzyme called Dicer cleaves the dsRNA into small fragments that are 21 nucleotides long, called small interfering RNAs (siRNAs). Dicer has a pocket that is exactly 65 Å in size, the distance that equals to 21 nucleotides. This pocket serves as molecular ruler. The siRNA duplex is dissociated into two strands by Dicer and one of the two strands (the sense or passenger strand) is degraded by the same enzyme. The second strand (the antisense or guide strand) gets incorporated into an enzymatic complex called the RNA Induced Silencing Complex, or RISC. RISC uses the antisense strand as template to find and hybridize with RNA with a complementary nucleotide sequence (more viral RNA), and to degrade it. In this way, cells are able to find and destroy viral RNA and to distinguish it from other cellular messenger RNA (mRNA).

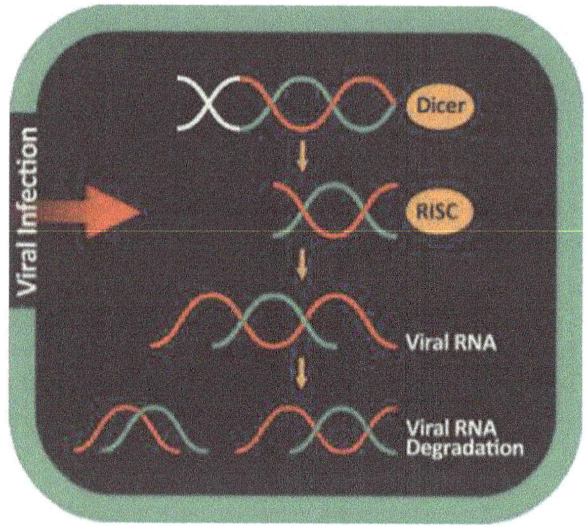

After viral infection, viral dsRNA is produced in eukaryotic cells. The dsRNA is recognized by the cell enzyme Dicer and processed into 21 nucleotide long siRNA molecules that are incorporated into the cell RISC complex and used to search for complementary viral RNA sequences for their degradation

While animals have evolved other defenses such as an interferon based signals to alert healthy cells about a virus attack, and antibodies to recognize specific viruses and other microbes, plants especially, but also insects, rely heavily on the RNAi pathway to defend themselves against viruses.

11.6 How Can We Manipulate RNAi to Induce Virus Resistance in Plants?

If we genetically transform plants to express double-stranded RNAs, the plant RNAi machinery will recognize the dsRNA and initiate a response. If the engineered plant expresses a plant virus double-stranded RNA sequence, the plant RNAi response will recognize the viral RNA, this strategy has been used to develop virus "immune" plants.

So, going back to the example of the transgenic plants expressing virus sequences such as those encoding for the viral coat protein, those plants are very efficient in recognizing and destroying the invading virus RNAs. If we analyze the genome of those transgenic plants, we can find the inserted viral sequence, and the corresponding 21 nucleotide siRNAs derived from the inserted sequence but resulting from the Dicer activity. Other transgenic plants produced in laboratory trials where RNAi against viruses has been exploited are: walnut, ryegrass, tomato, tobacco sweet potato, soybean, potato, rice, poplar, opium, maize, ornamental crops, apple.

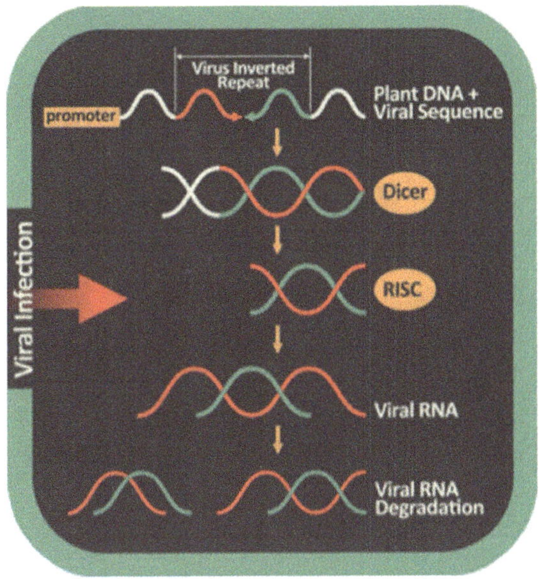

Transgenic plants can be made inserting a part of a viral sequence and its complementary sequence, under the control of a plant or virus promoter. The plant will express a double stranded form of the viral sequence that will trigger the plant RNAi pathway against those specific viral sequence. If the plant is challenged by the target virus also containing that particular sequence, the plant will use the primed RNAi pathway to destroy the viral RNA and halt infection

11.7 Does RNAi ALWAYS Involve the Use of Transgenic Plants?

'Viral induced gene silencing', or VIGS is used to introduce specific nucleotide sequences that can induce RNAi effects in plants via non-lethal recombinant viruses. These viruses are used as carriers to express the nucleotide sequence complementary to the one that we wish to target via RNAi, but since these viruses do not integrate into the plant genome, the resulting plants are not transgenic. These viruses are usually inoculated (or transferred) mechanically into their host plants. A type of VIGS is for example used in Brazil, Australia, South Africa and Japan to fight *Citrus tristeza virus* (CTV) in citrus orchards. There, a mild strain of CTV is inoculated in the trees, and serves as vaccine against severe CTV strains (Costa and Muller 1980).

Plant on the far left is a healthy citrus plant. Plant on the far right is infected with a severe CTV strain (depicted by a blue leaf) and plant in the middle left is infected with a mild CTV strain (red leaf). If a plant is inoculated first with a mild CTV strain (middle right) and then by a severe CTV strain (red and blue leaf respectively), the plant will be partially protected and will grow better than plants infected by the severe CTV strain

In 2020, the United States Department of Agriculture (USDA)'s Animal and Plant Health Inspection Service (APHIS) issued an environmental release permit to Southern Gardens LLC (Florida, US) to allow the release of 'CTV-SoD' within the state of Florida. This permit would allow the release of CTV clones genetically engineered to express defensin proteins derived from spinach to control Huanglongbing (HLB, citrus greening) disease, one of the most devastating diseases of citrus in the world. While these clones of CTV are transgenic, the plants inoculated with them are not, posing an interesting twist to the adoption of VIGS in agriculture.

11.7.1 Issues Linked with VIGS and RNAi

There are a few issues associated with the use of RNAi against plant viruses, due to the ability of plant viruses to evade the plants RNAi response. First, plant viruses evolve rapidly and can mutate their nucleotide sequences. If their nucleotide sequence changes compared to the one targeted by RNAi, it may not be recognized anymore by the RISC complex. Second, new strains of viruses with differences in the RNAi target region are always emerging, and would not be affected by RNAi. Third, plant viruses encode for proteins that can allow the virus to evade the RNAi machinery. These proteins are called 'suppressors of gene silencing' (Qu and Morris 2005), and other viruses expressing potent suppressors of gene silencing can sometimes protect the viruses that are the RNAi target, if the co-infect the transgenic plant. Mixed infections, where multiple viruses infect the same plants, are common in nature.

11.7.2 Modification of the RNAi Strategy: RNAi, or Gene Silencing, Can Be Used, for Instance, to Affect Insect Vector Performance

Viruses in nature can be either vertically transmitted via infected pollen and seeds, or horizontally transmitted by virus vectors, most commonly insects (Nault 1997). Some viruses can be transmitted in both ways, but often agriculturally significant viruses are transmitted mainly by their insect vectors. Viruses are picked up by insects when they feed on virus-infected plants and then are introduced into healthy plants when the insects move from plant to plant and eject saliva containing viruses in the newly encountered plants, while feeding.

Novel research focuses on provoking the plant to control insect vectors and their associated viral diseases using RNAi, either through genetic transformation or VIGS. Many studies suggest that RNAi effects can be induced in insect cells, and even in whole insects that feed on such plants (plants expressing the RNAi sequences by transformation or VIGS). Artificial dsRNAs can be used to trigger the RNAi pathway. If the artificial dsRNA nucleotide sequence is synthesized (artificially assembled) to be identical to a specific insect mRNA, then that mRNA becomes the target of the RNAi machinery for destruction, effectively 'silencing' the corresponding gene and stopping protein translation. For instance, if a plant is transformed to produce a dsRNA molecule whose sequence corresponds to an insect gene, the plant will produce siRNAs that will target the insect RNA target, and the plant might become insect and virus resistant. How? When insects feed on the transgenic plants, they ingest the plant-produced siRNAs, and since insects also have a RNAi machinery, their defense system will use the ingested siRNAs to find and destroy the corresponding target RNA sequence, in this case an insect RNA! Clever, eh?

A recent study (Baum et al. 2007) has reported the use of RNAi in corn roots to control the western corn rootworm. The transgenic corn plants express in its roots the dsRNAs against the western corn rootworm ATP-ase mRNA, (ATP-ase is expressed in the insect gut and necessary for many vital processes) and these plants have been shown to be highly resistant to rootworm damage. This technology can be used, for instance to increase the durability of transgenic corn using the Bt (*Bacillus thuringiensis*) resistance.

Aside from targeting insects, RNAi can also be used to target endogenous plant genes that produce unfavorable traits in crops. For instance, Non-Browning Arctic® Apple in which the apple enzyme responsible for fruit browning has been silenced through RNAi, has been deregulated in the US and is commercially available.

11.8　Future Perspectives

Transgenesis in plants and the use of RNAi technologies is a subject of hot debates. Recently, scientists have discredited a study published by Zhang et al. in 2012 where the authors have found that the plant microRNA168 can be found in blood of mammalians (humans and mice) fed on rice, and that this microRNA can regulate mammalian gene expression in the liver. MicroRNAs belong to a class of small RNA molecules very similar to the one composed by siRNAs. MicroRNAs are produced by every organism and are used to regulate gene expression, especially during the organism's growth and development. Since the structure of miRNAs is similar to the one of siRNAs, this study indirectly poses a question mark on the stability across the mammalian digestive system of siRNA generated by transgenic plants and, not surprisingly, is the subject of intense debate. SiRNAs seem to be stable in the digestive tract of arthropods and siRNAs generated by plants and ingested by insects in some cases have been shown to affect distant organs, but no study has proven the same stability in mammalians. At the same time, we are exposed every day to our own miRNAs and ingest miRNAs produced by plants, other animals and even by microorganisms in large amounts. Further studies are needed to examine the stability and potential effects of miRNAs in the mammalian digestive system.

A petition to deregulate a transgenic American chestnut (*Castanea dentata*) line (called 'Darling 58') has been submitted in January 2020 to the United States Department of Agriculture (USDA) by researchers at the State University of New York (SUNY) American Chestnut Research and Restoration Center (USDA-APHIS (2020), Petitions for determination of nonregulated status). These trees have been engineered to tolerate the destructive chestnut blight, with the intent to repopulate the American forests where the American chestnut have been decimated by the blight. The inserted gene, derived from wheat, produces an oxalate oxidase enzyme that inhibits canker formation and forces the fungus that is the causal agent of the blight to remain saprophytic (Powell et al. 2019). These transgenic trees have not been approved (as of summer 2020), yet, and a strong opposition has mounted in the US to avoid the release of these trees, since they are meant to cross with the native chestnut trees in the wild.

A new technology that is rapidly taking foot and that should be monitored in the future, is the use of gene or genome-editing, in which the genomic DNA of living organisms is modified by insertion deletion, modification or replacement (Maeder and Gersbach, 2016). This technology exploits the use of sequence-specific nucleases such as meganucleases, zinc finger nucleases (ZFNs), transcription activator-like effector nucleases (TALENs), and the clustered regularly interspaced short palindromic repeat (CRISPR)/CRISPR-associated protein (Cas). Of these nucleases, CRISPR (Cas) is probably the most flexible and popular, and it has been already applied under laboratory settings to confer resistance to plant viruses (Ji et al. 2019, Mahas et al. 2019; Zhang et al. 2019; Ahmad et al. 2020).

References

Ahmad S, Wei X, Sheng Z et al (2020) CRISPR/Cas9 for development of disease resistance in plants: recent progress, limitations and future prospects. Briefings Funct Genomics 19(1):26–39

Baum JA, Bogaert T, Clinton W et al (2007) Control of coleopteran insect pests through RNA interference. Nature Biotechnol 25:1322–1326

Costa AS, Muller GW (1980) Tristeza control by cross protection: a U.S.-Brazil cooperative success. Plant Dis 64:538–541

Dangl JL, Jones JDG (2001) Plant pathogens and integrated defence responses to infection. Nature 411:826–833

Gonsalves C, Cai W, Tennant P, Gonsalves D (1997) Efficient production of virus resistant transgenic papaya plants containing the untranslatable coat protein gene of papaya ringspot virus. Phytopathology 87:S34

Howe GA, Jander G (2008) Plant immunity to insect herbivores. Annu Rev Plant Biol 59:41–66

Ji X, Wang D, Gao C (2019) CRISPR editing-mediated antiviral immunity: a versatile source of resistance to combat plant virus infections. Sci China Life Sci 62:1246–1249

Kaniweski WK, Thomas PE (2004) The potato story. Agric Biol Forum 7:41–46

Mahas A, Aman R, Mahfouz M (2019) CRISPR-Cas13d mediates robust RNA virus interference in plants. Genome Biol 20(1):1–16

Márquez LM, Redman RS, Rodriguez R, Roossinck MJ (2007) A virus in a fungus in a plant: three-way symbiosis required for thermal tolerance. Science 315:513–515

Nault LR (1997) Arthropod transmission of plant viruses: a new synthesis. Ann Entomol Soc Am 90:521–541

Powell-Abel P, Nelson RS, De B et al (1986) Delay of disease development in transgenic plants that express the tobacco mosaic virus coat protein gene. Science 232:738–743

Powell WA, Newhouse AE, Coffey V (2019) Developing blight-tolerant American chestnut trees. Cold Spring Harb Perspect Biol 11(7):a034587

Qu F, Morris TJ (2005) Suppressors of RNA silencing encoded by plant viruses and their role in viral infections. FEBS Lett 579:5958–5964

Scorza SR, Callahan A, Dardick C et al (2013) Genetic engineering of Plum pox virus resistance: 'HoneySweet' plum—from concept to product. Plant Cell Tissue Organ Cult (PCTOC) 115:1–12

Tricoli DM, Carney KJ, Russell PF et al (1995) Field evaluation of transgenic squash containing single or multiple virus coat protein gene constructs for resistance to cucumber mosaic virus, watermelon mosaic virus 2, and zucchini yellow mosaic virus. Bio/Technology 13:1458–1465

USDA-APHIS (2020) Petitions for determination of nonregulated status. USDA-APHIS petitions for determination of nonregulated status. https://www.aphis.usda.gov/aphis/ourfocus/biotechnology/permits-notifications-petitions/petitions/petition-status

Voinnet O (2001) RNA silencing as a plant immune system against viruses. Trends Genet 17:449–459

Waterhouse PM, Wang MB, Lough T (2001) Gene silencing as an adaptive defence against viruses. Nature 411:834–842

Xu P, Chen F, Mannas JP et al (2008) Virus infection improves drought tolerance. New Phytol 180:911–921

Zhang L, Hou D, Chen X et al (2012) Exogenous plant MIR168a specifically targets mammalian LDLRAP1: evidence of cross-kingdom regulation by microRNA. Cell Res 22:107–126

Zhang T, Zhao Y, Ye J et al (2019) Establishing CRISPR/Cas13a immune system conferring RNA virus resistance in both dicot and monocot plants. Plant Biotechnol J 17(7):1185

Cristina Rosa is Associated Professor at the Plant Pathology and Environmental Microbiology Department at The Pennsylvania State University (USA). She worked on the use of RNA interference to control the glassy winged sharpshooter. The purpose of her research is to study the complex interaction between plant viruses, plants, and insect vectors.

Part IV
Sustainable Environment

Chapter 12
Root Traits for Improving N Acquisition Efficiency

Hannah M. Schneider and Jonathan P. Lynch

Abstract Global agriculture requires the development of nutrient-efficient crops to improve food security while reducing environmental pollution. In developing nations, low soil nitrogen (N) availability and limited fertilizer usage is a primary limitation to crop production, while in developed nations, intensive N fertilization is a primary economic, energy, and environmental cost to crop production. In order to mitigate these risks, the development of crops with superior root traits enhancing N acquisition is essential. The development of crops with enhanced N capture would increase crop productivity, enhance sustainability, and reduce environmental pollution. There is substantial genetic variation for root phenes that have potential to improve N capture and reduce the N requirement of crops. In this chapter, we explore root phenes that enhance N acquisition that are urgently needed in global agriculture. Root ideotypes for enhanced N capture are predicted to pave the way for more productive and sustainable cropping systems.

Keywords Agriculture · Root · Phenotype · N

12.1 N-Efficient Crops are Needed in Global Agriculture

A major challenge for global agriculture is to improve crop productivity to feed a growing population while reducing the environmental impact. Sustainable crop production is an ever-growing challenge as the global population and food demand continue to rise. In agroecosystems, nutrient efficient crops are important for crop productivity as suboptimal nitrogen (N) availability is a primary limitation to crop growth. In low-input agricultural systems, nutrient deficiency is a primary limitation to crop productivity and therefore food security. In high-input agroecosystems, the energy and economic cost and pollution caused by intensive N fertilization is unsustainable. Chemical fertilizers are widely used to enrich soils with nutrients and improve plant production, however, add a significant cost to crop production and are

H. M. Schneider · J. P. Lynch (✉)
Department of Plant Science, The Pennsylvania State University, University Park, PA, USA
e-mail: jpl4@psu.edu

A. Ricroch et al. (eds.), *Plant Biotechnology*,
https://doi.org/10.1007/978-3-030-68345-0_12

163

a major source of environmental pollution. In addition, as little as 50% of applied fertilizer is captured by crop roots (Pask et al. 2012). The production of fertilizer requires finite fossil fuels and mineral reserves and therefore the dependency of fertilizers must be minimized to achieve food security. The key to development of nutrient efficient crops is harnessing genetic variation for root phenes (phene is to phenotype as gene is to genotype) (see Box 1) that enhance soil resource acquisition. Understanding the functional utility and genetic architecture of plant phenes that enhance soil resource capture is urgently needed for global food security.

Nitrogen, after carbon, oxygen, and hydrogen, is the most abundant mineral element in plants and is an important part of proteins, nucleic acids, and chlorophyll. Nitrogen is one of the most important inputs to agroecosystems as it is one of the primary limiting resources in agriculture production. Therefore, N acquisition is one of the most important crop breeding targets. Generally, soil N exists in three forms: organic N compounds and inorganic ammonium (NH_4^+) and nitrate ions (NO_3^-). Most of the potentially available soil N at any given time is in the form of organic N as plant and animal residues and soil organisms. Organic N compounds are typically not directly available to the plant but may be converted to plant available N forms (e.g. nitrate) by microorganisms. In agricultural systems, nitrate is generally the most abundant form of available N and acquired by crops in the greatest amounts. In modern agriculture, inorganic N fertilizers have become indispensable for crop production. Over the past 50 years, food production has increased nearly twofold which can largely be attributed to the use of N fertilizer. In high-input systems, the use of intensive N fertilization has the potential to increase crop production, however it comes at an environmental cost. Most of the applied N fertilizer is not taken up by the crop and the remainder is incorporated into soil organic matter or lost through erosion, leaching, surface runoff, or volatilization causing environmental pollution. In addition, the production of inorganic N fertilizers through the Haber-Bosch process is extremely energy intensive and has a large carbon footprint. Driven mainly by increases in fossil fuel prices, the price of inorganic N fertilizer has increased reducing profit margins in rich nations. In the developing world, where the majority of the global population lives, access to fertilizer is limited and farmers are often poor and cannot afford to buy fertilizers. The development of crop cultivars with enhanced soil resource acquisition is an important goal for global agriculture.

Improving N use efficiency (NUE) is an important strategy for boosting plant performance and yield in both high and low input agroecosystems. NUE, or the grain weight produced per unit of soil N is a result of N uptake and utilization processes including absorption, assimilation, and remobilization. The development of plants with improvement in both uptake and utilization efficiency is ideal. Efficient uptake of N is important as over half of applied N is lost from farmlands due to leaching and other factors. Roots are the interface between the soil and plant and provide important functions for the plant including resource uptake, storage, and soil anchorage. Therefore, roots have a substantial potential to improve N uptake. The great potential in improvement of NUE through plant uptake highlights the potential of root ideotype breeding to improve NUE. In this chapter, we discuss root phenes that enhance N acquisition, their genetic architecture, and potential deployment in crop breeding programs.

12.2 Physiological Mechanisms of N Uptake

Nitrogen is considered to be a mobile soil nutrient and moves through the soil primarily through mass flow. The process by which nutrients are transported to the root surface through water movement (e.g. percolation, transpiration, evaporation) is mass flow and the rate of flow determines the amount of nutrients transported to the surface of the root. Nitrate ions have a high solubility and therefore have greater uptake at peak plant transpiration (Lebot and Kirkby 1992). Nitrate solubility also influences its soil mobility as N is rapidly leached into deeper soil domains and groundwater, lost through surface runoff, or temporarily immobilized during periods of drought. Generally, N is more available in deeper soil domains, especially later in the growth season, due to crop uptake and leaching throughout the growth season.

Activity of high-affinity transport systems (HATS) and low-affinity transport systems (LATS) help crop plants acquire soil N, mainly in the form of nitrate. HATS activity may enhance N acquisition in suboptimal N through the upregulation of HATS-N in low-input agroecosystems. LATS activity is primarily constrained to high N concentrations and it may enhance N acquisition if the plant encounters patches and pulses of high N. However, the contribution of LATS to N uptake is typically negligible as in most agricultural systems the concentration of nitrate in the bulk soil solution is low. In the majority of agroecosystems, HATS and its kinetic parameters is the most meaningful component of the root nutrient uptake system as most systems have suboptimal N and fluctuating availability of N due to leaching, mineralization, and denitrification. The Michaelis Menten equation quantitatively describes the dynamics of N uptake. The Michaelis Menten equation has two key parameters: Vmax (the maximum velocity of uptake and measurement of the maximum uptake rate) and Km (the substrate concentration at which half of the maximum velocity is attained and measurement of the affinity of the uptake sites for the nutrient) (Fig. 12.1) (Griffiths and York 2020). Presumably, the type and number of transporters, assimilation machinery, and anatomical phenes influence uptake kinetics.

12.3 Root Ideotype for Improved Acquisition of N

Soil resources are spatially and temporally heterogeneous therefore adaptations to root systems are important for soil resource acquisition and therefore plant fitness. Root phenes have the potential to improve plant performance in suboptimal N availability by improving the metabolic efficiency of soil exploration. Metabolic efficient soil exploration can be achieved by optimizing the allocation of resources and thus

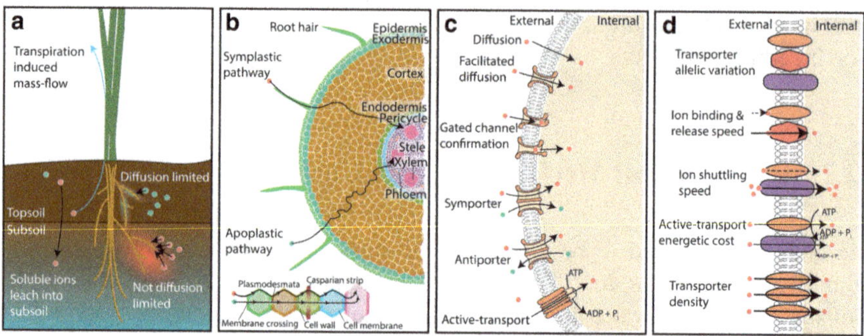

Fig. 12.1 Root uptake kinetics from the soil environment to transporter-level dynamics. **a** Mobility and bioavailability of soil nutrients determine the interception of nutrient ions with the root surface. **b** Nutrient ions may travel across the root through the apoplastic pathway (through the cell wall) or the symplastic pathway (through the cytoplasm) to reach the xylem for transport. **c** A variety of mechanisms exist for ions to enter the root through the soil solution. **d** Properties of transporters are presumably influenced by genetic variation and influence nutrient acquisition efficiency. From Griffiths and York (2020)

root foraging in specific soil domains. The construction and maintenance of root tissue requires an investment of resources (e.g. carbon and nutrients) by the plant. The reduction or elimination of unnecessary root tissue or investment in root tissues that are more metabolically efficient create roots with more efficient soil resource capture and greater plant performance in N stress. The metabolic cost (i.e. carbon and nutrient cost) of the root system is substantial and can exceed 50% of the daily photosynthesis (Lambers et al. 2002). The metabolic cost of soil exploration can significantly influence plant performance and yield under edaphic (soil-related) stress. Every unit of root growth requires each unit of leaf area to sustain relatively more non-photosynthetic tissue. Plants that are able to acquire N at a reduced metabolic cost will enable more metabolic resources to be available for further root growth and thus greater soil exploration, nutrient capture, and plant productivity. Strategic investment of plant resources results in more efficient root system function and resource capture.

An important ideotype in maize for the capture of soil N in both high input and low input agroecosystems is the 'steep, cheap, and deep' ideotype developed by (Lynch 2013) (Figs. 12.2 and 12.3). The 'steep, cheap, and deep' ideotype consists of several root architectural, anatomical, and morphological phenes that enhance N and water acquisition in maize by improving deep soil exploration at a reduced metabolic cost. The 'steep' component of the ideotype refers to the steep growth angle of nodal roots that increase root foraging in deep soil domains. The 'cheap' component of the ideotype refers to root anatomical and architectural phenes that reduce the metabolic cost of soil exploration therefore permitting further root growth in deeper soil domains. Significant natural genetic variation for 'steep' and 'cheap' root phenes that allow roots to explore 'deep' soil domains. Genetic variation for root phenes coupled with high throughput field phenotyping methods should enable

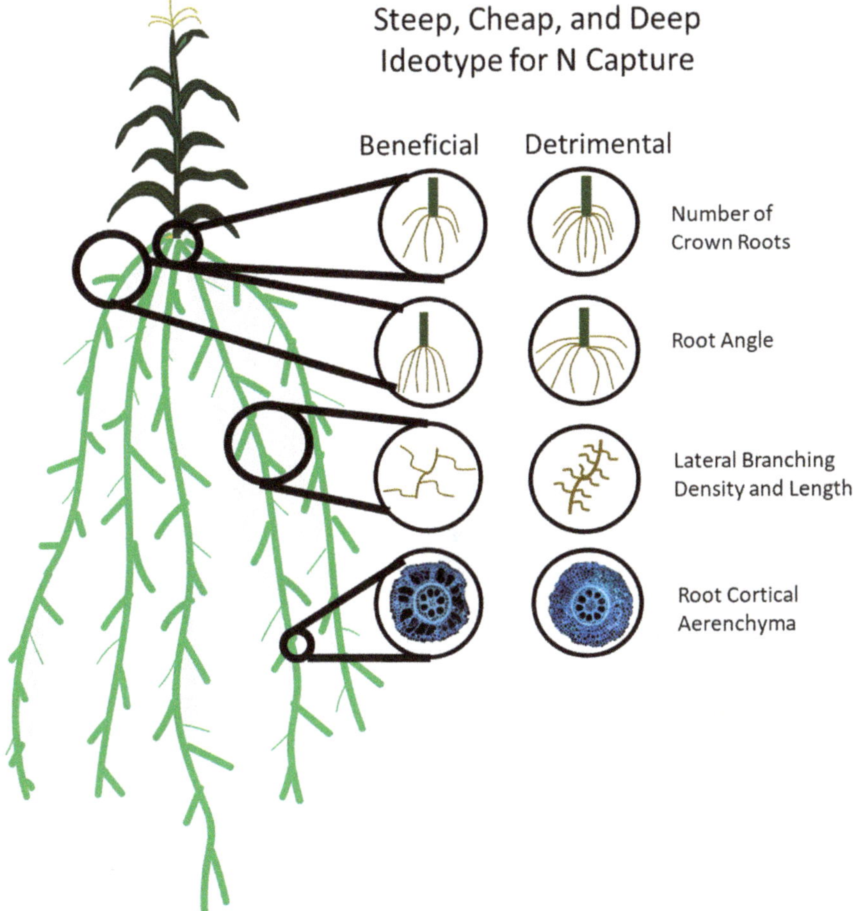

Fig. 12.2 The 'steep, cheap, and deep' ideotype for enhanced plant performance in N stress. In N stress, few nodal roots with a steep angle, few long lateral branches, and root cortical aerenchyma formation are adaptive responses for stress tolerance. Modified from Schneider and Lynch 2020

the understanding and deployment of improved root ideotypes into crop improvement programs. Understanding the fitness landscape of phenes, interactions between phenes, plastic responses of phenes, and their genetic control will lead to enhanced N acquisition and therefore more productive crops in both low and high-input systems. In the sections below we will briefly discuss some of the phene components of the 'steep, cheap, and deep' ideotype.

Fig. 12.3 The 'steep, cheap, and deep' and 'topsoil foraging' ideotypes in maize (upper panel) and common bean (lower panel). Root systems are simulated to 42 d after germination by OPENSIM-ROOT. Root systems in the center represent standard maize and common bean phenotypes. The maize root system represents a non-tillering monocot root architecture and the common bean root system represents an annual dicot root architecture. The 'steep, cheap, and deep' ideotype enhances subsoil resource acquisition and is beneficial for the capture of N. In contrast, the topsoil foraging phenotype is useful for the capture of topsoil resources including phosphorus, potassium, calcium, magnesium, and recently mineralized nitrate. From Lynch 2019

12.4   Root Architecture Ideotype for Improved Acquisition of N

Root architecture is the spatial configuration of the root system within the soil matrix. Root phenes have important roles in soil resource capture, particularly in environments with low nutrient availability. Root phenes determine the spatial and temporal placement of roots in specific soil domains and thus the capture of nutrient resources. As N is more available in deeper soil domains, root phenes that place roots deep in the soil profile at a reduce metabolic cost to the plant will be beneficial for N acquisition.

Root system architecture in monocots and dicots differs significantly (see Box 2), however many root phenes that confer enhanced N acquisition are similar. For example, in maize, root angle has an important role in N acquisition (Fig. 12.4). The angle which roots penetrate the soil influences plant depth and a steeper growth angle results in deeper rooted plants. Deeper roots enable the capture of N in deep soil domains and therefore had enhanced plant performance (Trachsel et al. 2013). Similarly in dicots, plants with a steep or fanned angle phenotype performed better under N stress when compared to plants with a shallow angle phenotype (Rangarajan et al. 2018). In maize, nodal root growth angles commonly are shallow in older nodes and steeper in younger nodes in maize. The emergence of roots with progressively steeper growth angles establishes a root architecture that initially is shallow, which coincides with greater availability of N in the topsoil during seedling establishment. The emergence of steeper growth angles in younger nodal roots over time coincides with the greater availability of N in deeper soil strata as the season progresses through leaching and crop uptake (York and Lynch 2015).

The production of fine roots (e.g. lateral roots) also plays a role in soil resource acquisition. Greater root length can be achieved with a fixed proportion of assimilates through reducing root diameter i.e. specific root length or the length of root per unit root mass. Reducing the construction and maintenance costs per unit root length enables a larger volume of soil exploration at a reduced metabolic cost. The length of lateral roots and the density at which they emerge from the main root axes plays a role in N capture and plant performance. In maize, fewer but longer lateral roots emerging from crown root axes are beneficial in N stress (Zhan and Lynch 2015) (Fig. 12.5). Similarly in common bean, a reduction in lateral root branching density enabled plants to accumulate a greater shoot biomass in low N environments (Rangarajan et al. 2018). Fewer lateral roots reduce competition for nutrients between different roots of the same plant (i.e. intra-root competition) and between roots of different plants (i.e. inter-root competition). Fewer, longer lateral roots reduce the metabolic cost of soil exploration and enable further root growth in deep soil domains and therefore improve plant performance and yield in low N environments.

The number of axial roots also influence soil N acquisition. In a field study, maize genotypes that had fewer crown roots had greater rooting depth, greater capture of deep soil N, and therefore greater plant growth and yield in low N environments (Fig. 12.6a). The emergence of fewer crown roots enables deeper soil exploration and the capture of deep N that has been leached into deeper soil domains throughout the growth season. Maize plants with fewer nodal roots, increase N capture and therefore plant performance through reallocation of carbon and resources to the growth of lateral roots, embryonic roots, and first node crown roots that grow in deep soil domains to increase soil foraging efficiency (Saengwilai et al. 2014b). Similarly in dicots, a smaller number of basal root whorls and fewer hypocotyl-borne roots increased root depth and enabled better N capture (Rangarajan et al. 2018). The

Phenotypic variation for crown root angle in the field

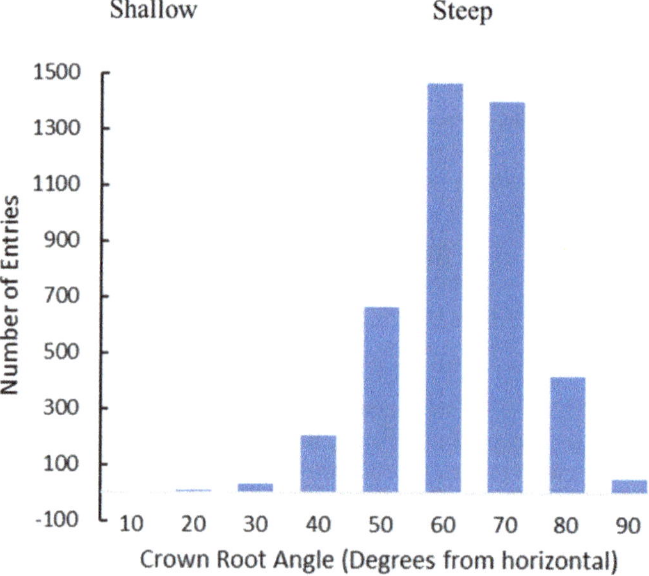

Fig. 12.4 Root growth angle influences the capture of N. Crown root angle in maize has a wide range of natural genetic variation in the field and greenhouse. Maize plants with steep crown root angles have greater root depth and plant growth in N stress

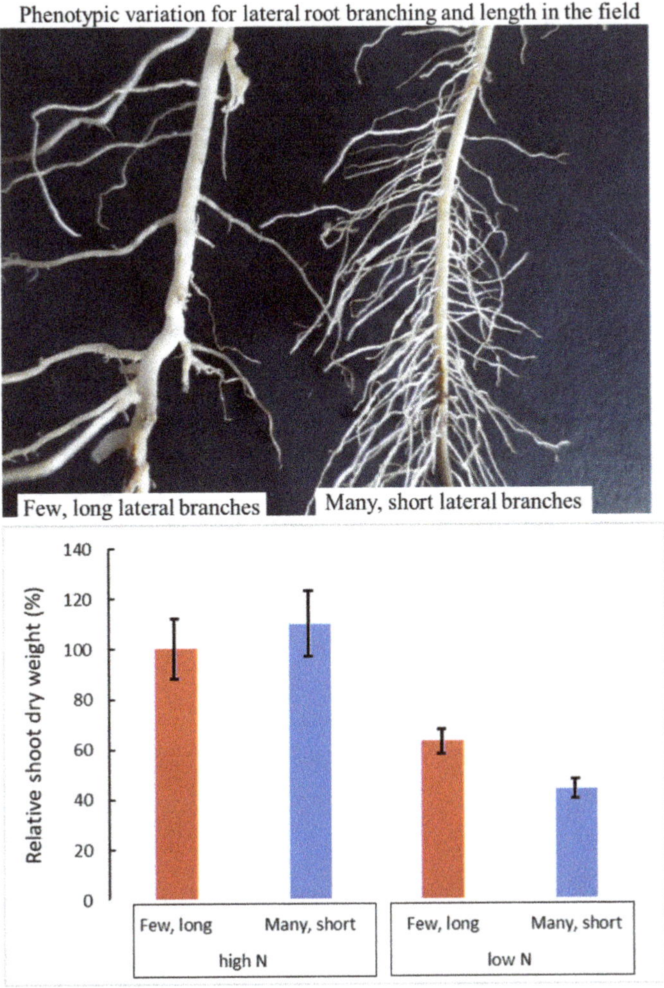

Fig. 12.5 Lateral root branching density and length is associated with N stress tolerance. Maize displays genetic variation for the density and length of lateral roots. In N stress in the field and greenhouse, maize lines with few, long lateral branches had 35% greater growth compared to lines with many short lateral branches Modified from Zhan and Lynch 2015

construction and maintenance of fewer axial roots enable further root growth of established axial roots into deeper soil domains.

Phenotypic variation for crown root number and root
cortical aerenchyma in the field

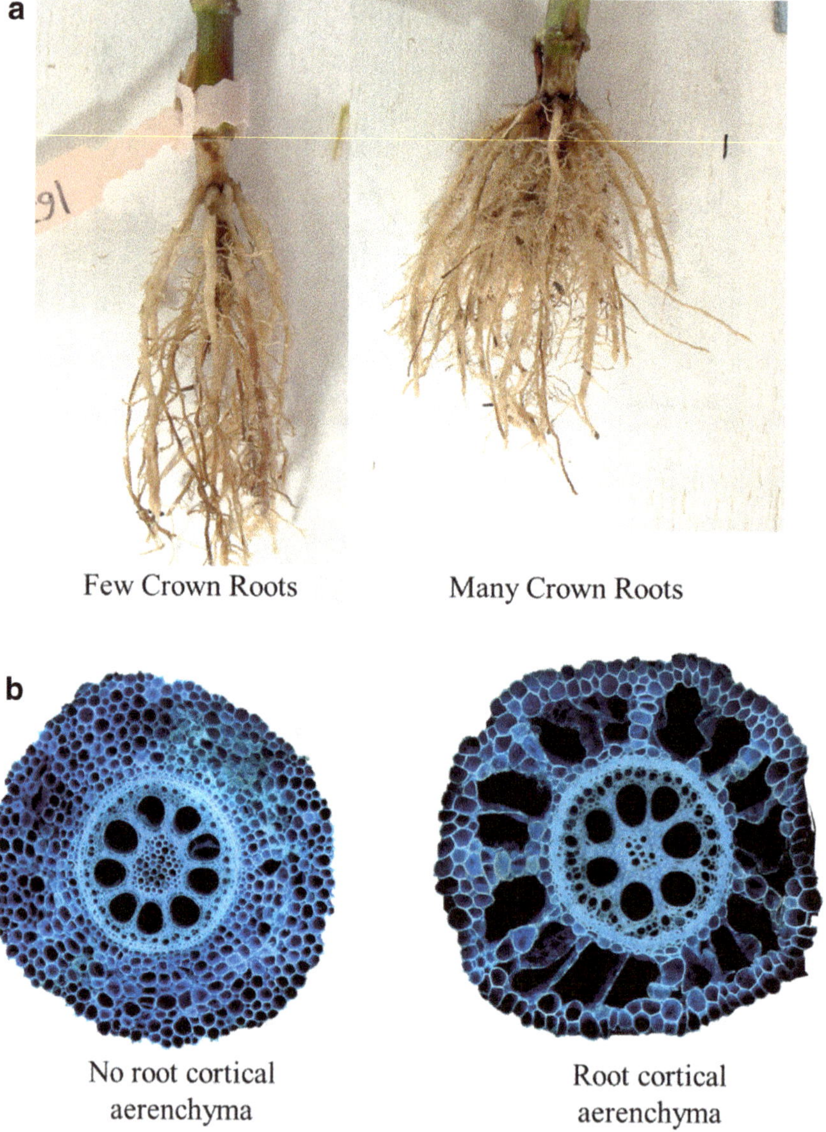

Fig. 12.6 Natural genetic variation exists for the formation of **a** the number of crown roots and **b** root cortical aerenchyma (RCA). Fewer crown roots and root cortical aerenchyma formation improve N capture

12.5 Root Anatomical Ideotype for Improved Acquisition of N

Root anatomical phenes can influence the metabolic cost of root tissue. Several anatomical phenes reduce the amount of living cortical tissue. A reduction in living cortical tissue therefore reduces root respiration and nutrient content, thereby permitting greater resource allocation to other plant processes including growth and reproduction. The change in ratio between respiring and non-respiring tissue reported in various anatomical phenes have been shown to have large effects on the metabolic cost of the root. For example, the formation of root cortical aerenchyma (RCA) replaces living cortical aerenchyma with air spaces (i.e. lacunae). Morphological, anatomical, and architectural phenes that reduce the metabolic cost of root tissue have the potential to enhance N acquisition.

Root cortical aerenchyma is primarily known for increased oxygen transport in hypoxic conditions, however RCA formation is induced by suboptimal availability of N, phosphorus (P), sulfur, and water. The formation of RCA converts living cortical parenchyma tissue into large intercellular spaces (i.e. lacunae) by programmed cell death. In low N environments, the formation of RCA in maize increased rooting depth in deep soil domains, increased leaf N and chlorophyll content, and increased plant biomass and grain yield through a reduction in the metabolic cost of root tissue (Fig. 12.6b). Similarly to RCA, root cortical senescence (RCS) involves the senescence of root cortical cells in temperate small grains through programmed cell death. RCS reduces the respiration and nutrient content of root tissue and the reduced metabolic costs associated with RCS can be attributed to enhance plant performance in low N environments.

Over the past 100 years, agricultural inputs and management practices have dramatically changed in rich nations from low fertilizer inputs and low planting densities to intensive fertilization and dense plant populations. Modern maize lines have a shallower root growth angle, fewer nodal roots, and a greater distance between nodal root emergence to lateral root emergence when compared to older maize lines. RCA formation increased in greater population densities. These changes in root architecture and anatomy have the potential to increase maize shoot growth by 16% in high N environments with a high planting density (York et al. 2015). The evolution of maize root anatomical and architectural phenes over the past century is consistent with greater N capture.

12.6 Root Phene Synergisms for N Capture

Root architectural and anatomical phenes do not function in isolation. Ideotype, or trait-based breeding, is a strategy to combine phenes that each would contribute to increased yield. Ideotype breeding not only considers phenes in isolation, but also the relationship between phenes as the integration of phenes determines how the whole plant functions. Phene synergisms (i.e. when the plant performance response of the interaction of multiple phenes is greater than the expected additive effect) are most effective between root phenes that affect the placement of roots in the soil profile and those that reduce the metabolic cost of root tissue. For example, in common bean the utility of root hairs is influenced by the root growth angle as the root angle determines the placement of root hairs in the soil profile (Miguel et al. 2015). Phene synergisms exist between RCA and the formation of lateral roots. Greater lateral root length branching density permits greater soil exploration and therefore the capture of soil resources. However, increased lateral branching length and density increases the metabolic demand of the root and due to competing sinks this could influence the growth of other root classes. This trade-off can be alleviated by decreasing the metabolic demand of the root through the formation of RCA. At an intermediate level of N, plants with RCA in lateral roots had 220% greater shoot biomass that was attributed to RCA formation (Postma and Lynch 2011). RCS had greater utility for resource capture and plant performance in plants with fewer tillers. Plants with RCS and few tillers had 20% greater shoot growth in suboptimal N availability. The growth of tillers coincides with the growth of nodal roots associated with tillers. In suboptimal N availability, nodal roots from tillers depleted limited soil nutrient resources faster and subsequently plant growth was reduced as soil resources became limiting in later growth stages. Therefore, a reduction in the number of tillers, and therefore nodal root development, reduced intra-root competition and enhanced plant growth in suboptimal N. Similarly, in low N environments, plants with RCS had 12% greater shoot growth when they also had a fewer lateral branches due to decreased intra-root and inter-root competition (Schneider et al. 2017). Understanding the fitness landscape of phene interactions is an important consideration for ideotype breeding.

12.7 Dimorphic Root Phenotypes for N Capture

In the field, plants may be exposed to dynamic or multiple, simultaneous stresses. For example, in low-input agroecosystems, availability of deep (i.e. N and water) and shallow (i.e. P and potassium (K)) resources are often co-limiting in crop production. The 'steep, cheap, and deep' ideotype may optimize the capture of mobile resources like N, however, may be maladaptive for the capture of shallow resources. For example, in common bean, plants with a steep or fanned root growth angle and fewer nodes of basal roots had enhanced N uptake, however the phenotype was maladaptive for phosphorus uptake. A root system optimized for phosphorus uptake has increased root length in shallow soil domains achieved through a shallow growth angle and a greater number of basal and hypocotyl-borne roots to enhance topsoil exploration (Rangarajan et al. 2018) (Fig. 12.3). An increased number of roots in shallow soil domains results in an increased metabolic cost of the root system and therefore a reduced root depth and trade-offs for N acquisition.

Dimorphic root phenotypes capable of effectively acquiring both deep and shallow soil resources would be beneficial in co-optimizing the capture of N, P, K, and water in environments where the availability of all soil resources is suboptimal (e.g. low-input systems). For example, a successful dimorphic root phenotype in common bean is a plant with an increased number of basal root whorls. A greater number of basal root whorls results in a greater range of growth angles and therefore a greater vertical range of root distribution and soil exploration which improves foraging in both the topsoil and deep soil domains (Rangarajan et al. 2018). Simulation modeling studies have demonstrated that numerous combinations of anatomical and architectural phenes could generate ideotypes for the capture of multiple, dynamic stresses. Integrated phenotypes optimized for the capture of soil resources are likely to differ between monocots and dicots.

12.8 Root Phenes Have Plastic Responses to N

Phenotypic plasticity is the ability of an organism to alter its phenotype in response to the environment (see Box 3). Soil is often heterogeneous and resource availability is spatially and temporally dynamic which often alters the expression of root phene states. For example, in suboptimal N environments, maize root angles became steeper enhancing soil exploration in deep soil domains and subsequently N capture (Trachsel et al. 2013). The formation of RCA increased in N deficit, enabling metabolic resources to be directed towards deeper root length, biomass, and reproduction (Saengwilai et al. 2014a). Similarly, nutrient deficiency, including suboptimal N availability, accelerated the rate of RCS formation. Enhanced RCS formation and the subsequent reduction in the metabolic cost of the root, enabled additional root growth in deep soil domains and enhanced the capture of N (Schneider et al. 2017).

In variable environments, phenotypic plasticity may be advantageous. However, in high input environments, with intensive fertilization and monoculture growth systems, phenotypic plasticity may be maladaptive. The evolution of crops took place in environments with biotic and abiotic stress and root growth and function evolved strategies for soil resource capture. However, in high input environments, constraints for soil resource acquisition are mitigated as crops are fertilized and irrigated. In most agricultural systems, root phenes, whether plastic or not, that contribute to the exploration of deep soil domains enhance N capture (Lynch 2019). The fitness landscape of phenotypic plasticity is dependent on specific environments and management practices. The genetic architecture of phenotypic plasticity is complex and highly quantitative. Before we integrate phenotypic plasticity into breeding programs, we first need to understand the utility of root phenes and interactions between root phenes in order to better understand adaptive or maladaptive plasticity under specific edaphic stresses.

12.9 Genetic Variation for Root Phenes and Breeding Strategies

Few genetic loci have been identified for anatomical phenes including root stele and xylem vessel diameter in rice (Uga et al. 2008, 2010), xylem vessel phenes in wheat (Sharma et al. 2010), and RCA in *Zea* species (Mano et al. 2005, 2007). In maize, genetic loci for the areas of cross section, stele, cortex, aerenchyma, and cortical cells, root cortical aerenchyma, and cortical cell file number have been identified through quantitative trait loci (QTL) mapping of recombinant inbred lines (Burton et al. 2014). In addition, genetic loci associated with root cross-sectional area, area of the stele and cortex, root cortical aerenchyma, lateral branching density and length, root angle, size of cortical cells and the number of cortical cell files in maize have been identified through genome-wide association mapping (Schneider et al. 2020a, b). However, the integration of these phenes into breeding programs is complex due to the highly quantitative nature of root phenes.

A large number of genetic loci each contribute small effects to the expression of a single phene. The genetic control of root phenes is further complicated by the distinct genetic control and expression of root phenes in abiotic stress and their plastic response to those stresses. Highly quantitative traits can pose a challenge for conventional breeding programs that use single-trait breeding strategies and marker assisted selection. The use of conventional tools would require the stacking of hundreds of genes for the development of desirable root ideotypes. Modern breeding methods, including genomic selection enable the selection of multiple loci. Genomic selection methods should include phenes and integrated phenotypes, not just selection for yield. Selection for individual root phenes has advantages compared to brute-force yield selection for edaphic stress. Landraces and wild germplasm may be an important source of phenotypic variation for ideotype development and in crop improvement programs. Genomic selection training sets must also consider wild and landrace germplasm, which presumably express more combinations and greater extremes of root phenes. Crop breeding programs must not only include expertise in genetics, plant pathology, and agronomy, but also soil science, plant nutrition, and ecophysiology in order to guide the development of crop ideotypes with enhanced N capture. We need to make a concerted effort to train and support scientists capable of working across disciplines to develop more N efficient crops.

Although the integration of plant phenes in breeding programs has been limited, there are a few notable examples. In wheat, breeding for smaller diameter xylem vessels has been employed as a means for improving water use efficiency. Plants with narrow xylem vessels had 3–11% greater grain yield when compared to the control in drought environments (Richards et al. 1989). In addition, the suberin content in the cortical exodermis has also been examined as a potential breeding target for Phytophthora resistance in soybean (Ranathunge et al. 2008). The 'Topsoil foraging' ideotype for improved phosphorus acquisition, which includes longer, denser root hairs and several architectural phenes that position more root foraging in shallow soil domains with greater phosphorus bioavailability (Lynch 2019), has been used to breed new varieties of common bean with substantially improved productivity in low fertility soils of Africa (Burridge et al. 2019). The continued development of root phenotyping platforms and the identification of genes associated with root phenes should facilitate the integration of root phenes in crop breeding programs.

12.10 Future Perspectives

The development of nutrient efficient crops has the potential to address urgent global challenges. Understanding plant evolution and biology requires the understanding of plant adaptation to nutrient constraints. However, this topic receives very little attention in basic and applied plant biology. The scientific community has made great progress in understanding and manipulating the plant genome, however most root phenes are genetically and physiologically complex and do not readily align with the gene-centric paradigm that dominates plant biology. Phenotypic plasticity and heterogenous soil environments complicate the identification of highly quantitative genes associated with root phenes through genome-wide association mapping. Future work must focus on understanding the fitness landscape (i.e. how root phenes affect cop performance in an array of environments and phene combinations). The fitness landscape of root phenes and their plastic responses is poorly understood. Differences in growth between controlled and field environments are often overlooked. Controlled environments (e.g. greenhouse, growth chamber) often do not represent the heterogeneous field environment (e.g. planting density, light, temperature, soil bulk density, nutrient distribution) and therefore are challenging to identify and understand plant phenes and their responses. Field phenotyping is often a bottleneck in crop physiology and breeding programs and high-throughput phenotyping often does not allow for the measurement of complex phene states. Field environments are usually heterogeneous and dynamic and difficult to replicate, measure, and monitor. In silico biology enables the study of complex interactions of the root fitness landscape that are not possible empirically and many environment and phenotype combinations that do not exist in nature. The study of the utility of root phenes for enhanced N capture requires many disciplines including ecology, physiology, development morphology, genetics and in silico biology.

12.11 Conclusion

In summary, there is great potential for the development of N efficient crops. Several high-throughput methods can be used to rapidly phenotype important root phenes for N capture. As discussed above, many root phenes for N capture have been well characterized and genetic loci have been identified enabling ideotype breeding. Root phenes important for N capture express significant genetic variation. Multiple phenes, each under distinct genetic control, interact with each other and the environment to determine the fitness of the root system. However, further research is needed to understand phene synergisms, trade-offs associated with other plant functions, and how phenes interact to influence plant fitness in different environments and management practices. The development of N efficient crops may allow farmers in low-input systems to increase crop productivity and climb out of the poverty trap of low input and low yields. N efficient crops may be easily adoptable in low-input systems as they have relatively few barriers for implementation and have the potential to have large effects on crop yields. In high-input systems, the development of N efficient crops may reduce environmental pollution and increase agricultural sustainability.

Box 1. What is a phene?

A phene is to phenotype as gene is to genotype. A phene is a distinct element of an organism's phenotype. A collection of distinct phenes comprise the phenotype, as a collection of distinct genes comprise the genotype. Genes have variants called alleles and similarly, phenes have variants called phene states. An example of a root phene is root growth angle. Root angle has at least two phene states: 'steep' and 'shallow'. The utility of soil resource acquisition is dependent on the phene state. For example, the 'steep' phene state is important for N acquisition in leaching environments.

Box 2. Root Classes

In annual crops, generally the root system of monocots consists of three major root classes: primary, seminal (or seed-borne) and nodal (shoot-borne) roots which all produce lateral (root-borne) roots of the first and second order. Nodal roots that emerge below ground are referred to as crown roots and nodal roots that emerge above ground are called brace roots. The root system of a dicot is distinct from that of a monocot. For example, a bean root system consists of a primary root, hypocotyl-borne roots, and basal roots which all produce lateral roots.

Box 3. Phenotypic Plasticity

Phenotypic plasticity is the ability of an organism to change its phenotype in response to the environment. Phenotypic plasticity may involve changes in physiology, morphology, anatomy, development, or resource allocation and it is phene specific, not a characteristic of an organism as a whole. The plastic response of an organism may be adaptive, maladaptive, or neutral in regard to its fitness. An example of an adaptive plastic response is the response of root angle in suboptimal N. In low N environments, maize plants had steeper growth angles that were able to explore deep soil domains and enhance N capture.

References

Burridge JD, Findeis JL, Jochua CN, Mubichi-Kut F, Miguel MA, Quinhentos ML, Xerinda SA, Lynch JP (2019) A case study on the efficacy of root phenotypic selection for edaphic stress tolerance in low-input agriculture: common bean breeding in Mozambique. Field Crops Res 244:107612

Burton AL, Johnson J, Foerster J, Hanlon MT, Kaeppler SM, Lynch JP, Brown KM (2014) QTL mapping and phenotypic variation of root anatomical traits in maize (Zea mays L.). Theor Appl Genet 128:93–106

Griffiths M, York LM (2020) Targeting root ion uptake kinetics to increase plant productivity and nutrient use efficiency. Plant Physiol 182:1854–1868

Lambers H, Atkin OK, Millenaar FF (2002) Respiratory patterns in roots in relation to their functioning. In: Waisel Y, Eshel A, Kafkaki K (eds) Plant roots, hidden half, third edit. Marcel Dekker Inc, New York, pp 521–552

Lebot J, Kirkby E (1992) Diurnal uptake of nitrate and potassium during the vegetative growth of tomato plants. J Plant Nutr 15:247–264

Lynch JP (2019) Root phenotypes for improved nutrient capture: an underexploited opportunity for global agriculture. New Phytol 223:548–564

Lynch JP (2013) Steep, cheap and deep: an ideotype to optimize water and N acquisition by maize root systems. Ann Bot 112:347–357

Mano Y, Omori F, Muraki M, Takamizo T (2005) QTL mapping of adventitious root formation under flooding conditions in tropical maize (Zea mays L.) seedlings. Breed Sci 55:343–347

Mano Y, Omori F, Takamizo T, Kindiger B, Bird RM, Loaisiga CH, Takahashi H (2007) QTL mapping of root aerenchyma formation in seedlings of a maize × rare teosinte "Zea nicaraguensis" cross. Plant Soil 295:103–113

Miguel MA, Postma JA, Lynch J (2015) Phene synergism between root hair length and basal root growth angle for phosphorus acquisition. Plant Physiol 167:1430–1439

Pask AJD, Sylvester-bradley R, Jamieson PD, Foulkes MJ (2012) Field crops research quantifying how winter wheat crops accumulate and use nitrogen reserves during growth. F Crop Res 126:104–118

Postma JA, Lynch JP (2011) Root cortical aerenchyma enhances the growth of maize on soils with suboptimal availability of nitrogen, phosphorus, and potassium. Plant Physiol 156:1190–1201

Ranathunge K, Thomas RH, Fang X, Peterson CA, Gijzen M, Bernards MA (2008) Soybean root suberin and partial resistance to root rot caused by phytophthora sojae. Biochem Cell Biol 98:1179–1189

Rangarajan H, Postma JA, Lynch JJP (2018) Co-optimisation of axial root phenotypes for nitrogen and phosphorus acquisition in common bean. Ann Bot 122:485–499

Richards RA, Passioura J (1989) A breeding program to reduce the diameter of the major xylem vessel in the seminal roots of wheat and its effect on grain yield in rain-fed environments. Crop Pasture Sci 40:943–950

Saengwilai P, Nord EA, Chimungu JG, Brown KM, Lynch JP (2014a) Root cortical aerenchyma enhances nitrogen acquisition from low-nitrogen soils in maize. Plant Physiol 166:726–735

Saengwilai P, Tian X, Lynch JP (2014b) Low crown root number enhances nitrogen acquisition from low nitrogen soils in maize (Zea mays L.). Plant Physiol 166:581–589

Schneider HM, Lynch JP (2020) Should root plasticity be a crop breeding target? Front Plant Sci 11:546. https://doi.org/10.3389/fpls.2020.00546

Schneider H, Klein S, Hanlon M, Brown K, Kaeppler S, Lynch J (2020a) Genetic control of root anatomical plasticity in maize. Plant Genome In Press

Schneider H, Klein S, Hanlon M, Nord E, Kaeppler S, Brown K, Warry A, Bhosale R, Lynch J (2020b) Genetic control of root architectural plasticity in maize. J Exp Bot In Press

Schneider H, Postma JA, Wojciechowski T, Kuppe C, Lynch JP (2017) Root cortical senescence improves growth under suboptimal availability of N, P, and K. Plant Physiol 174:2333–2347

Sharma S, Demason DA, Ehdaie B, Lukaszewski AJ, Waines JG (2010) Dosage effect of the short arm of chromosome 1 of rye on root morphology and anatomy in bread wheat 61, 2623–2633

Trachsel S, Kaeppler SM, Brown KM, Lynch JP (2013) Maize root growth angles become steeper under low N conditions. F Crop Res 140:18–31

Uga Y, Okuno K, Yano M (2008) QTLs underlying natural variation in stele and xylem structures of rice root. 14:7–14

Uga Y, Okuno K, Yano M (2010) Fine mapping of Sta1, a quantitative trait locus determining stele transversal area, on rice chromosome 9 9, 533–538

York LM, Galindo-Castaneda T, Schussler JR, Lynch JP (2015) Evolution of US maize (Zea mays L.) root architectural and anatomical phenes over the past 100 years corresponds to increased tolerance of nitrogen stress. J Exp Bot 66:2347–2358

York LM, Lynch JP (2015) Intensive field phenotyping of maize (*Zea mays* L.) root crowns identifies phenes and phene integration associated with plant growth and nitrogen acquisition. J Exp Bot 66:5493–5505

Zhan A, Lynch J (2015) Reduced frequency of lateral root branching improves N capture from low N soils in maize. J Exp Bot 66:2055–2065

Hannah M. Schneider is a Postdoctoral Scholar in the Department of Plant Science at The Pennsylvania State University, USA. She holds a Ph.D. degree in Horticulture from The Pennsylvania State University. Her research focuses on identifying and understanding the utility of root anatomical traits for improved soil water and nutrient acquisition. Using both genetic and physiological approaches her work aims to understand plant adaptations to drought and low fertility environments.

Jonathan P. Lynch is a Distinguished Professor in the Department of Plant Science at the Pennsylvania State University, USA.

Chapter 13
Sustainable Soil Health

Mary Ann Bruns and Estelle Couradeau

Abstract Soil is Nature's support medium for plant growth, but soils on only 12% of Earth's land area possess the inherent physical requirements to function for long periods as cropland. One-fourth of this land is now moderately to severely degraded, and efforts are now underway to determine the best means to increase their productivity by improving soil health. *Soil health* is a measure of how well soil functions in retaining water and recycling nutrients to support robust plant growth. Biotechnologies for crop improvement will have reduced effects if crops continue to be grown on degraded soils. Even when water and fertilizer are available, agricultural use of degraded soils often results in inefficient resource use and off-site pollution. *Soil quality*, a term sometimes used interchangeably with soil health, applies specifically to observable or measurable soil properties that indicate soil health. Although arable soils around the world differ in inherent properties that determine potential productivity, each soil has alterable properties that can be managed to sustain high productivity into the future. Alterable properties include soil organic matter content, root and microbial density, and macroporosity, all of which are highly dependent on maintaining biological diversity and activity in the soil. Because alterable properties undergo drastic changes when native vegetation is removed and land is disturbed for crop production, sustainable soil health involves restoring biological integrity through proper management. This chapter describes how soils are formed, why soils vary in productivity, and how soil quality can be evaluated in the field, in laboratories, and by advanced research facilities. The chapter concludes with a discussion of how soil health can be improved through diversified cropping, use of microbial technologies, and soil management practices that promote beneficial root-microbe interactions.

Keywords Soil quality · Organic matter · Soil aggregation · Rhizosphere · Microbial biomass · Microbiome

M. A. Bruns (✉) · E. Couradeau
Department of Ecosystem Science and Management, The Pennsylvania State University, University Park, PA 16802, USA
e-mail: mvb10@psu.edu

© The Author(s), under exclusive license to Springer Nature Switzerland AG 2021
A. Ricroch et al. (eds.), *Plant Biotechnology*,
https://doi.org/10.1007/978-3-030-68345-0_13

13.1 Introduction

Soil is a dynamic natural body composed of minerals, organic matter, and organisms lying between the Earth's crust and the atmosphere. Soil forms slowly as a function of root and microbial activity, downward water percolation, and mineral weathering. Soil formation rates are estimated at about 0.04–0.08 mm per year, resulting in annual natural accumulations of 0.5–1 ton of soil per hectare (Brady and Weil 2007). Soils which can support food production have taken centuries to develop and must not be confused with displaced "dirt" or "dust."

Soils on only 12% of Earth's land area possess the requirements to function as cropland over multiple human generations (e.g., sufficient soil depth, adequate moisture, permissive temperatures, moderate slope), while steeper and shallower soils on 26% more land permit function as pasture or rangeland (FAO 2013). Our soil resource is threatened, however, by erosion, deforestation, overgrazing and mismanagement, with soil loss rates estimated to be 10–30 times faster than rates of natural soil formation. Global averages of 5–40 tons of soil are lost per hectare per year, with wind- and water-borne particles ending up in rivers, reservoirs, and oceans (Pimentel 2006). With a projected world population of 9 billion people by 2050, food security depends on sustaining and intensifying agricultural productivity of these soils.

This chapter discusses soil as a critical food-producing resource and explains "soil health" as an integrative management objective for improving agricultural productivity. Despite decades of government-supported soil conservation programs, agricultural soils in developed countries continue to undergo serious degradation. Although developed countries can partly compensate for soil loss by increased use of fertilizer and irrigation, such resources are not available to most farmers in the developing world, where native soils are often less suited for agriculture. Soil conservation practices that focus only on keeping erosion losses to "tolerable Levels" will not provide food security. Management practices must instead aim to restore the biophysical integrity and biological diversity that characterize soil health. To that end, this chapter also describes methods that farmers, consultants, and scientists can implement to evaluate soil health.

13.2 Definition of Soil Health

Soil health is defined as a specific soil's capacity to provide requirements for vigorous plant growth while protecting the environment from off-site losses of soil, water, and nutrients. Healthy soils support beneficial levels of biological activity, making them resilient and capable of providing a self-regulating, low-stress habitat for life. While the terms "soil health" and "soil quality" have sometimes been used interchangeably, soil quality more precisely refers to individual soil properties that can be measured and which may or may not be changeable with management (Brady and Weil 2007). Observable or measurable properties include soil color, organic matter content, aggregation, porosity, and biological activity, all of which are used as *soil health indicators*. Soil health is a relative assessment that depends on how the soil is used (e.g., for row crops, pastures, orchards). Health is assessed on the basis of specific soils, because natural soils around the world (and even within a given region) vary greatly in their inherent capabilities to support growth of plants.

Removal of native vegetation and disruption of natural soil structure makes agricultural soils more susceptible to degradation. Soil degradation, whether it occurs through organic matter depletion, exhaustive cropping, or overgrazing, is the opposite of soil health building. Degraded soils cannot provide the same range of "ecosystem services" as biologically active, healthy soils can. Ecosystem services performed by healthy soils include the soil's ability to take up and store water, aerate and facilitate root growth, retain and recycle nutrients, support diverse biota to outcompete pests and pathogens, and prevent water runoff and soil erosion.

13.3 The Soil Resource

The suitability of a native soil for agricultural use is determined largely by inherent properties that arise from the integrated effects of five soil-forming factors: climate (temperature, precipitation); plants and other living organisms (above- and below-ground); parent material (bedrock or other substrata); relief (topographic or landscape position); and age (length of development). Professional evaluation of a soil to determine appropriate use involves assessment of its surrounding landscape, as well as excavation of the soil to a depth of at least one meter. This enables observation and measurement of soil layers ("horizons") that differ in color, thickness, permeability, and other properties.

The "soil profile" is a two-dimensional description of the horizons from the surface to the bottom of the excavation, while the "pedon" is the actual three-dimensional assemblage of horizons for a given soil. The pedon represents the smallest volume that can be called a soil and is used as a reference when depicting soil classes as polygons or units in soil mapping. Native topsoil horizons (termed "epipedons") are most directly influenced by plant cover, and because they are enriched with humus from decomposed plant residues, they are typically darker and more cohesive than horizons beneath them. Subsoil horizons are less affected by vegetation, but their thicknesses and physicochemical characteristics reflect local topography, water flow, composition of the parent material, and rates of mineral weathering. Parent materials such as limestone or wind-blown dust provide much more "native" soil fertility than materials like sandstone, which is rich in quartz but lacks minerals needed by plants.

The relative age of a soil and the climate under which it develops strongly influence soil pH and the pool of "base-generating" and "acid-generating" minerals available to plants. Soils subjected for millennia to tropical temperatures may have had practically all essential, base-generating minerals (i.e., calcium, potassium, and magnesium) leached away by heavy rainfall, leaving acid-generating minerals (i.e., aluminum, iron) that have adverse effects on plant growth. Soils exposed to weaker rainfall under moderate climatic regimes, on the other hand, tend to retain more of the basic minerals, especially if nutrients from litter from overlying vegetation are returned to the soil. Climate thus dictates the types and amount of vegetation that contribute to soil development, while topography influences a soil's net accumulations and losses of mineral and organic materials over time. Differences in these factors and their relative contributions to soil formation account for great spatial variability of soils.

13.4 Global Soil Classification

Soils are grouped into classes that reflect wide-ranging differences in the five soil-forming factors. Two frameworks for international soil classification are the World Reference Database (WRB) for Soil Resources of the Food and Agriculture Organization (IUSS Working Group WRB 2006) and U.S. Soil Taxonomy (USDA-NRCS 1999a). The WRB system uses two tiers of classification, with the first tier comprising 32 Reference Soil Groups (RSGs) and the second tier consisting of RSGs modified by specific descriptors that can be measured or observed. The U.S. system comprises 12 Soil Orders further divided into more than 20,000 soil series.

Table 13.1 lists the RSGs, their approximately correspondent Soil Orders, and the global percentages of ice-free land classified in each Soil Order. Soils best suited for agriculture (bold font in Table 13.1) include Mollisols, Alfisols, and Inceptisols. These soils possess the greatest native fertilities but account for only 26.4% of global land area. The most widely distributed soils, Entisols (16.3%) and Aridisols (12.7%), have low native fertility and receive inadequate rainfall. Gelisols (8.6%) are subject to low temperatures, while highly weathered Ultisols (8.5%) and Oxisols (7.6%) are high in aluminum and support mainly acid-tolerant plants. Thus, the majority of soils around the world have at least one severe limitation to their use for food production.

World soil maps created at scales of 1:5 million (i.e., 1 cm on the map represents 50 km on the ground) provide a broad understanding of the global distribution of major soil groups (Bai et al. 2010). However, maps produced at finer-scale resolution are required to depict soil variability at landscape levels sufficient to inform regional decision makers. Maps at scales of 1:10,000 or finer, however, may be required to capture local variation in soil depth, slope, and drainage, all of which influence soil suitability for producing food. Although finer-scale maps accompanied by detailed soil descriptions are available in most countries, understanding the limitations of soil at a given location typically requires direct observation and handling.

13.5 Soil Degradation

The Global Assessment of Land Degradation and Improvement (GLADA) conducted by the United Nations Environment Program (UNEP) employed remotely-sensed NDVI data over a 25-year period to assess the vigor of plant cover around the world

Table 13.1 List of World Reference Base (WRB) Reference Soil Groups (RSGs) and approximately equivalent Soil Orders in U.S. Soil Taxonomy

RSG diagnostic characteristics (horizons, properties, or materials that are observable or measurable in the field)	RSG	Soil Order (% of global land area)[a]
Soils with thick organic layers	Histosols	Histosols (1.2%)
Soils with strong human influence		
Soils with long and intensive agricultural use	**Anthrosols**	
Soils containing many artefacts	Technosols	
Soils with limited rooting due to shallow permafrost or stoniness		
Ice-affected soils	Cryosols	Gelisols (8.6%)
Shallow or extremely gravelly soils	Leptosols	
Soils influenced by water		
Alternating wet-dry conditions, rich in swelling clays	Vertisols	Vertisols (2.4%)
Floodplains, tidal marshes	**Fluvisols**	
Alkaline soils	Solonetz	
Salt enrichment upon evaporation	Solonchaks	
Groundwater affected soils	Gleysols	
Soils set by Fe/Al chemistry		
Allophanes or Al-humus complexes	Andosols	Andisols (0.7%)
Cheluviation and chilluviation	Podzols	Spodosols (2.6%)
Accumulation of Fe under hydromorphic conditions	Plinthosols	
Low-activity clay, P fixation, strongly structured	Nitisols	Oxisols (7.6%)
Dominance of kaolinite and sesquioxides	Ferralsols	
Soils with stagnating water		
Abrupt textural discontinuity	Planosols	
Structural or moderate textural discontinuity	Stagnosols	
Accumulation of organic matter, high base status		
Typically mollic	**Chernozems**	**Mollisols (6.9%)**
Transition to drier climate	**Kastanozems**	
Transition to more humid climate	**Phaeozems**	

(continued)

Table 13.1 (continued)

RSG diagnostic characteristics (horizons, properties, or materials that are observable or measurable in the field)	RSG	Soil Order (% of global land area)[a]
Accumulation of less soluble salts or non-saline substances		
Gypsum	Gypsisols	
Silica	Durisols	
Calcium carbonate	Calcisols	
Soils with a clay-enriched subsoil		
Albeluvic tonguing	Albeluvisols	Ultisols (8.5%)
Low base status, high-activity clay	Alisols	
Low base status, low-activity clay	Acrisols	
High base status, high-activity clay	**Luvisols**	**Alfisols (9.6%)**
High base status, low-activity clay	**Lixisols**	
Relatively young soils or soils with little or no profile development		
With an acidic dark topsoil	Umbrisols	Aridisols (12.7%)
Sandy soils	Arenosols	
Moderately developed soils	**Cambisols**	**Inceptisols (9.9%)**
Soils with no significant profile development	Regosols	Entisols (16.3%)

[a]Percentages of land area not included are either rock- or ice-covered

Bold font indicates groups that are inherently best suited for agriculture. Other groups have some type of limitation (described in column one) which must be addressed prior to agricultural use or which poses management challenges

(Bai et al. 2008, 2010). The NDVI (Normalized Difference Vegetation Index) is a measure of primary plant productivity based on absorbance of infrared radiation by chlorophyll. It was notable that many *degraded lands* identified by GLADA as undergoing degradation did not overlap with lands having *degraded soils*, which had been identified in an earlier assessment conducted by UNEP in 1988–91, the Global Survey of Human-Induced Soil Degradation (GLASOD, Fig. 13.1). The latter assessment defined soil degradation as a "process which lowers the current and/or future capacity of the soils to produce goods or services." The GLASOD study concluded that soils in 38% of the world's agricultural lands had been degraded either by wind or water erosion or other processes such as salinization and chemical pollution. The lack of spatial overlap between GLADA and GLASOD demonstrates how soil degradation estimates can be influenced by measurement method and reflects the challenges facing decision makers who must interpret available information.

Soils are most resistant to degradation when they serve as biologically intact foundations for "permanent," dense vegetative cover (i.e., native grassland, forest). It can be argued that converting such soils to agricultural use is an inherently degradative process. Plant removal and soil disturbance destroy intact root-microbial networks

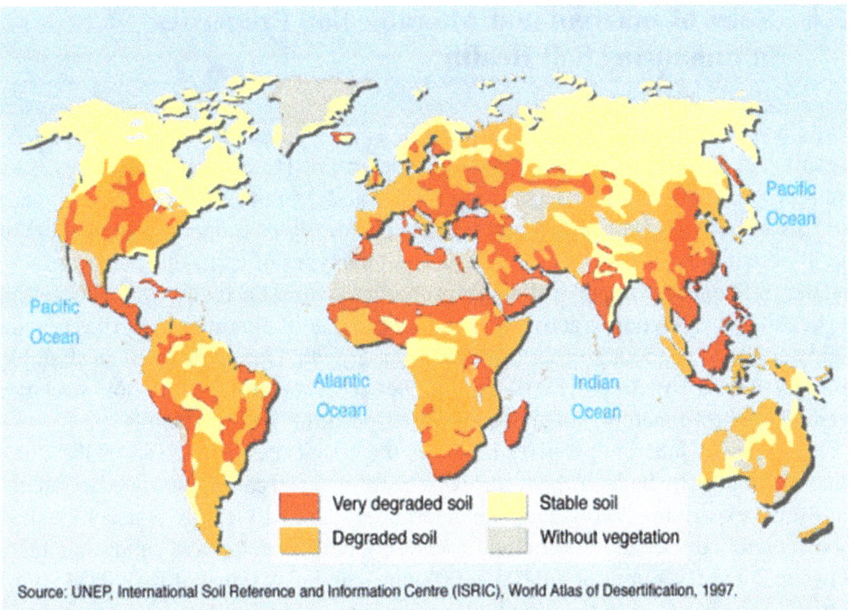

Fig. 13.1 World map produced with data from the Global Survey of Human-Induced Soil Degradation, Food and Agricultural Organization (*Source* Philippe Rekacewicz, UNEP/GRID-Arendal, World Map of Degraded Soils, https://www.grida.no/resources/6338, used with permission)

belowground that may have taken many years to become established but which are not recognized as being important to agricultural productivity. Because native plant-soil systems are co-adapted to resist the destructive forces of local climate, their disruption renders topsoils more vulnerable to drought and displacement by wind or water. The loss of surface soil, which is more nutrient-rich than deeper soil, rapidly reduces the soil's native fertility and water-holding capacity.

High NDVI readings in the GLADA study may have been obtained from plant cover growing on degraded soils managed with high inputs of fertilizer and irrigation. Because the negative effects of reduced soil health (i.e., poor soil structure, reduced water-holding capacity, low nutrient-use efficiency) on productivity can be masked by increased expenditures of nonrenewable resources, NDVI data may need to be interpreted in a more comprehensive way to account for all resources used in agricultural production. Similarly, soil conservation programs in North America are shifting from a sole focus on erosion reduction to practices that increase soil carbon (soil organic matter), which helps improve soil structure and increases efficiencies of nutrient and water use (USDA-SARE 2010).

13.6 Roles of Inherent and Alterable Soil Properties in Enhancing Soil Health

While a soil's suitability for agricultural use is determined by inherent properties, the soil's relative health and productivity are determined by alterable properties. Two examples of inherent properties are soil texture and depth to bedrock. Both of these are "givens" for the farmer, because it is not physically or economically feasible to modify them. Inherent soil properties can limit the types of crops that can be grown, and they influence the range of alterable properties achievable for that soil. Sustaining the health of a soil requires active management so that its alterable properties remain highly conducive to root proliferation and plant growth. One example of an alterable property is soil structure. The distinction between soil texture and soil structure exemplifies how inherent and alterable soil properties affect soil health.

Texture is an inherent property based on the coarseness or fineness of the soil's mineral fraction. In the field, the texture of a handful of soil can be estimated manually with some experience. More accurate determination of soil texture is accomplished by completely dispersing the soil and measuring the percentages of sand (diameters between 2 and 0.05 mm), silt (0.05–0.002 mm), and clay (smaller than 0.002 mm) particles. Since only particles smaller than 2 mm are considered soil, rock fragments larger than 2 mm are not considered in soil textural classifications. A specific soil can be grouped into one of 12 textural classes according to the relative distribution of particle sizes. Sands (>90% sand) and clays (>60% clay) are the two textural classes least suitable for agricultural use because they lie at the extremes of water and air permeability. Medium-textured soils, classified as loams (sandy loam, silt loam, silt, loam, sandy clay loam, clay loam, and silty clay loam), provide best conditions for air and water exchange in growing agricultural crops. Soils in the remaining textural classes (loamy sand, sandy clay, and silty clay) are intermediate.

Soil structure, on the other hand, is an alterable property based on organic matter as well as mineral fractions. Soil structure is the three-dimensional arrangement of mineral particles and organic matter into soil aggregates and pores of varying sizes. Soil structure reflects the amount and type of vegetation grown in the soil and the degree of physical disturbance to which the soil has been subjected. Imagine yourself standing in a well-managed garden or agricultural field—then envision the cubic meter (1 m^3) of soil immediately beneath your feet. Incredibly, about half that volume of soil is void space. The other half consists mainly of weathered minerals (sand, silt, clay, pebbles) and a relatively small, but functionally important, proportion (1–10%) of soil organic matter derived mainly from decomposing plant litter and root-microbe debris. To a great extent, biological activity controls this spatial arrangement of soil voids and solids.

Soil bacteria use organic carbon from living roots and decaying organic matter to obtain energy for growth and production of "extracellular polymeric substances" (EPS). Bacterial EPS facilitates adherence of cells to soil particles and results in the formation of *microaggregates*. In the presence of degradable organic matter and adequate moisture, bacterial and fungal activity act to bind smaller aggregates into larger ones. Experiments employing microbial inhibitors in soils demonstrate that *macroaggregates* (at least 0.25 mm in diameter) are formed only in the presence of active microorganisms (Bossuyt et al. 2001). Microbial activity in undisturbed soil also stabilizes organic matter as aggregates break down and reassemble (Six et al. 2004). In the field, soil structure is evidenced by its "friability," or ease with which the soil is broken apart into crumbs, which facilitate water infiltration and oxygen availability. Soils which contain more organic matter and which are infrequently disturbed (e.g., soils under long-term perennial vegetation) have more large pores and aggregates than heavily tilled agricultural soils in the same locale. The most effective way to enhance soil structure is to increase the soil's organic matter content (e.g., by tilling in young cover crops or by amending the soil with composts or manures).

Differences in soil structure are reflected in laboratory measurements of soil bulk density, which is determined from the dry mass of soil solids within a known volume. Although bulk density does not quantify soil structure per se, it can be used to assess the efficacy of management practices aimed to enhance organic matter content and improve soil health. For a given soil, the higher the bulk density, the less porosity it has for air and water movement. Bulk densities range from 0.8 g cm^{-3} for uncultivated soils under perennial vegetation, to 2.2 g cm^{-3} for heavily compacted soil. Bulk densities of 1.0–1.4 g cm^{-3} provide highly favorable conditions for plant growth in most soils. However, soil texture affects the value of a soil's "ideal" bulk density for agricultural use, as well as the degree to which bulk density can be increased before root growth is restricted (Table 13.2).

Soil organic matter is a heterogeneous component of soil, consisting of freshly added plant material, living and decaying roots and microorganisms, partially decomposed plant, animal, and microbial materials, and fully decomposed humus. Soil organic matter content, like other alterable soil properties, can be measured in the analytical laboratory, although increasing darkness of topsoil color is a good field

Table 13.2 Relationship between soil texture classes and bulk density values

Soil texture	Ideal soil bulk densities and root growth limiting bulk densities for soils of different textures		
	Ideal bulk densities	Bulk densities that can affect root growth	Bulk densities that can restrict root growth
	g/cm^3 (expressed as Mg m^{-3} in International Scientific Units)		
Sand, loamy sand	<1.60	1.70	>1.80
Sandy loam, loam, sandy clay loam, clay loam, silt, silt loam, silty clay loam	<1.40	1.60	>1.75
Sandy clay, silty clay, clay	<1.10	1.50	>1.60

Table from the USDA-NRCS (1999b) Soil quality test kit guide indicator of higher organic matter levels. Other alterable properties include water-extractable carbon, soil aggregate stability, soil pore size distribution, water-holding capacity, water infiltration rate, carbon and nitrogen availability, microbial biomass content, pH, and salt content. Resources are available for farmers and landowners to learn about soil health indicators, field assessment methods, and management practices that are effective in improving soil health (Moebius-Clune et al. 2016; USDA-SARE 2010).

13.7 Why Organic Matter Enhances Soil Health

Food security and sustainable soil health depend on minimizing erosive soil losses. To achieve this, vegetative cover on arable lands must be maintained as continuously as possible to help rebuild root-microbial networks that keep soil in place. Living vegetative cover sustains plant roots that exude and secrete organic compounds. Topsoil under living vegetation has an increased proportion of rhizosphere soil, which is defined as soil immediately adjacent to plant roots (typically within 2 mm of root surfaces). Rhizosphere soil contains greater concentrations of organic carbon and denser populations of active soil microorganisms than non-vegetated soils. Because living roots "pump" carbon into soils, vegetated soils are well-aggregated and more erosion-resistant than non-vegetated soils.

Root-enriched soils harbor diverse communities of bacteria, archaea, and fungi. These three groups comprise the "soil microbial biomass," which is responsible for organic residue decomposition and the release of such inorganic nutrients as ammonium, phosphates, and sulfates. Macro- and micronutrients, when bound within organic compounds in decaying plant tissue, remain unavailable to roots until microbes and their degradative enzymes break these tissues down. Slightly larger organisms, the protozoans and nematodes, prey upon the microbial biomass. Upon consumption of microbial cells, these "grazers" release inorganic nutrients in their wastes that once again become available for plant uptake. Along with other groups of soil organisms, including microarthropods and earthworms, a soil's entire biological assemblage is sometimes referred to as the "edaphon," from the Greek word for "ground" or "soil." Because soil biology is a comparatively young science, much remains to be learned about how the soil influences its edaphon and vice versa. It is widely recognized, however, that the soil edaphon is highly correlated with the amount and quality of soil organic matter, as well as organic matter quality and frequency of addition.

Just as plant roots require adequate nutrients, water and air to thrive, so does the soil microbial biomass. Although most of the organic carbon taken up and decomposed by the soil microbial biomass is released by aerobic respiration as carbon dioxide, some of the carbon becomes stabilized when microbial exopolysaccharides bind to clays to facilitate aggregate formation. Soil aggregation acts as positive feedback to improve soil porosity and the microbes' own habitat. As organic compounds become "humified" (i.e., resistant to further microbial degradation), intimate chemical interactions between humus carbon and soil minerals help protect and stabilize the carbon, particularly within microaggregates. Continual additions to topsoil of fresh organic matter sources (i.e., legume cover crops, composts, manures) feed a self-regulating, stabilizing activity by soil microorganisms. If organic matter is not added continually, either from living plant roots or with soil amendments, net soil loss will occur.

13.8 Bio-Based Tools for Evaluating Soil Health

Every soil possesses a site-specific combination of inherent and alterable properties that support a range of biological activities contributing to soil health. Arable but never-cultivated soils, for example, contain biota that have become conditioned to native vegetation and local conditions, and they exhibit adapted levels of biological activity which are disrupted upon cultivation. After soils have been tilled intensively and crops harvested year after year, the failure to replenish the soil with plant or other organic residues results in depleted soil organic matter, loss of porous structure, low soil microbial biomass and reduced biological activity. Even degraded soils can contain millions of microorganisms and hundreds of species in each gram, and scientists still have much to learn about how microbial diversity and activity contribute to soil health status. Three kinds of bio-based tools for evaluating soil health are discussed here, starting with simple field observations, followed by methods performed in commercial and research laboratories, and finally advanced biotechnologies being explored in specialized research facilities.

13.8.1 Visual and Manual Assessments in the Field

Farmers and other landowners familiar with their soils can rely on experience and sensory cues to evaluate soil health, particularly using sight and touch. Darker soil colors signify greater organic matter content, which in turn signifies greater water-holding capacity and nutrient availability. Softer, less-compacted soil structure denotes easier water infiltration and root penetrability. Earthworms and other soil fauna indicate greater abundance of a microbial biomass that can support more diverse and larger organisms in soil communities. Cohesive soil crumbs, in contrast to dusty surface crusts, signal greater resistance to wind and water erosion and better capacity for soil to remain in place. Finally, achieving good growth of disease-free crops without excess fertilizer, water, and pesticides represents the clearest hallmark of soil health. While on-site evaluation can be assisted with simple tools like spades, buckets, and penetrometers, probably the most valuable tool for evaluating soil health is knowledge of its inherent characteristics and its cropping and management history.

13.8.2   Wet Laboratory Methods

Because the soil microbial biomass responsible for beneficial biological services cannot be observed directly in the field, wet laboratory methods have been developed to obtain quantifiable measurements of microbial biomass and its general or specific activities. In laboratory testing, heavily cultivated, continuously cropped soils are expected to exhibit lower microbial biomass and biological activity than native soils. On the other hand, intentional replenishment of organic matter from living roots of diverse crops and other organic amendments, combined with minimal physical disturbance, should lead to increases in soil microbial biomass and biological activity exceeding steady-state levels. Reliable assessment of soil health status using one type of laboratory test, however, is elusive because there is so much spatial variability in soils, climates, and management histories. Therefore, the relationships between soil health status and laboratory test results are better correlated when comparing similar soils over time, as when a grower uses tests to evaluate effects of management changes on soils in the same field over several years. The Cornell Soil Health Assessment exemplifies a program that employs a variety of biochemical, physical and process-based tests to generate overall soil health scores (Moebius-Clune et al. 2016). Three types of laboratory testing for soil health-biochemical properties, microbial biomass, and biological activity-are discussed here.

13.8.2.1   Biochemical properties

Soil fertility (macronutrients, micronutrients, pH) is tested routinely by commercial laboratories, but such tests are done mainly to provide chemical fertilizer and liming recommendations for specific crops. When it comes to determining a soil's biological status, the only test offered by most commercial laboratories is measurement of soil organic matter. While a soil's organic matter content is a particularly useful characteristic for assessing biological health, organic matter content can take many years to undergo a significantly measurable change and depends on soil type and texture. The partitioning of total organic matter into easily degradable, intermediate, and difficult-to-degrade fractions, on the other hand, does undergo more rapid change and can indicate how soils are responding to in management. Procedures for assessing soil organic matter fractions (e.g., extractable, labile carbon sources, fresh organic matter particles) are typically conducted in academic research laboratories, where soil microbial biomass carbon, an especially critical fraction, is also measured.

13.8.2.2   Microbial biomass estimates

Bacteria and fungi are the most abundant microorganisms driving soil organic matter turnover and nutrient release to plant roots and microbiota. Typical cell widths of bacteria and fungi (~0.5 μm and 5 μm, respectively) make it impossible to observe them directly, and even under the microscope, soil microorganisms are difficult to discern when they are enmeshed in clays and organic debris. Historically, soil microorganisms and their physiologies have been studied in the laboratory by dispersing them in soil slurries, diluting them out, and cultivating in nutrient medium in petri plates. When culturable microorganisms are enumerated using aerobic plate counts, it has long been known that only 1–10% of the viable microorganisms initially present in the soil sample are actually recovered in plates. Such low recovery has been confirmed consistently after examining stained smears of the same soils under the microscope, a tedious and often subjective technique. Since microbial counting is subject to high variability, a more quantifiable approach is to estimate total microbial biomass carbon by fumigating a soil sample with chloroform and measuring the microbial carbon released upon cell lysis. Another quantifiable approach is to extract and measure microbial phospholipid fatty acids (PLFA), which can also generate information on diversity of total, bacterial, and fungal fatty acids. A few commercial labs conduct PLFA testing, but the correlations between PLFA data and soil health status are not yet clearly established.

13.8.2.3 Biological activities

The capacity for soil microorganisms to recycle inorganic nutrients from organic matter for plant root uptake is reflected in the rate of CO_2 released during respiration from moist soil samples incubated in the laboratory under controlled conditions (Franzluebbers 2018). Outcomes of respiration and other activity measurements involving added substrates or various enzymes are all highly sensitive to procedures used for sampling, storing, preparing, and incubating the soil samples. The highly variable physicochemical properties of soils and the potential for small deviations during laboratory procedures has made it challenging to develop universal, standardized tests for soil biological activity as indicators of soil health. Laboratory test results have the most reliable and satisfactory correlations with soil health status when similar soils are compared over time and consistent testing procedures are scrupulously followed.

13.8.3 Advanced Biotechnologies

Soils represent some the most diverse microbiomes on the planet with billions of cells and tens of thousands of species in each gram (Fierer 2017). More advanced technologies are being developed in research facilities to understand the nature of soil organic matter and the microorganisms that contribute to its formation and transformations. Fundamental knowledge of what microorganisms can do has been facilitated by studying purified laboratory cultures, which has made it easy to test their metabolic properties and sequence the DNA in their genomes. This approach has been critical for understanding microbial life, and it is still a route that is being pursued using advanced, high-throughput cultivation methods. However, despite all the efforts invested, only a small fraction (~1%) of extant microbial species have been cultured in the laboratory. Besides our failure to culture most microorganisms, the strains growing in cultures are not being studied in a physiological context that accurately depicts their natural environments. This is especially true for soils, where the physicochemical nature of soils is not reproduced during growth on a petri plate, rendering it difficult to apply knowledge learned from the lab to real soil systems. Three advanced technologies, which do not rely on cultivation (i.e., "cultivation-independent"), will be discussed here: microbiome analyses; stable isotope probing, and cell tagging and capture.

13.8.3.1 Microbiome analyses

The direct study of microbial function in situ remains a challenge and we are just starting to understand how microbial processes integrate through space and time to confer to the soil its emergent health indicator properties (Baveye et al. 2018). The study of soil microbiomes (i.e. the total assemblage of microorganisms in a given soil) suffers the same challenges as other microbiome sciences (such as human and food microbiomes) and more. These limitations have existed ever since cultivation-independent methods were developed, where soil microbiomes could be analyzed by extracting their nucleic acids (DNA, RNA) or other cellular components (PLFA) from the soil. Although cultivation-independent techniques also bear a fair set of their own biases, they undoubtedly shed much light on microbiome composition and function, leading to recent and rapid increases in their development and use. The flagship of these methods consists of total DNA extraction from soils followed by PCR amplification and next generation sequencing of the marker genes for 16S rRNA (for Bacteria and Archaea) and 18S rRNA (for Eukaryotes) that are good taxonomic markers. Producing large sets of sequences (rRNA libraries) from soil extracts has helped us recognize the tremendous diversity of soil microorganisms, typically comprising dozens of major phyla, some with no cultivated organisms. The development of next generation sequencing, especially the Illumina platform, has permitted the analysis of millions of rRNA genes per sample, thus informing us about which microorganisms are present in soils.

Knowledge of which microbes are present, however, does not provide sufficient information about which functions they can carry out. It is commonly hypothesized that a healthy soil will host different microbial species having redundant functions, so that each critical microbial function will be carried out by at least one species as environmental conditions vary. To understand which microbial functions are possible in soils, the description of community diversity using rRNA gene amplicon libraries is not sufficient, and we need to employ methods that inform us about genes encoding functional enzymes. An example of studying a specific function, such as N_2 fixation, is to extract soil DNA (or soil RNA to be converted to copy DNA) and use quantitative PCR (qPCR) to estimate the numbers of *nifH* genes or transcripts encoding a portion of the nitrogenase enzyme. Since nitrogenase is just one of thousands of microbial enzymes present in soil, a more comprehensive approach involves using next generation sequencing to copy sequences from as many genes or transcripts as possible, in a process called metagenomic or metatranscriptomic sequencing of total DNA or RNA pools, respectively. Bioinformatics tools are then used to assemble sequences and assign them using multiple databases to produce a list of genes with or without an assigned function. That list can be used to assess the different categories of functions present in the community. More recently, binning algorithms have been developed that allow the assembly of genomes from metagenomic data. To date tens of thousands of metagenome-assembled genomes (MAGs) have been produced, permitting gene functions to be studied in their genomic context even for organisms that cannot be cultivated (Parks et al. 2017). Being able to obtain a genome without having to cultivate an organism is clearly a revolution in the field of microbiome research, but it comes with the caveat that a genome assembled in silico may not in fact exist as such in the community, and studies of cultured organisms are still needed to understand microbial physiology. Thus, it appears critical to continue developing both cultivation-based and cultivation-independent approaches so that knowledge from both approaches can advance our understanding of microbial communities.

Another caveat associated with soil microbiome research is that up to 40% of total DNA extracted from soil may actually come from free DNA, which is DNA released from dead microbial cells and becomes stabilized when adsorbed to soil colloids (Carini et al. 2016). Because sequencing of DNA from dead cells is clearly not informative about active functions, total RNA can be extracted instead, because it has a much shorter half-life outside the cell. Nevertheless, up to 95% of cells in soil can be dormant at any one time (Blagodatskaya and Kuzyakov 2013) so that much of the RNA in bulk extracts may come from less active cells. Thus, methods that label nucleic acids from active microorganisms prior to separation from nucleic acids from inactive microbes would have more value in assessing actual contribution to microbiome function.

13.8.3.2 Stable isotope probing (SIP)

This method involves incubating a soil sample after adding a substrate labelled with a heavy isotope such as ^{13}C or ^{15}N. The labelled substrate will be incorporated into newly synthesized biomass, including nucleic acids. Post incubation, DNA is extracted, and the "heavy" DNA fraction belonging to organisms having incorporated the labeled substrate is separated from the rest of the DNA via gradient ultracentrifugation. This DNA can then be collected for sequencing using metagenomics or amplicon sequencing as described above. This technique has been successfully used in soil and has helped us understand the elemental turnover in total communities, applying it for instance to the identification of nitrogen fixers (Pepe-Ranney et al. 2016) or species able to degrade polyaromatic carbon substrates. More recently this method was employed with deuterium water (D_2O), which in theory would be an ideal label since the added substrate (water) should not modify cell physiology. However, separation of D_2O-labeled organisms has been difficult and required the use of optical tweezers coupled to Raman spectroscopy at a throughput of a few hundred cells per day (Berry et al. 2015). Considering the millions to billions of cells in a single gram of soil, this approach is not realistic for characterizing soil microbiome function.

13.8.3.3 Cell tagging and capture

Other methods have been used to probe active microbes in soil, but none has gained momentum in the scientific community as they are usually difficult to perform and require specialized equipment. It is however worth noting that a novel approach called Bioorthogonal Non-Canonical Amino-Acid Tagging (BONCAT) could be game-changing. This method relies on the incorporation of a modified water-soluble amino acid into newly made proteins. When the modified amino acid is added to a soil sample, it is taken up by active microbes during incubation. Following uptake and incorporation of the amino acid into proteins, soil suspensions are prepared and microbes are detached from soil particles for collection on a filter. When the cells on filters are treated with "click chemistry" reagents, fluorescent probes bind to the modified amino acids in new proteins, thus causing active cells to be fluorescent, while no binding occurs in inactive cells. Fluorescence-activated cell sorting (FACS) by flow cytometry can then be used to separate the groups of cells for counting. Further downstream analyses of separated cells can include any high throughput sequencing technique discussed above. Interestingly, the labelled proteins can also be captured themselves by conjugating them to a column for analysis by proteomics. Unlike SIP, this method does not rely on cell division for labelling and can therefore label slow-growing organisms, which are often important in soils. This method was recently shown to be robust and reproducible for the study of soil microbiomes, opening new avenues for characterization of microbial processes in situ and understanding how processes are integrated at the soil scale (Couradeau et al. 2019).

Field-based and lab-based methods of evaluating soils are respectively carried out by farmers and field consultants, wet labs, and research facilities, but to be informative, they all need to be able to relate test results to knowledge of soil management and productivity. If results of advanced research techniques are consistent with predictions from less costly methods, this would simplify soil health evaluation and make it more accessible to all.

13.9   Management Practices to Improve Soil Health

13.9.1   Crop Diversification and Nitrogen Nutrition

Agricultural policies exert pressures on farmers to produce one of a few commodity crops such as maize, wheat, soybean, and rice. Subsidies and disaster payments based on commodity crop acreage are powerful disincentives for farmers to include non-commodity crops in soil-building rotations. Growing the same crop in the same field year after year results in the introduction of a narrow range of organic carbon compounds into the soil, leading to low diversity of soil biota and selection of pathogens or pests that thrive on the predictable food source. More importantly, continuous "monocropping" precludes the use of soil-building crops, the roots and residues of which are especially important sources of organic carbon to enhance soil health. Crop diversification not only helps improve soil diversity, it can improve nitrogen (N) use efficiency, a critical problem for today's agricultural systems which lose on average 50% of applied N to the environment.

Legumes represent an important group of soil-building crops, because their symbiotic relationship with rhizobia bacteria enables them to incorporate N_2 from the atmosphere. This evolutionary relationship, in which N_2-fixing rhizobia proliferate in tumor-like root nodules, occurs only within the legume family, which is particularly fortuitous for these plants, because prokaryotes (bacteria and archaea) are the only organisms on Earth that can reduce gaseous N_2 to NH_3 (ammonia). Biologically fixed N differs from industrial N fertilizer (i.e., NH_4NO_3, or ammonium nitrate), in that it is fixed directly within the plant.

Another distinction is that biologically fixed N is less likely to be lost from soils than industrially fixed N, because it is coupled immediately to carbon in such intracellular organic molecules as proteins and amino acids. Organic N is recycled in the soil more slowly than inorganic N, because it first must undergo microbial decomposition before NH_4^+ is released into the soil. Moreover, the majority of ion exchange sites on soil particles are negatively charged, so that the NH_4^+ tends to be held in soils by ionic forces. As a tightly held cationic nutrient, NH_4^+ is less mobile and therefore less subject to leaching losses after heavy rainfall than the NO_3^- anion.

The carbon added to soils in decaying legume residues can also stimulate new growth of microbial biomass, leading to cellular re-incorporation of inorganic N. The shuttling of inorganic N in and out of microbial biomass helps to prolong "N soil residence time." The majority of soil microorganisms are heterotrophs (dependent on organic carbon for energy and cell material). When heterotrophic competition for NH_4^+ is sustained with new supplies of organic carbon, soil NH_4^+ is less available for oxidation to NO_3^- by the specialist microbes known as "nitrifiers."

Nitrifying bacteria and archaea do not need organic carbon because they are "lithoautotrophs", i.e., they depend on NH_4^+ or NO_2^- oxidation for energy generation and CO_2 fixation for cell material. By enhancing organic carbon inputs, heterotrophic uptake of NH_4^+ makes it less available to nitrifiers, preventing or delaying autotrophic conversion to the more mobile form of NO_3^-. In addition to being more leachable,

NO_3^- also can be denitrified to N_2O or N_2 in wetter soils and lost to the atmosphere. Thus, lower N losses which have been observed for legume-based cropping systems (Drinkwater et al. 1996) may be explained in part by heterotrophic suppression of nitrifier activity and the predominance of reduced over oxidized forms of soil N.

13.9.2   Reduced Tillage

When permanent vegetation is removed and soils are disturbed, soil organic matter levels begin to decline unless organic carbon is added back in the form of living roots, cover crop residues, composts, or animal manures. If organic matter is not returned to the soil, reductions in soil organic matter occur rapidly, particularly in systems that employ conventional (inversion) tillage. In this type of tillage, a moldboard plow is used to cut into and turn over topsoil. Additional equipment such as cultivators and diskers are used to break up the soil further so that the seedbed can be smoothed with a harrow. Repeated physical disruption promotes soil aeration and mixing of soil microbial biomass with crop residues, resulting in rapid oxidation of organic matter, reducing the amount of carbon that will be processed slowly into stable humus.

As soil organic matter declines, the soil becomes more susceptible to erosion, reinforcing a feedback cycle that leads to even more soil carbon being lost. Because moldboard plowing of extensive areas of prairie sod in the Midwestern United States led to the Dust Bowl in the 1920s and 30s, the United States Soil Conservation Service and cooperating farmers began to investigate tillage methods that entailed less physical disturbance and left more crop residue on the soil (Montgomery 2007). These first forms of "conservation tillage" employed chisel plows with narrow points spaced to create furrows for seed introduction without the need to invert the entire soil or bury all crop residues. Subsequent types of conservation tillage (i.e., no-till planting, mulch tillage, ridge tillage) all result in less soil disturbance than conventional tillage.

Although conservation tillage has focused on the amount of crop residues left on the soil surface (must be at least 30%), an additional benefit of reduced tillage is the slowed rate of organic matter oxidation within the soil. In addition, reduced tillage alters the composition of the soil microbial biomass, resulting in fungi making up a greater proportion of total biomass. No-till methods reduce physical breakage of fungal hyphae, thus favoring fungal proliferation and enhancing formation of macroaggregates, which improves soil porosity. Reduced tillage also is less damaging to endomycorrhizal fungi (i.e., root "endophytes"), which produce fine networks of nutrient-scavenging hyphae extending beyond roots to distances of several cm. Tips of these hyphae can tap into water and nutrient supplies otherwise unavailable to plant roots and bring nutrient-laden water back to the plant. Reduced tillage methods thus improve soil biological integrity by slowing organic matter loss and preserving root-microbe symbioses.

13.9.3 Microbial Augmentation

In addition to maintaining the biophysical integrity of soils, many microorganisms interact with plant roots in positive ways. "Plant growth-promoting rhizobacteria" (PGPR) comprise a diverse array of beneficial bacteria that have been recovered from plant roots and studied for several decades. Various functions of PGPR have been observed under laboratory conditions, but the extent to which these functions can be replicated in the field is uncertain. Beneficial functions provided by PGPR include production of phytohormones (e.g., indoleacetic acid), which influence root growth, and release of phosphate-solubilizing enzymes that make the phosphorus in soil minerals more available to the plant. Recognition of PGPR has spurred interest in increasing their presence in the rhizosphere by developing their use as inoculants.

Inoculation as a plant-growth-promoting technology has been used for decades, mainly with rhizobia and legume seeds to promote the establishment of the N_2-fixing symbiosis. Most inoculant products contain a mixture of rhizobial strains that have been proven to be "rhizosphere-competent," or capable of surviving in rhizosphere soil, entering host plant roots, and inducing the host to develop "effective" nodules. Many steps involving chemical signaling and recognition between the legume and the symbiont are required for an effective symbiosis to be established. Effective nodulation, which results in N_2 fixation, can be determined in the field by observing the red interiors of nodules when cut open. The red coloration is due to the presence of leghemoglobin, a protein produced by the plant to prevent O_2 from interfering with rhizobial enzymes responsible for N_2 fixation.

The problems and pitfalls that have been encountered with rhizobial inoculants should be taken into account when developing novel inoculants. Failure to use appropriate rhizobial strains which are compatible with a specific legume variety can make inoculation futile. Since specificity is clearly observed between legume varieties and their compatible rhizobial strains (i.e., "cross-inoculation groups"), similar relationships may exist between other plant hosts and microbial associates. Other factors can interfere with effective symbiosis, even when the correct rhizobial strain is present. For example, high levels of ammonium and nitrate in the soil render it unnecessary for plants to invest in the energy and carbohydrate delivery needed to support effective nodules. Soil conditions such as phosphorus or trace element deficiencies can also result in ineffective nodulation.

It is reasonable to expect that any beneficial microorganisms introduced through inoculation or soil augmentation will interact with and encounter competition from the indigenous microbial community (i.e., the edaphon). Inoculant technologies must be based on a recognition that introduced organisms will face competition from resident microorganisms which are already adapted to soil conditions. Resident soil rhizobia from previous legume crops, for example, can also persist in soil and compete with inoculated strains.

Considering the investment required to develop new inoculants, a more feasible alternative to promoting beneficial root-microbe relationships is to enhance generalized activity of the entire edaphon by adding organic matter and using management practices that promote biophysical integrity. The concept of the "plant microbiome"—the array of microorganisms living in and on plant tissues—is analogous to the "human microbiome" but is likely to be strongly influenced by the "soil microbiome" (Chaparro et al. 2012). The same biotechnological tools that inform us about how our own health is affected by microorganisms residing on and in the human body are therefore applicable to elucidating how soil- and plant-associated microorganisms contribute to crop health.

13.10　Conclusions

Agricultural management goals often focus on attaining maximal yields of the most profitable crops in the short-term, rather than on sustaining long-term soil productivity. As much as 25% of the world's agricultural lands are considered moderately to severely degraded (Bai et al. 2008, 2010), but soil damage can be masked with added fertilizer and irrigation. Such inputs are often unavailable in developing countries, but even where water and fertilizer are abundant, agricultural use of degraded soils leads to increasing resource inefficiencies. Improving soil health depends on maintaining or increasing organic matter content and minimizing disturbance to promote beneficial biological processes known to occur in plant-soil systems. A key challenge for farmers, researchers, and policy makers is to determine how crop production practices can complement or accommodate these processes in soils. Policy innovations for sustainable soil health will be based on explicit recognition that soil organic matter and biota are critical to erosion control and keeping soils in place.

Acknowledgements　We thank Nina Camillone and Claudia Rojas-Alvarado for their reviews and helpful comments during the preparation of this chapter. This work was supported by the USDA National Institute of Food and Agriculture Federal Appropriations under Hatch Project PEN04710.

References

Bai ZG, Dent DL, Olsson L, Schaepman ME (2008) Global assessment of land degradation and improvement. 1. Identification by remote sensing. Report 2008/01. ISRIC—World Soil Information, Wageningen

Bai ZG, de Jong R, van Lynden GWJ (2010) An update of GLADA—global assessment of land degradation and improvement. ISRIC report 2010/08. ISRIC—World Soil Information, Wageningen

Baveye PC, Otten W, Kravchenko A, Balseiro Romero M, Beckers É, Chalhoub M, Darnault C, Eickhorst T, Garnier P, Hapca S, Monga O, Müller C, Nunan N, Pot V, Schlüter S, Schmidt H,

Vogel H-J (2018) Emergent properties of microbial activity in heterogeneous soil microenvironments: different research approaches are slowly converging, yet major challenges remain. Front Microbiol. https://doi.org/10.3389/fmicb.2018.01929

Berry D, Mader E, Lee TK, Woebken D, Wang Y, Zhu D, Palatinszky M, Schintlmeister A, Schmid MC, Hanson BT, Shterzer N, Mizrahi I, Rauch I, Decker T, Bocklitz T, Popp J, Gibson CM, Fowler PW, Huang WE, Wagner M (2015) Tracking heavy water (D_2O) incorporation for identifying and sorting active microbial cells. Proc Natl Acad Sci 112(2):E194–E203. https://doi.org/10.1073/pnas.1420406112

Blagodatskaya E, Kuzyakov Y (2013) Active microorganisms in soil: critical review of estimation criteria and approaches. Soil Biol Biochem 67:192–211. https://doi.org/10.1016/j.soilbio.2013.08.024

Bossuyt H, Denef K, Six J, Frey SD, Merckx R, Paustian K (2001) Influence of microbial populations and residue quality on aggregate stability. Appl Soil Ecol 16:195–208

Brady NC, Weil RR (2007) The nature and properties of soils, 14th edn. Prentice-Hall, Upper Saddle River, p 980

Carini P, Marsden PJ, Leff JW, Morgan EE, Strickland MS, Fierer N (2016) Relic DNA is abundant in soil and obscures estimates of soil microbial diversity. Nat Microbiol 2:1–6. https://doi.org/10.1038/nmicrobiol.2016.242

Chaparro JM, Sheflin A, Manter DK, Vivanco JM (2012) Manipulating the soil microbiome to increase soil health and plant fertility. Biol Fertil Soils 48:489–499

Couradeau E, Sasse J, Goudeau D, Nath N, Hazen TC, Bowen BP, Malmstrom RR, Northen TR (2019) Probing the active fraction of soil microbiomes using BONCAT-FACS. Nat Commun 10(2770). https://doi.org/10.1038/s41467-019-10542-0

Drinkwater LE, Wagoner P, Sarrantonio M (1996) Legume-based cropping systems have reduced carbon and nitrogen losses. Nature 396:262–265

Fierer N (2017) Embracing the unknown: disentangling the complexities of the soil microbiome. Nat Rev Microbiol 15(10):579–590. https://doi.org/10.1038/nrmicro.2017.87

Food and Agriculture Organization of the United Nations (2013) FAO statistical yearbook 2013. World Food and Agriculture, Rome, 289 pp

Franzluebbers AJ (2018) Short-term C mineralization (aka the flush of CO_2) as an indicator of soil biological health. CAB Rev 13, No. 017. https://doi.org/10.1079/PAVSNNR201813017

IUSS Working Group WRB (2006) World reference base for soil resources 2006. World Soil Resources Reports No. 103. FAO, Rome. ISBN 92-5-105511-4

Moebius-Clune BN, Moebius-Clune, DJ, Gugino BK, Idowu OJ, Schindelbeck RR, Ristow AJ, van Es HM, Thies JE, Shayler HA, McBride MB, Kurtz KSM, Wolfe DW, Abawi GS (2016) Comprehensive assessment of soil health—the Cornell framework, edition 3.2. Cornell University, Geneva. https://soilhealth.cals.cornell.edu/training-manual/

Montgomery DR (2007) Dirt—the erosion of civilizations. University of California Press, Berkeley, 285 pp

Parks DH, Rinke C, Chuvochina M, Chaumeil P-A, Woodcroft BJ, Evans PN, Hugenholtz P, Tyson GW (2017) Recovery of nearly 8,000 metagenome-assembled genomes substantially expands the tree of life. Nat Microbiol 903:1–10. https://doi.org/10.1038/s41564-017-0012-7

Pepe-Ranney C, Koechli C, Potrafka R, Andam C, Eggleston E, Garcia-Pichel F, Buckley DH (2016) Non-cyanobacterial diazotrophs dominate dinitrogen fixation in biological soil crusts during early crust formation. ISME J 10(2):287–298. https://doi.org/10.1038/ismej.2015.106

Pimentel D (2006) Soil erosion: a food and environmental threat. Environ Dev Sustain 8:119–137

Six J, Bossuyt H, Degryze S, Denef K (2004) A history of research on the link between (micro)aggregates, soil biota, and soil organic matter dynamics. Soil Tillage Res 79:7–31

United States Department of Agriculture, Natural Resources Conservation Service (1999a) Soil taxonomy: a basic system of soil classification for making and interpreting soil surveys. NRCS, Washington, 871 pp

United States Department of Agriculture, Natural Resources Conservation Service (1999b) Soil quality test kit guide. USDA, Washington, 82 pp

United States Department of Agriculture, Sustainable Agriculture Research and Education, Sustainable Agriculture Research and Education (2010) Building soils for better crops, 3rd ed. Available online https://www.sare.org/Learning-Center/Books/Building-Soils-for-Better-Crops-3rd-Edition

Mary Ann Bruns Ph.D. is Professor of Soil Microbiology at the Pennsylvania State University. She advises graduate students and teaches courses in soil ecology and soil microbiology. Her current research focuses on the microbial ecology of soil nitrogen cycling and the roles of biological crusts to stabilize soil surfaces. Bruns has served as Chair of the Soil Biology and Biochemistry Division of the Soil Science Society of America and as Associate Editor of the Soil Science Society of America Journal, Applied Soil Ecology, and the Canadian Journal of Soil Science.

Estelle Couradeau is Assistant Professor of Soils and Environmental Microbiology in the Department of Ecosystem Science and Management and a member of the Huck Institutes of the Life Sciences at The Pennsylvania State University-University Park, PA, U.S.A. Her lab focuses on the study of soil microbiomes using various techniques including omics tools and advanced imagery. She teaches environmental sustainability and advanced methods in microbial ecology. Current research includes simulating and understanding the fate of soil microbiomes in the face of climate change.

Chapter 14
Environmental Phytoremediation and Analytical Technologies for Heavy Metal Removal and Assessment

Ephraim M. Govere

Abstract Biotic or abiotic environmental systems do not function well under chemical stresses. The realization that heavy metal stressors are a major threat to the environment and its ecosystems and the organisms that depend on them, have created an urgent need for novel ways to take effectively, efficiently, economically, and eco-friendly sustainable bioremediation interventions to decontaminate and protect the environment from heavy metal stressors. One very promising green technology intervention is phytoremediation. Phytoremediation technologies presented in this chapter include phytoaccumulation (phytoextraction), phytodegradation (phytotransformation), phytostabilization, phytovolatilization, phytofiltration. Because the effectiveness of these technologies is evaluated by assessing accumulated, degraded, immobilized, volatilized, precipitated, removed, extracted, adsorbed, desorbed, or leached heavy metals, this chapter also covers atomic spectroscopic techniques, the commonly used analytical techniques for the assessment of heavy metals in the environment. The analytical techniques presented in the chapter include atomic absorption spectroscopy, atomic absorption spectroscopy, inorganic mass spectrometry, atomic fluorescence spectroscopy, and X-ray fluorescence. Results from analytical techniques are critical to chemical hazard identification, exposure assessment, dose-response assessment, and nature and the magnitude of risk to biotic and abiotic environmental constituents. The goal is to protect, restore, and promote sustainable use of the environment to meet the needs of present and future generations.

Keywords Heavy metals · Phytoremediation · Atomic spectroscopy

E. M. Govere (✉)
Department of Ecosystem Science and Management, The Pennsylvania State University, 457 Agricultural Sciences and Industries Building, University Park, PA 16802, USA
e-mail: emg900@psu.edu

© The Author(s), under exclusive license to Springer Nature Switzerland AG 2021
A. Ricroch et al. (eds.), *Plant Biotechnology*,
https://doi.org/10.1007/978-3-030-68345-0_14

14.1 Introduction

Environmental chemical stressors are chemical pollutants at concentration levels that are toxic to organisms and threaten environmental sustainability. When the concentration of chemicals in the environment is at low, non-toxic levels, they are merely chemical contaminants rather than stressors. Heavy metal stressors are metallic elements and metalloids (elements that contain properties of both metals and nonmetals). They have a relatively high density compared to water (Tchounwou et al. 2012). They become toxic when they exceed their threshold concentration in an organism. Some metals are essential to sustaining animal and plant life as nutrients. They include cobalt (Co), copper (Cu), chromium (Cr), iron (Fe), magnesium (Mg), manganese (Mn), molybdenum (Mo), nickel (Ni), selenium (Se), and zinc (Zn). The other metals: aluminum (Al), antinomy (Sb), arsenic (As), barium (Ba), beryllium (Be), bismuth (Bi), cadmium (Cd), gallium (Ga), germanium (Ge), gold (Au), indium (In), lead (Pb), lithium (Li), mercury (Hg), nickel (Ni), platinum (Pt), silver (Ag), strontium (Sr), tellurium (Te), thallium (Tl), tin (Sn), titanium (Ti), vanadium (V) and uranium (U) are not essential to sustaining the life of organisms. Of these metals, Cr, As, Cd, Hg, and Pb have the highest environmental impact because of their widespread industrial, domestic, agricultural, medical, and technological use. They are referred to as systemic toxicants because they are known to induce multiple organ damage, even at lower levels of exposure (Tchounwou et al. 2012). Major natural sources of heavy metal stressors include natural and anthropogenic. Natural sources include decomposition (rock weathering and plant and animal tissue decomposition), precipitation, leaching, volatilization, rainfall and runoff, snowmelt, erosion (water surface runoff and windblown dust), natural fires, and volcanic eruption. Anthropogenic sources are a result of industrialization and urbanization. They include industries (mine and thermal, nuclear, petroleum, textile, electroplating, battery, plastic, smelter, tannery, fuel, paper, electronics, weaponry), sewage sludge, stormwater runoff, and plant growing inputs (fertilizers and manure; insecticides, herbicides, and fungicides; irrigation water, industrial byproducts), domestic and industrial usages and generated wastes (Srivastava et al. 2017). The presence of heavy metals in soil and water and the subsequent uptake of the heavy metals by plants and aquatic and terrestrial animals pose high animal and human health risks. The risk is greatest with humans and animals due to biomagnification—an increase in the concentration of metals with an increase in the food chain. Besides the direct adverse on human health, heavy metals decrease the diversity and activity of soil and aquatic microorganisms, and plant communities. A decrease in soil microorganisms reduces the recycling of plant nutrients, maintenance of soil structure, detoxification of noxious chemicals, and the control of plant pests (Singh and Kalamdhad 2011). Uptake of heavy metals by plants and animals contaminate the food chain and threaten the integrity and productivity of an ecosystem and the benefits derived from it. There should be global efforts to develop and implement effective, efficient, economic, eco-friendly, and sustainable corrective and preventative remedies to environmental heavy metal stressors. Modern phytoremediation techniques are among the most promising interventions for removing/reducing heavy metals from the environment.

14.2 Phytoremediation Techniques for Heavy Metal Removal

Phytoremediation is a bioremediation technology aimed at purifying and reclaiming environmental systems contaminated with heavy metals using natural and genetically modified plants. Phytoremediation processes are augmented with natural and genetically modified microbes to improve their effectiveness and efficiency. Major phytoremediation technologies are phytoaccumulation, phytodegradation (phytotransformation), phytostabilization, phytovolatilization, phytofiltration, and rhizofiltration.

The phytoaccumulation, also known as phytoextraction, involves extraction, uptake, and accumulation of heavy metal stressors by plants such as aquatic macrophytic species *Eichhornia crassipes* and *Centella asiatica* that hyper-accumulate heavy metals 100,000 times more than dissolved in aquatic ecosystems (Muthusaravanan et al. 2018). *Cyperus rotundus* accumulates Cr, and Cd; and *Brassica juncea* is a sink for Pb, Cu, and Ni. *Eichhornia crassipes* and *Centella asiatica* can remove 97.3 and 99.6% Cu, respectively (Muthusaravanan et al. 2018).

Phytoremediation technology is even more attractive because some plants can compartmentalize the heavy metals and restrict their accumulation to only roots and rhizosphere. For example, grasses do not accumulate metals in their shoots. That they can be grown on heavy metal contaminated areas and still have grazers feed on them without risk from heavy metal toxicity. This process is called phytostabilization and is complementary to phytoaccumulation. Thus, plants can "arrest" (accumulation) heavy metals and keep them in "jail" (immobilization), so to speak. Phytostabilization in combination with phytoaccumulation constitutes a double-action bioremediation technique. These technologies are greatly enhanced by the addition of natural and genetically modified microbes, a process called bioaugmentation. The bioaugmentation process not only improves the accumulation and stabilization of heavy metals but also reduces their toxic effects. Soil microbes enhance plant growth while at the same time they produce organic functional groups that fix (immobilize) heavy metals, reduce their bioavailability, and transform them into non-toxic chemical forms. For example, inoculating tomato plants with endophytic bacteria *Magna-porthe oryzae* and *Burkholderia sp.*, increases the plant growth plus limit Ni and Cd accumulation in roots and shoots of the tomato. The Bioaugmentation process can be added to bioremediation processes to create hybrid biotechnology to remove heavy metals from the environment. Lee and Kim (2010) reported the use of hybrid bioleaching technology to remove heavy metals from soil. For active bioaugmentation, they first bleached the soil with acidophilic *Acidithiobacillus thiooxidans* (bioleaching). The bacteria get their energy from the oxidation of sulfur minerals and in the process create acidic conditions that free metal ions and make them active in soil solution. To ensure the heavy metals are not re-adsorbed or fixed to soil surfaces, they created a heavy metal electromigration process by adding anionic ethylenediaminetetraacetic acid (EDTA) in the cathode compartment. The EDTA acts as a sink for heavy metals.

This new novel bioelectrokinetic technology has great potential and was found to have a high removal efficiency of 92.7% for Pb.

While most heavy metal stressors in the environment exist in non-volatile forms, three very toxic ones—Se, As and Hg can be converted to volatile forms such as dimethyl selenide, Trimethyl arsine, and mercuric oxide, respectively. They become atmospheric pollutants through the evaporation or the volatilization processes. Phytovolatilization technology is applied to minimize the toxic effects of volatile organometallic forms of Se, As and Hg generated mostly from landfills. The phyto-volatilization process involves selecting natural or genetically modified plants that can take up the metals and volatilize them in less toxic forms via transpiration. Plants like Indian mustard (*Brassica juncea*) and musk-grass (*Chara canescens*) are capable of degrading the toxic compounds and free them to the atmosphere. The use of enzymes improves the phytovolatilization technology. For example, geneti-cally engineering *Arabidopsis thaliana L.* and *Nicotiana tabacum L.* with mercuric reductase and bacterial organomercurial lyase enhances absorption of Hg(II) and methyl mercury from the soil and volatilizing them as Hg(0) (Muthusaravanan et al. 2018). When plants metabolically or in combination with enzymes are used to uptake, metabolize, and degrade metal stressors into less toxic forms, the process is called phytodegradation or phytotransformation (Ahmadpour et al. 2012). The enzymes commonly associated with this process are peroxidase, nitroreductase, nitrilase dehalogenase, and laccase (Muthusaravanan et al. 2018).

Two hydraulic phytoremediation techniques, phytofiltration and rhizofiltration, have the greatest potential in removing trace metals and radioactive elements in aquatic environments such as municipal wastewaters, industrial wastes, effusions of arable lands, exudates from waste dumps and landfills, and contaminated groundwa-ters. Both techniques use natural or engineered plants to sorb and accumulate metals. The difference between them is that with the rhizofiltration technique, contaminants accumulate in the vacuoles of the root cell cortex or penetrate directly into the root tissues and then get exudated onto the surface of the roots as precipitates (Hanus-Fajerska and Kozminska 2016). This purification technology is most appropriate for purifying hydroponic and aeroponic cultures. *Eleocharis acicularis* and *Eichhornia crassipes* are examples of hyper-accumulating plant species used in these aquatic phytoremediation technologies.

Another plant and microbial mediated bioremediation mechanism is the use of biosurfactants. Surfactants are chemical compounds generally known as amphiphilic. These compounds are both hydrophilic (water-loving) and lipophilic (fat-loving). The extent of their hydrophilicity and lipophilicity (HLB) is a very important char-acteristic in their use for heavy metal decontamination. The HLB is an indicator of the emulsifying, foaming, and dispersing capacities of the surfactant. A surfactant with high HLB is effective in forming an oil-in-water emulsion, while a low HLB likely forms a water-in-oil emulsion (Halecký and Kozliak 2020). In other words, a surfactant with high HLB is more effective in improving the surface-surface inter-actions (mixing) between immiscible liquid/air, liquid/liquid, and liquid/solid inter-face systems by reducing surface-surface repulsive forces between those systems (Halecký and Kozliak 2020). Adding more surfactant in the system reduces the surface-surface tension. Surfactants can be grouped into two groups: synthetic surfactants and biosurfactants. Biosurfactants are naturally produced by bacteria, fungi, yeast, or plants. For example, *Bacillus subtilis*, *Pseudomonas aeruginosa*, *Acinetobacter calcoaceticus*, and *Acinetobacter radioresistens* produce rhamnoplids, high-molecular-weight bioemulsifier lipopolysaccaharide, and bioemulsifier complex Alasan, respectively (Bustamante et al. 2012). Biosurfactants produced by microorganisms increase cell wall hydrophobicity and this further enhances solubility, mobility, and bioavailability of the heavy metals, making it easy to remove them from the environment. Plants produce biosurfactants, for example, saponins from soapberry, and lecithins from soyabeans. These phytogenic surfactants are released from decaying plant roots. With today's technology, biosurfactants can be synthesized from microorganisms and plants. Unlike synthetic surfactants, biosurfactants are eco-friendly because they are biodegradable, less toxic, more effective, and efficient, more heavy metal-specific, and work under harsh environmental conditions such as extreme temperature, pH, and salinity. They are being used to increase the binding ability of zeolites and clays used as adsorbent materials for the removal of heavy metal stressors (Jiménez-Castañeda and Medina 2017).

14.3 Atomic Spectroscopic Techniques for Assessing Heavy Metal Stressors

The environmental risk from heavy metals that are present in the environment before, during, and after phytoremediation interventions need to be assessed. The environmental risk from a heavy metal stressor is the probability that its measured concentration in the environment would exceed the toxic effect concentrations derived from the affected organism's sensitivity distributions (Fedorenkova et al. 2012). Atomic spectroscopy is one of the commonly used techniques for the assessment of heavy metals in the environment.

Atomic spectroscopic techniques are commonly used to determine the elemental composition of heavy metal stressors in solid, gaseous, and aqueous environmental samples by their electromagnetic or mass spectra. The underlying characteristic of atomic spectroscopic techniques is the absorption or emission of energy by a given atom. That means there must be a source of energy. With no energy source, the atom stays at the ground state, which is the level at which the atom's electrons are at the lowest possible energy level. The ground state is considered the normal state of an atom. When subjected to an energy source, the atom absorbs the energy to what is referred to as, excited state, at which its electrons are at a higher energy level than the ground state. To go back to the ground state from the excited state, the atom must emit energy. The amount of energy absorbed or emitted corresponds exactly to the difference between the ground state and excited energy levels. The energy absorbed or emitted is in form of light (photons = visible light displaying wavelike properties) when outer-valence electrons are involved or X-ray photons when inner-shell electrons of the atom collide with high-energy charged particles. Thus, the atomic spectroscopic techniques for assessing heavy metals are named based on these underlying principles and include atomic absorption spectroscopy (AAS), atomic emission spectroscopy (AES), atomic fluorescence spectroscopy (AFS), and X-ray fluorescence (XRF). With these techniques, the composition of heavy metals in an environmental sample is determined by its electromagnetic spectrum using a spectroscope. In addition to these techniques is the inorganic mass spectroscopy (MS) technique in which heavy metal atoms are separated in a magnetic field according to their mass to charge (m/z) ratio (mass spectrum) rather than the intensity of absorbed or emitted photons. Flame atomic absorption spectroscopy (FAA), graphite furnace atomic absorption spectroscopy (GFAA), inductively coupled plasma optical emission spectroscopy (ICP-OES), inductively coupled plasma mass spectrometry (ICP-MS), atomic fluorescence spectroscopy (AFS), and X-ray fluorescence (XRF) are specific analytical techniques operating under the atomic spectroscopy principles (Akash and Rehman 2020; Van Loon 2012).

14.3.1 Flame Atomic Absorption Spectroscopy (FAA)

Flame atomic spectroscopy utilizes a flame generated from a mixture of air and acetylene or nitrous oxide and acetylene to heat an aqueous environmental sample. Environmental samples such as animal and plant tissue, soil, slurries, or sediment are solubilized (make more soluble or soluble) and digested, typically with a concentrated acid such as nitric, hydrochloric, or sulfuric acid. A sample aliquot is aspirated into the flame. The heat from the flame dissociates the sample atoms into an atomic vapor, a process called vaporization. The atoms are at their ground-state. Radiation from a selected excited heavy metal in a hollow cathode lamp or electrodeless discharge lamp is passed through the vapor containing ground-state atoms. Optical devices called monochromators separate polychromatic light into monochromatic light or individual (mono) wavelengths of selected heavy metal. The radiation energy from the selected metal is absorbed by the ground-state atoms. The absorption causes the intensity of the transmitted radiation to decrease in proportion to the amount the metal atoms in the sample. High absorbance means low transmittance of light which in term means a high concentration of the selected metal (radiation source). The absorbed radiation is quantified by a spectrophotometer, a photosensitive device that measures the attenuated transmitted radiation. A photomultiplier tube is used to enable the measurement of the absorption of the photons in a sample with very low concentrations of heavy metals.

14.3.2 Graphite Furnace Atomic Absorption Spectroscopy (GFAA)

The GFAA technique is like the FAA but differs in the sample introduction and atomization process. Instead of a flame, the atomizer in the GFAA technique is a small graphite furnace (tube) into which an environmental sample is placed. The tube is electrically heated to selected temperatures that can go up to 2700 °C. A small representative solid or aqueous sample is placed into the graphite tube. By pre-programming the increase in temperature, the sample is the tube is evaporated to dryness, charred, and atomized. The atomic vapor is trapped in the tube. Radiation from a selected excited heavy metal in a hollow cathode lamp or electrodeless discharge lamp is passed through the vapor containing ground-state atoms of that metal. The metal atoms absorb the energy to get excited. Like FAA, the amount of radiation absorbed at the specific wavelength corresponds to the number of atoms of the heavy metal in the sample and is quantified by a spectrophotometer. The GFAA has more advantages over the FAA technique. It has high sensitivity and low detection limits due to the absence of flame gases and the extended period that the radiation passes throw the atomic vapor; and it allows for the analysis of small amounts of both solid and aqueous samples with less preparation than with the FAA technique.

14.3.3 Inductively Coupled Plasma Optical Emission Spectroscopy (ICP-OES)

Inductively Coupled Plasma Optical Emission Spectroscopy (ICP-OES) is the measurement of the light emitted by the elements in a liquid or gas samples directly introduced into the inductively coupled plasma (ICP). Unlike the FAA and GFAA, the ICP-OES technique exploits the fact that excited electrons emit energy at a given wavelength as they return to the ground state after excitation by high temperature. While the source for FAA is flame and that of GFAA is an electric current, the ICP-OES uses a heating source called inductively coupled plasma. The plasma source can generate temperatures as high as 9726.85 °C compared to 2700 °C for FAA and GFAA, and to 6000 °C on the surface of the sun. The capacity to generate high temperatures enables the ICP instruments to not only atomize but also ionize all heavy metals with ease. How is ICP generated? Plasma is a cloud of protons, neutrons, and electrons acting as a whole because they are loose from their atoms. To generate the plasma, a radio frequency power (RF) power and frequency typically of 0.2–2.0 kW and 27–40 MHz produces electric and magnetic fields that accelerate the argon ions and electrons by inductive coupling (inductively coupled plasma ICP). As they accelerate, they collide with other argon atoms, causing further ionization and in the process create a very intense, brilliant white, teardrop-shaped, high-temperature plasma. The plasma created high temperatures are capable of exciting any heavy metal element, unlike FAA and GFAA in which the low temperatures fall to free atoms from molecules and hinder radiation absorption, a problem referred to as chemical interferences.

In the ICP-OES technique, a solution is pumped into the instrument, a nebulizer sprays it into a spray chamber as an aerosol. The argon gas carries the sample mist into the plasma in the torch. The sample is subjected to the plasma temperatures starting at 10,000 °K and ending around about 600 °K. The following are the stages the sample goes through as it passes through the plasma:

- Environmental Sample: biotic or abiotic in gaseous, solid (e.g. animal and plant tissue, soil, slurries, or sediment) or aqueous state.
- Solution: turning an environmental sample into a liquid. Animal and plant tissue, soil, slurries, or sediment are solubilized (make more soluble or soluble) and digested, typically with a concentrated acid such as nitric, hydrochloric, or sulfuric acid.
- Desolvation: drying of the sample by removing the solvent resulting in microscopic solid particulates, or a dry aerosol.
- Vaporization: converting solid particulates or a liquid into gaseous state molecules.
- Sublimation: directly converting a solid into gaseous state molecules without passage through a liquid stage.
- Atomization: breaking the gaseous molecules into atoms by breaking the chemical bonds to yield free atoms.
- Excitation: supplying the amount of energy (called excitation energy) to an atomic nucleus, an atom, or a molecule to move it from a condition of lowest energy (ground state) to one of higher energy (excited state).
- Ionization: supplying the amount of energy (called ionization energy) to move excited atoms from a condition of lowest ionization energy (ionization ground state) to one of higher ionization energy (excited ionization state).

The plasma heat excited and ionize the atoms. However, the analytical temperatures in the plasma are lower than initial excitation and ionization temperatures. The atoms lose energy in the form of light at specific wavelengths peculiar to its atomic character. The measured emission intensities are then compared to the intensities of standards of known concentration to obtain the elemental concentrations in the unknown sample. The ICP-OES is capable of simultaneous analysis of multiple heavy metal stressors. The use of detectors such as segmented array charge-coupled device detector (SCD) with over 200 small subarrays of 20–80 pixels each, make it is possible for simultaneous multi-element detection of 70 out of a total of 92 naturally occurring. Furthermore, ICP-OES has higher accuracy, precision, reproducibility, and resolution, and a wide dynamic range compared to FAA and GFAA and can tolerate a high amount of total dissolved solids in the sample.

14.3.4 Inductively Coupled Plasma Mass Spectrometry (ICP-MS)

Inductively Coupled Plasma Mass Spectrometry is a type of analytical mass spectrometry for the determination of elemental composition. An ICP-MS combines a high-temperature ICP source with a mass spectrometer. The operating principles for ICP-MS are very similar to that of ICP-OES. The difference major difference is that ICP-MS operates in the ionization zone. Instead of measuring the intensity of light emitted by the atoms, ICP-MS separates and detected ions by a mass spectrometer based on their mass to charge ratio (m/z). ICP-MS has many advantages over ICP-OES, GFAA, and FAA. The main advantage is its high sensitivity (parts per trillion (ppt) level) and the ability to determine isotopic composition. However, it requires "clean" samples and higher-level skills to operate and maintain.

14.3.5 Atomic Fluorescence Spectroscopy (AFS)

Atomic fluorescence spectroscopy (AFS) sometimes referred to as atomic spectrofluorometry or atomic fluorimetry operates under the atomic emission principles. In this technique, the UV is the excitation energy source. The excited atoms become unstable and quickly reemit the radiation energy in the form of fluorescence (light). A monochromator separates the polychromatic light into monochromatic light whose intensity detected by a spectrometer, is proportional to the concentration of the metal. Another technique like AFS but differing in sample preparation to enable an analysis of volatile heavy metals at room temperature is the cold vapor atomic fluorescence spectroscopy (CVAFS). It is capable of sub-trace level detection of hydride-forming elements (As, Sb, Se, Sn) and Hg in environmental samples. An ICP-MS instrument with flow injection (FI) coupled with a hydride generation system (HG) can also be used to measure hydride-forming elements and Hg.

14.3.6 X-Ray Fluorescence Spectroscopy (XRF)

X-ray fluorescence spectroscopy (XRF) and ICP-MS are the most applied analytical spectroscopy techniques for heavy metals. Unlike FAA, GFAA, ICP-OES, ICP-MS, and AFS, the X-ray fluorescence technique is non-destructive. The sample is ground to a fine powder and analyzed directly for most heavy metals. However, the ground sample can be mixed with a chemical flux and use a furnace or gas burner to melt the powdered sample to create a homogenous glass that can be analyzed to compute the abundance of the metal (Wirth and Barth 2012). The sample is illuminated with a primary X-ray beam (high-energy, short-wavelength radiation) that excite heavy metal atoms in the sample. The atoms then absorb X-ray energy by ionizing (losing electrons). This high energy dislodges tightly held inner electrons making the atom unstable and resulting in outer electrons replacing the missing dislodged ones. During the excitation process, atoms emit X-rays along a spectrum of wavelengths characteristic of the types of atoms present in the sample. The emitted radiation is fluorescent. It is of lower energy than the primary incident X-rays. The fluorescent radiation is measured using a wavelength-dispersive spectrometer (Wirth and Barth 2012).

14.4 Conclusion

The quality of our physical environment and its major components (soil, water, and air) and the health of its ecosystems, define and sustain the wellbeing of the current and future generations. Poisoning our environment with toxic substances such as heavy metals and failing to monitor and take corrective and preventative action is unethical. Phytoremediation techniques are the most promising green technology to purify environmental components and ecosystems contaminated with heavy metal stressors. Analytical techniques such as atomic absorption and emission spectroscopy play a big role in assessing the levels and impact of heavy metals on ecosystems and human and animal health and the effectiveness of phytoremediation technologies. Therefore, there should be continuous efforts to improve both the phytoremediation technologies and analytical methods to assess their effectiveness in removing heavy metal from the environment.

References

Ahmadpour P, Ahmadpour F, Mahmud TMM, Abdu A, Soleimani M, Tayefeh FH (2012) Phytoremediation of heavy metals: a green technology. Afr J Biotechnol 11(76):14036–14043

Akash MSH, Rehman K (2020) Atomic spectroscopy. In: Essentials of pharmaceutical analysis. Springer, Singapore

Bustamante M, Duran N, Diez MC (2012) Biosurfactants are useful tools for the bioremediation of contaminated soil: a review. J Soil Sci Plant Nutr 12(4):667–687

Fedorenkova A, Vonk JA, Lenders HJ, Creemers RC, Breure AM, Hendriks AJ (2012) Ranking ecological risks of multiple chemical stressors on amphibians. Environ Toxicol Chem 31(6):1416–1421. https://doi.org/10.1002/etc.1831. Epub 2012 Apr 27

Halecký M, Kozliak E (2020) Modern bioremediation approaches: use of biosurfactants, emulsifiers, enzymes, biopesticides, GMOs. In: Filip J, Cajthaml T, Najmanová P, Černík M, Zbořil R (eds) Advanced nano-bio technologies for water and soil treatment. Springer International Publishing, Cham, pp 495–526

Hanus-Fajerska EJ, Kozminska A (2016) The possibilities of water purification using phytofiltration methods: a review of recent progress. BioTechnol J Biotechnol Comput Biol Bionanotechnol 97(4)

Jiménez-Castañeda ME, Medina DI (2017) Use of surfactant-modified zeolites and clays for the removal of heavy metals from water. Water 9(4):235

Lee K-Y, Kim K-W (2010) Heavy metal removal from shooting range soil by hybrid electrokinetics with bacteria and enhancing agents. Environ Sci Technol 44(24):9482–9487

Muthusaravanan S, Sivarajasekar N, Vivek JS, Paramasivan T, Naushad M, Prakashmaran J, Al-Duaij OK (2018) Phytoremediation of heavy metals: mechanisms, methods and enhancements. Environ Chem Lett 16(4):1339–1359

Singh DJ, Kalamdhad A (2011) Effects of heavy metals on soil, plants, human health and aquatic life. Int J Res Chem Environ 1:15–21

Srivastava V, Sarkar A, Singh S, Singh P, de Araujo ASF, Singh RP (2017) Agroecological responses of heavy metal pollution with special emphasis on soil health and plant performances. Front Environ Sci 5(64)

Tchounwou PB, Yedjou CG, Patlolla AK, Sutton DJ (2012) Heavy metal toxicity and the environment. Experientia Suppl 101:133–164

Van Loon AT (2012) Analytical atomic absorption spectroscopy: selected methods. Elsevier

Wirth K, Barth A (2012) X-ray fluorescence (XRF). Geochemical instrumentation and analysis

Ephraim Muchada Govere Ph.D., M.B.A., CPSS, is the Director of the Soil Research Cluster Laboratory (SRCL) and teaches a graduate course—Soil Ecosystems Analytical Techniques—in the Department of Ecosystem Science and Management, Pennsylvania State University, USA. SRCL is a multi-function, multi-user laboratory that provides leadership and expertise, and analytical instrumentation and equipment necessary for research and teaching in the areas of soil chemistry, soil fertility, soil physics, pedology, hydropedology, and other ecosystem areas of agricultural and environmental testing.

Part V

Contributions to Food, Feed, and Health

Chapter 15
Production of Medicines from Engineered Proteins in Plants

Kathleen Hefferon

Abstract Plants have always played a predominant role in both the nutritional and medical components of human health. Innovations in plant biology, in particular plant biotechnology, have helped these disciplines advance significantly. While some of our modern drugs have their origins in plant species, many pharmaceuticals that are commercially available today using conventional production systems such as yeast or mammalian cells can now be generated within the plant material itself, through a novel approach known as molecular pharming. This chapter describes the use of plants as a production platform to generate medicines to address infectious and chronic diseases. Examples of biologics produced in plants, from vaccines, monoclonal antibodies to therapeutic agents are described. The development of nanoparticles for cancer therapy based on plant viruses is also discussed. The chapter concludes with a projection of future applications for plant-based pharmaceuticals.

Keywords Molecular pharming · Developing countries · Pandemic · Vaccines · Monoclonal antibodies · Cancer · Nanoparticles · Drugs · Transgenic plants · Chloroplasts

15.1 Introduction

Since the dawn of man, plants have been central to our health requirements. Using modern technologies, plants are being engineered for improved nutritional benefits, ranging from biofortification (such as improved vitamin and mineral content) to enhanced nutraceutical properties (such as oil crops that produce the omega-3 fatty acids that are regularly found in seafood and are essential for human health). Plants are also being developed for their medicinal properties, through a process known as bioprospecting, and some known bioactive compounds can now be produced inexpensively in plant cell culture, including cancer fighting compounds such as

K. Hefferon (✉)
Department of Microbiology, Cornell University, Ithaca, NY 14850, USA
e-mail: klh22@cornell.edu

A. Ricroch et al. (eds.), *Plant Biotechnology*,
https://doi.org/10.1007/978-3-030-68345-0_15

taxol. Finally, plants can be engineered to act as production platforms for modern pharmaceuticals, such as vaccines and monoclonal antibodies.

There are many advantages to generating pharmaceutical proteins in plants. First and foremost, plant-derived biologics are inexpensive to produce, often less than a hundredth of the cost of their conventional counterparts. In addition to this, plant-made pharmaceuticals can be lyophilized so as to not require refrigeration, and thus can be stably stored at ambient temperatures for months, or even years (Twyman et al. 2005). The low cost and lack of cold chain requirements make plant-made pharmaceuticals amenable for storage and delivery throughout developing countries.

Plants do not carry human pathogens such as cytomegalovirus, and thus are safer than the use of mammalian production systems such as Chinese hamster (CHO) cells. Plant-produced proteins also contain similar post-translational modifications as do mammalian cells, but which are lacking in bacterial cell culture. The number of plants used can easily be scaled up to produce a larger amount of pharmaceutical product on demand. If the recombinant protein is made in a transgenic plant, the seed can be stored and provided to countries who could then grow their own (Hefferon 2013).

It should be noted that plants have their own N-glycosylation systems which may complicate results due to the introduction of molecular heterogeneity and subsequent difficulties in immune cell recognition. Therefore, plants have been modified via the knocking out of plant-specific N-glycosylation genes and the deliberate introduction of genes coding for mammalian sugar synthesis, so that they produce more 'humanized' proteins (Gomord et al. 2010).

Due to man's extensive history and background in plant breeding research, much is known about methods to extract plant protein, and this knowledge makes the extraction and purification of medicinal proteins from engineered plants facile. Moreover, in certain instances, only the crude extracts taken from edible tissue of plants are necessary; these can then be delivered to patients by oral consumption. Plant-made biologics ideally can both simplify the manufacturing process and preserve the biological and immunological activities of the plant derived antigen; the thick cell wall protects the antigen in question from the harsh environment of the gastrointestinal tract, so that it can be better recognized by the immune surveillance system (Azegami et al. 2014). Plants such as carrot, lettuce, maize, potato and rice have been used to generate such oral vaccines.

A previous impediment for the use of plant-based biopharmaceuticals (Thomas et al. 2011) has concerned variations in the amount of the target antigen produced between different species of plants or even from leaf to leaf or fruit to fruit of the same plant, which then complicates standardized delivery of the concerned antigen. This problem has largely been mitigated using various delivery techniques, which will be described in the next section.

15.2 Technologies Used to Produce Plant-Based Pharmaceuticals

Several technologies are routinely used to produce heterologous proteins in plants, the most popular of these include the generation of transgenic plants or the use of transient expression strategies. Transgenic plants are advantageous as they involve the incorporation of genes which can be transmitted stably from one generation to the next and also stored as seed. Both nuclear and transplastomic (chloroplast) transformation are used routinely by scientists. Chloroplasts have the advantage of producing multiple copies of a gene in a single plant cell, thus generating a much greater amount of protein than is possible for nuclear transformation. In addition to this, chloroplasts are not found in pollen and thus there would be no biocontainment concerns if these plants were grown in the open field (Oey et al. 2009). Caveats include the fact that transplastomic technology is rather new and thus not available for all crop types. Chloroplasts do not contain glycosylation machinery as is found in the cytoplasm of plant cells, and therefore could not produce some proteins that have highly specific folding requirements (Verma and Daniell 2007).

Transgenic plants are commonly generated using Agrobacterium-mediated transformation or else by biolistic delivery via a gene gun (Laere et al. 2016). Both are effective means of inserting foreign DNA into plant tissue. In vitro techniques such as protoplast, cell culture and culture of hairy root propagation have been attempted to introduce genes specifying biopharmaceuticals in alfalfa, carrot, lettuce, maize, peanut, potato, rice, tobacco, tomato and soybeans.

Transient expression strategies include the use of Agroinfiltration, plant virus recombinant expression vectors, or else a combination of the two. Transient expression has the advantage of being short-term and rapid, often providing high yields within a few days, as opposed to the months needed for the generation of transgenic plants. Plant viruses commonly used for the production of pharmaceutical proteins include Alfalfa mosaic virus, Cowpea mosaic virus, Potato virus X (PVX), Bamboo mosaic virus, Papaya mosaic virus, Tomato bushy stunt virus, Tobacco mosaic virus and Plum pox potyvirus (Salazar-Gonzalez et al. 2015). Either full-length proteins can be generated in a plant virus expression vector, or else short epitopes can be displayed on the surface of assembled plant virus nanoparticles. An advantage is that multiple epitopes or proteins can be expressed in this fashion in a single plant cell, thus making the rapid production of complex proteins with multiple subunits feasible. Infiltration technologies can also be used to produce pharmaceuticals transiently in plants. For example, a technique known as Magnifection involves vacuum infiltrating tobacco plants with Agrobacterium containing either binary vectors or second generation deconstructed viral vectors (Leuzinger et al. 2013). The Canadian biotechnology company Medicago was a pioneer in this agroinfiltration technology (Landry et al. 2010). Using the MagniCOM technology, the German plant biotechnology company Icon Genetics has generated several plant-derived vaccines, including yields of HBV surface antigen as great as 300 mg/kg of *Nicotiana benthamiana* fresh leaf weight (Huang et al. 2006). The next section provides other examples of vaccines, monoclonal antibodies and therapeutic agents produced in plants.

15.3 Vaccines Produced in Plants

15.3.1 Plant-Made Vaccines for Human Papillomavirus (HPV)

Human papillomavirus is a causative agent of cervical cancer. In 2018, 311,000 women died from cervical cancer resulting from HPV infection; more than 85% of these deaths occurring in low- and middle-income countries. A current vaccine based on virus-like particles exists but is expensive. It would be very beneficial to have an inexpensive plant made vaccine to HPV available, particularly for sub-Saharan Africa (Bicmelt et al. 2003). Several research projects have focused on this possibility (Scotti and Rybicki 2013). For example, the E7 protein of Human papillomavirus (HPV-16 E7) has been expressed using an expression vector based on Potato virus X (PVX) in *N. benthamiana* plants. E7 is a protein expressed by the virus that has been shown to play a role in oncogenesis. A crude plant extract containing this vaccine protein was shown to generate cytotoxic T cell (CTL) as well as T helper cell (Th1 and Th2) responses in mice. This recombinant virus was demonstrated to inhibit tumour growth in the mice by as much as 40% (Franconi et al. 2002). When E7 expression was increased fivefold by targeting its expression into the secretory pathway of plants, tumour growth was inhibited in as many as 80% of the mice tested (Franconi et al. 2006). It was also demonstrated in this study that the plant extracts themselves possessed adjuvant properties. An HPV-16 E7 mutant fused through homologous recombination to the chloroplast genome of the unicellular alga *Chlamydomonas reinhardtii* was used in a different study and showed that 60% of mice injected with the purified protein were protected from developing tumours, demonstrating that this vaccine retained its therapeutic potential in an algal system (Demurtas et al. 2013). This system is ideal for vaccine production as the E7 protein that was produced remained soluble and was thus more amenable to downstream processing that is cost-efficient. Chlamydomonas is also an attractive choice as it is easy to grow and transform and is completely amenable to GMP guidelines.

HPV virus-like particles (VLPs) have also been generated in plants. Paz De la Rosa et al. (2009) demonstrated immunogenicity in mice of HPV-16 VLPs generated in tobacco. Fernández-San Millán et al. (2010) generated VLPs in tobacco chloroplasts using the L1 capsid protein of HPV. The VLPs isolated from these plants were highly immunogenic. Professor Ed Rybicki's group in Cape Town, South Africa has explored the use of L1 and L2 to assemble highly immunogenic VLPs from plants using an expression vector that allows post translational targeting of the protein to plant chloroplasts (Hitzeroth et al. 2018). Further studies showed that epitopes derived from L2 could be incorporated into the loops of L1 to make a VLP that elicited cross neutralizing antibodies, essential for protection against multiple strains of HPV (Chabeda et al. 2019). More recently, Yazdani et al. (2019) expressed VLPs using a Grapevine fanleaf virus expression vector that expressed epitopes to L2 of HPV on the VLP surface.

15.3.2 Plant-Made Vaccine for Cholera

Cholera remains a prominent disease in the world today. Bacterial enterotoxins such as the B subunit of cholera toxin (CTB), has been used as part of a fusion protein with a poorly immunogenic protein to elicit a robust mucosal immune response for a number of multicomponent vaccines (Yu and Langridge 2001). For example, a tricomponent subunit vaccine produced in transgenic potato by fusion of cholera toxin B to the enterotoxin of rotavirus and *E. coli* adhesion protein has provided protection against cholera, rotavirus and enterotoxigenic *E. coli* (Yu and Langridge 2001). Matoba (2015) found that plant made CT-B was N-glycosylated, and this extra motif provided some advantages for its use as an epitope scaffold and vaccine platform.

It has long been considered that a plant-made vaccine against cholera would be efficacious, feasible and affordable for developing countries. Hamorsky et al. (2013a) demonstrated that a cholera vaccine produced from a plant production platform would circumvent the challenges of cost and scalability to facilitate mass distribution in developing countries. The authors added an endoplasmic reticulum retention signal and optimized secretory signal, improving the yield dramatically to >1 g per kg of fresh leaf material. The vaccine protein could then be efficiently purified by simple two-step chromatography, making it feasible for use in a low resource setting. Furthermore, Yuki et al. (2013) and Kashima et al. (2016) used transgenic rice as a platform to produce cholera vaccine. The authors produced and stored seed, grew plants under hydroponic conditions and rice was collected, polished and powdered to meet regulatory specifications.

15.3.3 Plant-Made Vaccines for Influenza Virus

In a fashion similar to the cholera example, plants have been intensively examined for their ability to produce inexpensive and efficacious influenza vaccines. Influenza vaccines are usually produced using chicken eggs, this is a process that is both slow and costly. Plant-made influenza vaccines, therefore, have proven to be attractive alternatives (Pillet et al. 2015, 2016; Shoji et al. 2012; Makarkov et al. 2017; Won et al. 2018). Recently, Hodgins et al. (2019) have demonstrated that plant-based influenza vaccines derived of influenza virus hemagglutinin can protect very old mice from death and frailty after challenge with virus, even aged mice who are associated with co-morbidities. Other studies demonstrated that vaccines based on plant-made virus-like particles (VLPs) displaying wild-type influenza hemagglutinin (HA) are unusually immunogenic, eliciting both humoral and cellular responses. These VLPs were stable for 12 months at 4 °C, are identical in appearance to wild type virus and are currently completing clinical trials.

15.4 Antibodies to One Health Diseases Produced in Plants

15.4.1 Ebola Virus

The first fully automated industry that can produce mass scale quantities of vaccines and therapeutics using tobacco plants and plant viral vector technology was developed to address Ebola virus. This came about through a collaboration between engineers at the Fraunhofer Center for Manufacturing Innovation at Boston University (CMI) and biologists at the Fraunhofer Center for Molecular Biology (CMB). At present, ZMapp, a cocktail of anti-Ebola virus antibodies, is manufactured in *N. benthamiana*. The plant-generated anti-Ebola antibodies showed more robust antibody-dependent cellular cytotoxicity (a mechanism in which an effector cell of the immune system lyses a target cell that is coated with highly specific antibodies) when compared to anti-Ebola antibodies generated in a Chinese hamster kidney mammalian cell line (Budzianowski 2015). This suggests that plants could be used as bioreactors for bulk production of ZMapp antibodies to meet the increasing demand for anti-EBOV vaccines. Plant-based AU:15 vectors have been generated using both DNA (Bean yellow dwarf virus) and RNA viruses to express heterologous proteins such as the EBOV GP, VP40 and NP. mAbs against EBOV have been generated through Agroinfiltration using plants such as lettuce (Laere et al. 2016). The protective and neutralizing anti-EBOV mAb6D8 has been expressed at levels as high as 0.5 mg/g of leaf mass. These antibodies were first tested as an experimental treatment on Americans who returned from West Africa with Ebola symptoms during the outbreak of 2015. An advantage is that they can be stockpiled and stored at room temperature and at low cost for future emergencies. ZMapp was undergoing advanced clinical trials but has recently been outperformed by newer treatments.

15.4.2 Plant-Made Antibodies to HIV

HIV continues to be a devastating infectious disease, with serious healthcare and economic repercussions for all countries, particularly Southeast Asia and sub-Saharan Africa. A vaccine to prevent HIV transmission has been particularly elusive, and only recently has some ground been gained in this area. The disease has been treated largely with a cocktail of antiviral proteins, but they are expensive and can be difficult to come by, particularly for the poor in remote regions. Plant-based versions of vaccines and antiviral proteins to prevent HIV infection are also under exploration.

One alternative solution has been the development of monoclonal antibodies in transgenic plants that block HIV transmission. Tobacco mosaic virus (TMV) has also been used to produce the broadly neutralizing antibody known as VRC01 in tobacco plants (Hamorsky et al. 2013b). Similarly, Sack et al. (2015) generated the HIV-neutralizing monoclonal antibody 2G12 in transgenic tobacco plants. These plants are being examined further for their ability to generate safe and efficacious mono-clonal antibodies in a double-blind, placebo-controlled clinical trial (Ma et al. 2015). Loos et al. (2015) produced broadly neutralizing monoclonal antibodies PG9 and its derivative RSH in transgenic tobacco plants. Besides neutralizing virus transmis-sion, these re-engineered plant-made antibodies are capable of inducing antibody-dependent cellular cytotoxicity, an activity not observed for PG9 produced in Chinese hamster cells. The authors thus expect that plant-produced anti-HIV-1 antibodies will be superior to their conventional counterparts.

15.4.3 Plant-Made Vaccines to Dengue and West Nile Virus

Antibodies have been produced in plants to combat other One Health Diseases, such as West Nile Virus and Dengue Fever Virus. Both viruses can cause antibody dependent enhancement (ADE) over multiple infections; as a result, a conventional vaccine to either of these viruses can create adverse immune responses. Similarly, ADE can cause problems if someone is pre-exposed to one and then later exposed to another type of flavivirus. When antibodies were generated against a plant made envelope protein from WNV, they did not elicit ADE in pre-immunized mice that were later infected with Dengue fever virus, indicating that plants may provide a means to circumvent the phenomenon of ADE entirely.

15.5 Plant-Made Therapeutic Agents

A plant-derived pharmaceutical for a genetic disorder has been generated in cultured transgenic tobacco and carrot cells by the Israeli company Protalix. In 2012, the US FDA granted approval to Protalix AU:5 and its partner, Pfizer for production of the drug taliglucerase alfa (a recombinant glucocerebrosidase) in plants to treat Gaucher disease. Gaucher disease is a genetic metabolic disorder that is largely found in Jewish populations and treatment is expensive. Carrot cells can produce the same drug but for a fraction of the cost. The glucocerebrosidase produced in carrot cells is correctly glycosylated and biologically functional. The drug can therefore easily be administered orally in the form of a juice.

Plant biotechnology has also been used to treat chronic diseases such as diabetes and hypertension. For example, Wakasa et al. (2011) used transgenic rice seed expressing an antihypertensive peptide derived from ovalbumin, known as novokinin. Transgenic rice seed expressing this peptide exhibited significant antihypertensive activity, and in a long-term administration for 5 weeks, even a smaller dose (0.0625 g/kg) of transgenic seeds could confer antihypertensive activity. Kawaka et al. (2015), examined the use of plants to produce γ-aminobutyric acid (GABA), a non-protein amino acid that functions as a major neurotransmitter and also as a blood-pressure lowering agent. The authors generated GABA-fortified transgenic rice by manipulating the GABA-shunt pathway in plants. A field trial of these transgenic plants was performed and milled rice was administered orally to rats on a daily basis for a period of 8 weeks. The study demonstrated an anti-hypertensive effect in spontaneous hypertensive rats, suggesting that GABA-fortified rice may be applicable as a staple food to control or prevent hypertension.

Daniell et al (2016) further explored the use of plants to produce drugs inexpensively using chloroplast engineering for other diseases including diabetes, Alzheimers disease, hemophilia and retinopathy. These noninfectious diseases all have significant impact on our health and healthcare systems. For example, in 2019, 463 million adults (20–79 years) worldwide were living with diabetes; by 2045 this will rise to 700 million (International Diabetes Federation). On average, people with diagnosed diabetes have medical expenditures approximately 2.3 times higher than what expenditures would be in the absence of diabetes (American Diabetes Association). According to the Harvard School of Public Health, the global cost of diabetes has now reached 825 billion dollars per year. Plant made insulin could reduce costs enormously and would eliminate the need for refrigeration or needles to administer this drug. Insulin could be produced in lettuce or tobacco; mice studies have shown that oral delivery of insulin from these plants can significantly lower blood glucose levels in a manner analogous to conventional injections. Daniell's group has shown that tobacco chloroplasts could easily yield up to 20 million daily insulin doses per acre of tobacco per year. This research group has also demonstrated that the glucagon-like peptide exendin-4 (EX4), when delivered orally, could lower glucose levels just as effectively as its conventional counterpart.

15.6   Plant Virus Nanoparticles to Combat Cancer

Nanoparticles are tiny assembled sets of particles between 1 and 100 nm in diameter and have a wide variety of uses. Plant viruses are increasingly being used as viral nanoparticles (VNPs) for immunotherapy to combat a variety of cancers. Plant viruses have been found to enter into human cells using vimentin as their receptor, a protein found in most cell types. One advantage of VNPs is that they can target and induce highly localized immune reactions to confront solid tumours. Plant virus nanoparticles can also act as drug delivery vehicles and can be utilized for tissue imaging (Steinmetz 2013). VNPs are also highly biocompatible as they are nontoxic, stable, amenable for genetic engineering, easy to scale-up and are less expensive than other nanomaterials (Steinmetz 2010). Plant viruses have a relatively short half life in the bloodstream of less than two weeks; however, this time can be extended by coating the nanoparticle with polyethylene glycol or other similar stabilizing compounds. Plant viruses can be directly injected into a solid tumor, or else can be administered parenterally, so that the virus particles will become affixed within the tumor itself. Thus, plant virus nanoparticles conjugated to fluoromeres can be used to screen and identify tumors in a patient using an MRI scan.

An example of a plant virus being used as a nanoparticle is the Potato virus X (PVX) nanofilament, carrying a breast cancer targeting monoclonal antibody. This VNP, known as trastuzumab, has been shown to cause apoptosis in breast cancer cell lines (Esfandiari et al. 2016). Tobacco mosaic virus (TMV) has also been used as a nanoparticle for drug delivery in cancer chemotherapy. In this instance, the drug phenanthriplatin has been shown to be delivered by these VNPs in breast cancer mouse models (Czapar et al. 2016). Furthermore, Cowpea mosaic virus (CPMV) nanoparticles have been shown to mediate anticancer effects by stimulating the immune system. In situ vaccination of tumours using empty eCPMV nanoparticles generated in *N. benthamiana* plants has been successful in cancer immunotherapy of breast, colon, melanoma and ovarian cancer models (Lizotte et al. 2016). Plant viruses are increasingly being used for the delivery of nucleic acid therapies (Lam and Steinmetz 2018) and have been found to be superior to mammalian vectors with respect to their high production yields, low cost, enhanced safety due to their inability to replicate in mammalian cells combined with the low risk of insertional mutagenesis.

15.7 Conclusions

Plants used as a production platform for pharmaceuticals have select advantages. The plant tissue itself, when orally consumed, becomes the delivery vehicle. Plant cell walls offer some resistance to degradation in the harsh environment of the gastrointestinal tract, thus enabling more vaccine protein to be recognized by the immune surveillance system and as a result, eliciting a more robust immune response than their conventional vaccine counterparts. Finally, by avoiding any sophisticated equipment required for protein purification, plant-made biologics become a feasible, affordable choice for developing countries.

Under today's current COVID-19 pandemic, several solutions, including vaccines and antiviral agents, are under exploration. Some of these solutions are being pursued under plant production platform. Medicago, a biopharmaceutical company based in Canada, has successfully developed a Virus-like particle (VLP) of the coronavirus 20 days after obtaining the SARS-CoV-2 genetic sequence. Instead of using egg-based methods to develop vaccines, this technology inserts a genetic sequence encoding the spike protein of COVID-19 into *Agrobacterium*, a common soil bacterium that is taken up by plants. The resulting plants that are developed produce a virus like particle that is composed of plant lipid membrane and COVID-19 spike protein. The VLPs are similar in size and shape to actual coronavirus but are lacking in nucleic acid and are thus noninfectious. Previously, Medicago has made VLPs composed of influenza virus haemagglutinin, and have demonstrated their safety and efficacy in animal models as well as in human clinical trials (Pillet et al. 2019). The cost of producing a plant-made vaccine based on VLPs is a small fraction compared to its conventional counterpart.

British American Tobacco, through its biotech subsidiary in the US, Kentucky BioProcessing (KBP), is developing a potential vaccine for COVID-19 and is currently undergoing pre-clinical testing (Gretler 2020). Experts at KBP cloned a part of the genetic sequence of SARS-CoV-2, which they used to develop a potential antigen that was inserted into *N. benthamiana* plants for production. The vaccine has elicited a positive immune response by pre-clinical testing and has moved onto Phase 1 human clinical trials, which could begin in late June. BAT could manufacture as much as 1–3 million doses of COVID-19 vaccine per week (they made 10 million vaccines of flu in a month as well as an Ebola vaccine using the same plant-based approach).

These two recent examples of the rapid installation of plant made vaccine production during the coronavirus pandemic illustrate their urgent need and enormous potential. Hopefully, as they enter the world stage, plant-made pharmaceuticals will remain a much-needed and permanent component of public health.

References

Azegami T, Yuki Y, Kiyono H (2014) Challenges in mucosal vaccines for the control of infectious diseases. Int Immunol 26:517–528

Biemelt S, Sonnewald U, Galmbacher P, Willmitzer L, Müller M (2003) Production of human papillomavirus type 16 virus-like particles in transgenic plants. J Virol 77(17):9211–9220

Budzianowski J (2015) Tobacco against Ebola virus disease. Przegl Lek 72(10):567–571

Chabeda A, van Zyl AR, Rybicki EP, Hitzeroth II (2019) Substitution of human papillomavirus type 16 L2 neutralizing epitopes into L1 surface loops: the effect on virus-like particle assembly and immunogenicity. Front Plant Sci 10:779

Czapar AE et al (2016) Tobacco mosaic virus delivery of phenanthriplatin for cancer therapy. ACS Nano 10(4):4119–4126

Daniell H, Chan HT, Pasoreck EK (2016) Vaccination via chloroplast genetics: affordable protein drugs for the prevention and treatment of inherited or infectious human diseases. Annu Rev Genet 50:595–618

Demurtas OC et al (2013) A Chlamydomonas-derived human papillomavirus 16 E7 vaccine induces specific tumor protection. PLoS ONE 8(4):e61473

Esfandiari N et al (2016) A new application of plant virus nanoparticles as drug delivery in breast cancer. Tumor Biol 37(1):1229–1236

Fernández-San Millán A, Ortigosa SM, Hervás-Stubbs S, Corral-Martínez P, Seguí-Simarro JM, Faye L, Gomord V (2010) Success stories in molecular farming—a brief overview. Plant Biotechnol J 8(5):525–528

Franconi R et al (2002) Plant-derived human papillomavirus 16 E7 oncoprotein induces immune response and specific tumor protection. Cancer Res 62(13):3654–3658

Franconi R et al (2006) Exploiting the plant secretory pathway to improve the anticancer activity of a plant-derived HPV16 E7 vaccine. Int J Immunopathol Pharmacol 19(1):187–197

Gomord V, Fitchette A, Menu-Bouaouiche L, Saint-Jore-Dupas C, Michaud D, Faye L (2010) Plant-specific glycoprotein patterns in the context of therapeutic protein production. Plant Biotechnol J 8:564–587

Gretler C (2020) Tobacco-based coronavirus vaccine poised for human tests. Bloomberg, 15 May. https://www.bloomberg.com/news/articles/2020-05-15/cigarette-maker-s-coronavirus-vaccine-poised-for-human-tests

Hamorsky KT, Kouokam JC, Bennett LJ, Baldauf KJ, Kajiura H, Fujiyama K, Matoba N (2013a) Rapid and scalable plant-based production of a cholera toxin B subunit variant to aid in mass vaccination against cholera outbreaks. PLoS Negl Trop Dis 7(3):e2046

Hamorsky KT, Grooms-Williams TW, Husk AS, Bennett LJ, Palmer KE, Matoba N (2013b) Efficient single tobamoviral vector-based bioproduction of broadly neutralizing anti-HIV-1 monoclonal antibody VRC01 in Nicotiana benthamiana plants and utility of VRC01 in combination microbicides. Antimicrob Agents Chemother 57(5):2076–2086

Hefferon K (2013) Plant-derived pharmaceuticals for the developing world. Biotechnol J 8:1193–1202

Hitzeroth II, Chabeda A, Whitehead MP, Graf M, Rybicki EP (2018) Optimizing a human papillomavirus type 16 L1-based chimaeric gene for expression in plants. Front Bioeng Biotechnol 6:101

Hodgins B, Pillet S, Landry N, Ward BJ (2019) Prime-pull vaccination with a plant-derived virus-like particle influenza vaccine elicits a broad immune response and protects aged mice from death and frailty after challenge. Immun Ageing 16:27

Huang Z, Santi L, LePore K, Kilbourne J, Arntzen C, Mason H (2006) Rapid, high-level production of hepatitis B core antigen in plant leaf and its immunogenicity in mice. Vaccine 24:2506–2513

Kashima K, Yuki Y, Mejima M, Kurokawa S, Suzuki Y, Minakawa S, Takeyama N, Fukuyama Y, Azegami T, Tanimoto T, Kuroda M, Tamura M, Gomi Y, Kiyono H (2016) Good manufacturing practices production of a purification-free oral cholera vaccine expressed in transgenic rice plants. Plant Cell Rep 35(3):667–679

Laere E, Ling APK, Wong YP, Koh RY, Lila MAM, Hussein S (2016) Plant-based vaccines: production and challenges. J Bot 1–11

Lam P, Steinmetz NF (2018) Plant viral and bacteriophage delivery of nucleic acid therapeutics. Nanomed Nanobiotechnol 10(1):e1487

Landry N, Ward B, Trepanier S, Montomoli E, Dargis M, Lapini G et al (2010) Preclinical and clinical development of plant-made virus-like particle vaccine against avian H5N1 influenza. PLoS ONE 5:e15559

Leuzinger K, Dent M, Hurtado J, Stahnke J, Lai H, Zhou X et al (2013) Efficient agroinfiltration of plants for high-level transient expression of recombinant proteins. J Vis Exp 77:e50521

Lizotte P et al (2016) In situ vaccination with cowpea mosaic virus nanoparticles suppresses metastatic cancer. Nat Nanotechnol 11(3):295–303

Loos A, Gach JS, Hackl T, Maresch D, Henkel T, Porodko A, Bui-Minh D, Sommeregger W, Wozniak-Knopp G, Forthal DN, Altmann F, Steinkellner H, Mach L (2015) Glycan modulation and sulfoengineering of anti-HIV-1 monoclonal antibody PG9 in plants. Proc Natl Acad Sci USA 112(41):12675–12680

Ma JK, Drossard J, Lewis D, Altmann F, Boyle J, Christou P, Cole T, Dale P, van Dolleweerd CJ, Isitt V, Katinger D, Lobedan M, Mertens H, Paul MJ, Rademacher T, Sack M, Hundleby PA, Stiegler G, Stoger E, Twyman RM, Vcelar B, Fischer R (2015) Regulatory approval and a first-in-human phase I clinical trial of a monoclonal antibody produced in transgenic tobacco plants. Plant Biotechnol J 13(8):1106–1120

Makarkov AI, Chierzi S, Pillet S, Murai KK, Landry N, Ward BJ (2017) Plant-made virus-like particles bearing influenza hemagglutinin (HA) recapitulate early interactions of native influenza virions with human monocytes/macrophages. Vaccine 35(35 Pt B):4629–4636

Matoba N (2015) N-glycosylation of cholera toxin B subunit: serendipity for novel plant-made vaccines? Front Plant Sci 6:1132

Oey M, Lohse M, Kreikemeyer B, Bock R (2009) Exhaustion of the chloroplast protein synthesis capacity by massive expression of a highly stable protein antibiotic. Plant J 7:436–445

Paz De la Rosa G, Monroy-García A, Mora-García Mde L, Peña CG, Hernández-Montes J, Weiss-Steider B, Gómez-Lim MA (2009) An HPV 16 L1-based chimeric human papilloma virus-like particles containing a string of epitopes produced in plants is able to elicit humoral and cytotoxic T-cell activity in mice. Virol J 6:2

Pillet S, Racine T, Nfon C, Di Lenardo TZ, Babiuk S, Ward J, Kobinger GP, Landry N (2015) Plant-derived H7 VLP vaccine elicits protective immune response against H7N9 influenza virus in mice and ferrets. Vaccine 33(46):6282–6289

Pillet S, Aubin É, Trépanier S, Bussière D, Dargis M, Poulin JF, Yassine-Diab B, Ward BJ, Landry N (2016) A plant-derived quadrivalent virus like particle influenza vaccine induces cross-reactive antibody and T cell response in healthy adults. Clin Immunol 168:72–87

Pillet S, Couillard J, Trépanier S, Poulin JF, Yassine-Diab B, Guy B, Ward BJ, Landry N (2019) Immunogenicity and safety of a quadrivalent plant-derived virus like particle influenza vaccine candidate—two randomized phase II clinical trials in 18 to 49 and ≥50 years old adults. PLoS One 14(6):e0216533

Sack M, Rademacher T, Spiegel H, Boes A, Hellwig S, Drossard J, Stoger E, Fischer R (2015) From gene to harvest: insights into upstream process development for the GMP production of a monoclonal antibody in transgenic tobacco plants. Plant Biotechnol J 13(8):1094–1105

Salazar-Gonzalez J, Banuelos-Hernandez B, Rosales-Mendoza S (2015) Current status of viral expression systems in plants and perspectives for oral vaccines development. Plant Mol Biol 87:203–217

Scotti N, Rybicki EP (2013) Virus-like particles produced in plants as potential vaccines. Expert Rev Vaccines 12(2):211–224

Shoji Y, Farrance CE, Bautista J, Bi H, Musiychuk K, Horsey A, Park H, Jaje J, Green BJ, Shamloul M, Sharma S, Chichester JA, Mett V, Yusibov V (2012) A plant-based system for rapid production of influenza vaccine antigens. Influenza Other Respir Viruses 6(3):204–210

Steinmetz NF (2010) Viral nanoparticles as platforms for next-generation therapeutics and imaging devices. Nanomed Nanotechnol Biol Med 6(5):634–641

Steinmetz NF (2013) Viral nanoparticles in drug delivery and imaging. Mol Pharm 10:1–2

Thomas DR, Penney CA, Majumder A, Walmsley AM (2011) Evolution of plant-made pharmaceuticals. Int J Mol Sci 12:3220–3236

Twyman R, Schillberg S, Fischer R (2005) Transgenic plants in the biopharmaceutical market. Expert Opin Emerg Drugs 10:185–218

Verma D, Daniell H (2007) Chloroplast vector systems for biotechnology applications. Plant Physiol 145:1129–1143

Wakasa Y, Zhao H, Hirose S, Yamauchi D, Yamada Y, Yang L, Ohinata K, Yoshikawa M, Takaiwa F (2011) Antihypertensive activity of transgenic rice seed containing an 18-repeat novokinin peptide localized in the nucleolus of endosperm cells. Plant Biotechnol J 9(7):729–735

Won SY, Hunt K, Guak H, Hasaj B, Charland N, Landry N, Ward BJ, Krawczyk CM (2018) Characterization of the innate stimulatory capacity of plant-derived virus-like particles bearing influenza hemagglutinin. Vaccine 36(52):8028–8038

Yazdani R, Shams-Bakhsh M, Hassani-Mehraban A, Arab SS, Thelen N, Thiry M, Crommen J, Fillet M, Jacobs N, Brans A, Servais AC (2019) Production and characterization of virus-like particles of grapevine fanleaf virus presenting L2 epitope of human papillomavirus minor capsid protein. BMC Biotechnol 19(1):81

Yu Y, Langridge WHR (2001) A plant-based multicomponent vaccine protects mice fromenteric diseases. Nat Biotechnol 19(6):548–552

Yuki Y, Mejima M, Kurokawa S, Hiroiwa T, Takahashi Y, Tokuhara D, Nochi T, Katakai Y, Kuroda M, Takeyama N, Kashima K, Abe M, Chen Y, Nakanishi U, Masumura T, Takeuchi Y, Kozuka-Hata H, Shibata H, Oyama M, Tanaka K, Kiyono H (2013) Induction of toxin-specific neutralizing immunity by molecularly uniform rice-based oral cholera toxin B subunit vaccine without plant-associated sugar modification. Plant Biotechnol J 11(7):799–808

Kathleen Hefferon received her Ph.D. from the Department of Medical Biophysics, University of Toronto (Canada) and currently teaches microbiology at Cornell University (USA). Kathleen has published multiple research papers, chapters and reviews, and has written three books. Kathleen is the Fulbright Canada Research Chair of Global Food Security and has been a visiting professor at the University of Toronto over the past year. Her research interests include the use of biotechnology to promote global health.

Chapter 16
Low Gluten and Coeliac-Safe Wheat Through Gene Editing

Luud J. W. J. Gilissen and Marinus J. M. Smulders

Abstract Cereal consumption by humans is older than agriculture. During preceding eras, humans gradually acquired the necessary knowledge and tools for processing and cultivation of cereals, making themselves unconsciously prepared to become the first farmers when climate improved at the end of the last Ice Age, some 12 thousand years ago, in the Near East (Fertile Crescent) region. Early wheat cultivation in fields surrounded by wild relatives facilitated the occurrence of inter-specific hybridizations; one of the resulting hybrids was bread wheat, with superior nutritional and food-technological qualities, largely due to the favourable composition of the gluten proteins in the grain. Bread wheat developed from that time into the commodity crop that it is nowadays.

Wheat-based foods and food with wheat ingredients can provoke coeliac disease in genetically predisposed susceptible individuals representing 1–2% of the population. Some specific digestion-stable fragments of wheat gluten are recognized by immune cells that become inflammatory in the small intestine, ultimately resulting in a variety of severe symptoms. The large number of gluten genes in wheat's complex genome hinder conventional breeding to produce a coeliac-safe wheat while maintaining good baking properties, although low gluten levels can be reached. CRISPR/Cas9 enables now to produce wheat line that potentially have a strongly reduced coeliac-immunogenicity, as will be explained from two model studies. An efficient screening pipeline to detect and select promising coeliac-safe(r) wheat lines will be described. Legal and societal aspects regarding the (non) GMO status of gene-edited plants, gluten-free food production chain management, and food labelling will be discussed.

Keywords Neolithic farming · Hybrid wheat plants · Polyploid · Gluten · Gliadin · Epitope · Advanced wheat breeding · Mutagenesis · Screening pipeline · Gene editing · GM · Gluten-free · Legislation · Genomic techniques · CCP (critical control points)

L. J. W. J. Gilissen · M. J. M. Smulders (✉)
Plant Breeding, Wageningen University & Research, Wageningen, The Netherlands
e-mail: rene.smulders@wur.nl

L. J. W. J. Gilissen
e-mail: luud.gilissen@wur.nl

A. Ricroch et al. (eds.), *Plant Biotechnology*,
https://doi.org/10.1007/978-3-030-68345-0_16

16.1 Wheat

16.1.1 History

The plant family *Triticeae* and the genus *Homo* share a long history that started several million years ago in East Africa when early humanoids gradually changed their lifestyle and moved from the forest into the savannah. There, grasses grew with ungulates feeding upon. The humanoids preferred both as food, as is reflected in fossil tooth enamel. It was some two million years ago that several populations of *Homo* species started to leave Africa in sequential waves. First, *Homo erectus*, who spread eastward over Asia. Then, *Homo neanderthalensis* went westward into the European peninsula and lived there between 0.5 million and 30 thousand years ago. More recently, about 100,000 years ago, *Homo sapiens* followed its predecessors. All travelled via the Near East, where *H. neanderthalensis* and *H. sapiens* not only shared the habitat for many millennia, but also generated progeny together. However, the reason that this Near East area is also known as the Fertile Crescent relates specifically to the wide variety of plant (several cereals and pulses) and animal (several ungulates) species that were abundantly present there and formed a complete food package. Evidently, *H. neanderthalensis* consumed various plants as food, including *Triticeae* seeds, already 50 thousand years ago. Remarkably, starch grain identification on their skeletal dental calculus revealed that their cereal food was cooked before consumption. Cereal seed processing is also known from *H. sapiens* living in South-Italy 32 thousand years ago: collected oat grains appeared to have been heated before milling, as was shown by an analysis of remaining starch grains from the surface of an excavated grinding stone. At the same time, the Cro-Magnons, known as hunters from their famous grotto paintings, made first steps towards horticulture by using tools to prepare the soil for plant cultivation. They brought plants to their living place: domestication. Further, a large camp site from almost 20 millennia ago has been excavated in the Fertile Crescent area; people lived there for twelve hundred years on a two-hectare area. Together, these facts demonstrate that during a long history *Homo* acquired necessary knowledge and tools, cultivation and processing technologies, and cultural practices to generate a new view on food and living. It is not surprising, therefore, that when twelve thousand years ago the global climate improved, the conditions were optimum to make the first steps towards modern agriculture just in the Fertile Crescent area: the first farmers came at stage. Cereals (wheat and barley) and pulses became the founder crops; sheep, goats and aurochs were the first domesticated animals. These plant and animal species formed the Neolithic food package of the first farmers. Farming became successful and caused overpopulation, resulting in migration waves during the following four thousand years from the Fertile Crescent area over the Mediterranean Basin by seafaring colonists as well as by north-west movements of farmers over land into the European continent, together with their domesticated cattle and their crops, including several wheat species (Gilissen and Smulders 2020, and references therein).

16.1.2 Diversity of Wheat Species

The genus *Triticum* includes several cultivated wheat species. Some of these species have originated from interspecific hybridization events in the wild, or, during early agriculture, in fields surrounded by related wild wheat species. As the genomes in such hybrids do not pair correctly during meiosis, the hybrids are sterile, but fertility can often be restored by spontaneous chromosome doubling. This, then, gives rise to new, allopolyploid species, i.e. with multiple sets of chromosomes originating from the original wheat species.

The individual genomes of the wheat hybrids are indicated with letters A, B, C, D etc. Some of the wild wheat species are traditionally classified under the genus name *Aegilops*. They can hybridize with cultivated wheat (the genus *Triticum*). In modern wheat breeding, the hybridization potential of different wheat species is still applied to produce artificial hybrids in which desired traits such as disease resistances are transferred from one species to the other. Artificial hybrids can be produced from species of different genera: *Tritordeum* (wheat × barley) and *Triticale* (wheat × rye) are well-known examples of recently created novel hybrid species that are currently cultivated.

Seeds of wild einkorn (*Triticum monococcum*, a diploid wheat species with the AA genome) and wild emmer (*T. turgidum*, a tetraploid species with the AABB genome) were already collected by humans for food before the onset of agriculture. These became the first cereal crops because of their high domestication potential. Domestication is de facto a mutualistically beneficial relationship. Regarding wheat, it includes a gradual process of natural and steady selection starting from heterogeneous (wild) plant populations into crop types with improved agricultural and food-quality traits. Regarding wheat, such traits that were initially selected for by early farmers, consciously and unconsciously, included larger grains, a non-brittle rachis (leaving the intact ear on the plant), and naked grains (also called 'free-threshing'), which enables the grains to separate easily from the chaff, free from the hulling glumes. This characteristic is the result of a single point mutation in the DNA changing the recessive *q* gene into dominant *Q*. This natural mutation occurred only once in the evolution of a tetraploid wheat species, and was transferred into hexaploid wheat species through hybridization and further introgression (that is the movement of a gene from one species into the gene pool of another species by repeated backcrossing of an interspecific hybrid with one of the parent species) (Matsuoka 2011).

The common bread wheat (*T. aestivum*, a hexaploid wheat species with the AABBDD genome) originated as a new hybrid species probably during early times of agricultural practice near the Caspian Sea. In this region, many wild *Aegilops* species are native. There, a cultivated free-threshing tetraploid wheat (AABB) in an agricultural field will have been hybridized with a local wild diploid *Aegilops tauschii* with its DD genome, which led to the hexaploid species *T. aestivum*. For centuries bread wheat was cultivated as part of a mixture with the tetraploid emmer and durum wheat (the latter had evolved from a free-threshing tetraploid wheat during ongoing domestication). It was only in Roman times that the milling technology was sufficiently advanced to produce fine flour from bread wheat enabling to bake a high-volume bread with a light crumb, superior to the flat pancake-like and compact breads from other wheat species and related grains such as rye and barley. Roman bread bakers commanded great respect from emperors and other wealthy people.

Presently, next to the diploid einkorn, eight tetraploid *T. turgidum* subspecies and five hexaploid *T. aestivum* subspecies are being cultivated, but almost 95% of the world's volume of 700–750 million metric tons is bread wheat (*T. aestivum* ssp *aestivum*); durum wheat (*T. turgidum* ssp. *durum*), used for pasta, makes up five per cent. In comparison, the other species (including the increasingly appreciated spelt wheat, *T. aestivum* ssp. *spelta*) are marginal crops.

Economically, world grain prices follow the current price of wheat. Why has especially bread wheat (and durum wheat to a lesser extent) become so popular? That is mainly because it contains a highly versatile protein component: Gluten.

16.2 Wheat Gluten

16.2.1 Complexity

A wheat grain contains three major parts, the bran (the outer layers of the grain), the germ (embryo) and the endosperm tissue containing starch and proteins (mainly gluten) as storage material that will support the seedling during the first growing phase after germination through rapid breakdown into glucose (for energy) and amino acids (to build new proteins from). Similar gluten proteins are also produced in barley and rye grains.

Gluten proteins belong to the protein superfamily of the prolamins. Wheat gluten are very diverse and comprise three protein classes (Table 16.1; Goryunova et al. 2012; Huo et al. 2018a, b; Shewry 2019; Altenbach et al. 2020):

1. HMW-glutenins of 65–90 kDa, which make up 6–10% of the total gluten fraction; the coding genes are located on the long arms of the homoeologous chromosomes 1 (1A, 1B, 1D);

Table 16.1 Gluten genes found in the genome of variety Chinese Spring

Gluten family	A genome	B genome	D genome	Total	Expressed
HMW-glutenins	2	2	2	6	4
LMW-glutenins	3	5	7	17	10
Alpha-gliadins	26	11	10	47	28
Gamma-gliadins	4	6	4	14	11
Delta-gliadins	2	1	2	5	2
Omega-gliadins	4	7	6	17	11

Based on Huo et al. 2018a, b

2. sulphur-rich alpha- and gamma-gliadins and B- and C-type LMW glutenins of 30–45 kDa, taking 70–80% of the total gluten fraction; also a minor group, the delta-gliadins, can be distinguished; the coding genes are located in clusters on the short arms of the chromosomes 1, except for the alpha-gliadin genes, which are located in tandem on the short arms of the chromosomes 6;
3. sulphur-poor omega-gliadins and D-type LMW-glutenin subunits of 30–75 kDa, representing 10–20% of the total gluten fraction, with their genes located on the short arms of the chromosomes 1.

Evolutionary, the gamma-gliadins are considered the oldest of the gliadin family. DNA sequence analysis revealed early duplications of the original gamma-gliadin gene, followed by further mutations, new duplications, pseudogenisations and deletions.

A single bread wheat variety genome (e.g. from the variety Chinese Spring) may contain as many as ~100 different gluten genes of which about sixty are expressed into proteins, as was shown by mass spectrometry analysis (Huo et al. 2018a, b; Shewry 2019; Altenbach et al. 2020; Table 16.1). Comparing different wheat varieties, variation exists in the number of expressed gluten genes, in the sequences of the encoded proteins, and in the amount of produced protein per gene. Additional variation in the gluten composition (quantitatively and qualitatively) is induced by environmental factors in the field during the growing season, such as the temperature during certain stages of crop development, and nutrient availability from the soil, in particular nitrogen and sulphur (Shewry 2019).

16.2.2 *Versatility and Functionality of Gluten in Food Products*

Gluten proteins are water-insoluble. After milling of the kernel and adding a certain, limited volume of water to the flour, its gluten together with the starch can form a dough. The elastic network of the dough is made up by the glutenins, while the gliadins provide viscosity to it. The quality of the glutenins thus determines the rheological characteristics of the dough and baking quality. The role of gliadins is supportive. Other gluten-related wheat prolamins may additionally contribute to some extent to baking quality: of these, the alpha-amylase/trypsin inhibitors (ATIs), the farinins, the purinins and the grain softness protein (GSP) have some relevance in pasta-making, dough mixing, viscosity and milling, respectively (Shewry 2019).

The overall baking quality of an individual wheat variety is determined by its total gluten quantity and its gluten quality, especially that of the HMW subunits fraction, and by the gliadin/glutenin ratio. These factors determine for which type of application a wheat variety is most suitable: for bread, cookies, or pasta. To guarantee and standardize the baking quality of flour, even though multiple varieties are cultivated and the growing conditions also vary strongly from year to year, several batches of grains from different locations, consisting of different varieties, are mixed in quantities such that the desired quality of the flour for a specific application is obtained. A lower gluten content in one variety will thus be compensated for by mixing with a variety with a higher gluten content from another region in the country, the continent or the world. Wheat is a commodity with a high travel standard.

Washing the dough with an excess of water will remove the starch, leaving a rubbery mass of gluten proteins, called vital wheat gluten (VWG). Extra VWG added to a dough will increase the volume of the bread. VWG can be added to flour to improve the rheological and technical properties to the specific levels of dough quality required by the baking industry for the production of a large range of different products: steam buns, toasted breads, crusty breads, sweetbreads, leavened and laminated sweet goods, laminated puff pastries, rolls and buns, crackers, cookies, sponge cakes, wafers, snacks, etc. The consumption of these products has increased during the last decades and has become an integrated part of the 'western lifestyle' (Igrejas and Branlard 2020). Accordingly, the intake of VWG has tripled since 1977 from 0.37 to 1.22 g per capita per day; this is a modest addition to the ~15 g of total daily gluten consumption per capita from bread (Kasarda 2013).

Wheat flour and wheat-derived ingredients (wheat starch, wheat glucose syrup, and VWG) are applied in many more food products than in bakery alone and can be found in highly processed foods such as sweets, frozen meals, packed soups and chips, and more unexpectedly in vinegars, popcorns, gourmet products, vitamins, ice-creams, coffee, nuts, rice crackers, soy sauce, canned vegetables, cheeses, sea foods, etc. Wheat ingredients were detected in 29.5% of 10,235 labelled food items in the supermarket (Atchison et al. 2010). Declaration on the product label of the origin of e.g. the glucose syrup is not always adequate. The presence of such additives or ingredients is no issue to healthy individuals, but it may make life complex and health-impairing to susceptible persons suffering from wheat-related diseases.

16.3 Human Health

16.3.1 Positive Effects of Wheat Consumption

The consumption of healthy foods may increase life expectancy and well-being, and it may help to substantially reduce health care costs. Whole grain wheat fits in the category of healthy foods. Whole grain foods contain the three main parts of the grain: the bran (rich in fibres), the starchy endosperm (rich in carbohydrates and proteins) and the germ (rich in vitamins and micronutrients). Several large cohort studies have clearly shown that consumption (the more the better) of whole grain products (including whole grain wheat) significantly reduces the risk of several 'western lifestyle'-related chronic diseases, including obesity and diabetes, heart and vascular diseases, immune-related diseases, and several forms of cancer. Governmental agencies of many countries therefore advise to consume whole grain foods (Gilissen and Van den Broeck 2018, and references therein).

16.3.2 Negative Effects of Wheat Consumption

Next to the nutritional and health benefits for all consumers, wheat may also cause allergic and intolerance reactions in certain people (Gilissen et al. 2014). True wheat allergies, mainly occurring in children, are relatively rare, with a prevalence of 0.25%. Another condition, called wheat or gluten sensitivity (more specifically 'non-coeliac wheat/gluten sensitivity', NCWGS) is not well understood. It has a self-diagnosed prevalence of ~10% and a medically estimated prevalence of 1%. The symptoms are generally mild. The causal food compounds are not yet determined, but gluten or wheat alone is not likely. In contrast, coeliac disease (CD), a chronic inflammation of the small intestine induced by the consumption of gluten proteins from wheat, barley and rye, has been elucidated largely after decades of thorough scientific studies. Genetic factors from both the plant (knowledge on coeliac-immunogenic gluten fragments, the epitopes) and human (composition of specific receptor proteins on immune system-related T cells) have been determined. CD has a prevalence of 1–2% of the general population worldwide, which means that in the EU alone at least 4.5 million people suffer from this disease.

In the absence of a cure for CD, prevention by following a strict, life-long gluten-free diet is the only remedy currently available. In practice this is a challenge for CD patients because of the presence of wheat and gluten in many food products as explained above (Gilissen et al. 2014; Rustgi et al. 2020). Indeed, compliance with the gluten-free diet is difficult and presently far from 100% (Scherf et al. 2020). Extensive research has led to EU regulation EC828/2014 stating that gluten-free products should not contain gluten above the 20 ppm threshold. This has raised interest in the development of food processing and breeding strategies for coeliac-safe and healthy wheat products (Jouanin et al. 2018a). Extensive knowledge gained during the last twenty years on the aetiology and on the causative gluten epitopes appeared helpful in such strategies. On the other hand, the development and marketing of gluten-free products (including many bakery products) has generated a billion euro/dollar market, not only for coeliac patients: gluten-free consumption has become a trend for several other (true or supposed) reasons.

16.3.3 Coeliac Disease Epitopes

Relevant for coeliac disease are two features of gluten proteins: they are rich in glutamine (Q) and proline (P) amino acids; and they contain protein domains that are repetitions of short sequences rich in these amino acids. Especially the high abundance of these two amino acids make gluten proteins partly resistant to the digestive proteases in the mouth, the stomach and the small intestine. This means that relatively long peptides can survive in the small intestine. In genetically predisposed individuals, certain fragments (containing a core sequence of nine amino acids, the so-called epitopes) may be recognized and bound by the HLA-DQ receptors

HLA-DQ2 (notably HLA-DQ2.5) or HLA-DQ8, present on specific immune cells, the CD4$^+$ T-cells, that then become activated. This activation leads to cascade of cellular and molecular immune reactions ultimately resulting in an inflammation of the intestinal mucosa causing degeneration of the villi (flattening) of the small intestine surface with serious consequences for adequate uptake of nutrients, minerals and vitamins. This disturbed food uptake leads to a variety of symptoms, ranging from bowel disorders to skin, bone, nerve, and muscle problems. Because of the variety of symptoms, the majority of the patients have not yet been (properly) diagnosed.

In wheat, three alpha-gliadin, eight gamma-gliadin, two omega-gliadin and glutenin epitopes have been detected related to HLA-DQ2.5 recognition. Next to these, four minor epitopes (one from alpha-gliadin, two from gamma-gliadin and one from glutenin) with a different T-cell recognition pattern, related to HLA-DQ8, have been identified (Sollid et al. 2012, 2020). Some gluten proteins may contain multiple (up to six, in case of the alpha-gliadin 33-mer peptide) overlapping epitopes. Importantly, all CD-immunogenic epitopes contain one or more glutamic acid (E) residues, whose charge is necessary for increased affinity (recognition) by the T-cell receptor. Several of these E residues are not present in the original gluten fragment but are formed in the intestine by deamidation of glutamine (Q) into glutamic acid (E) through tissue transglutaminase-2 (TG2), an enzyme naturally present in the human intestine. For example, the DQ2.5-glia-alpha 1a epitope has the amino acid sequence PFPQPQLPY in the natural alpha-gliadin protein, but it becomes immunogenically active only after deamidation of Q at amino acid position 6 into PFPQPELPY. More details of important epitopes are shown in Box 1.

Box 1 Dominant Epitopes

Most patients respond to multiple gluten epitopes. Some epitopes, listed below, are very frequently found (dominant epitopes) in HLA-DQ2.5 and HLA-DQ8 positive patients. Note the general presence of proline (P) at the amino acid positions one and eight in the DQ2.5 epitopes. For deamidation of Q into E by TG2, a QxP residue sequence is the optimum target site (Tye-Din et al. 2010; Sollid et al. 2012, 2020; Salentijn et al. 2012); these Q residues are bold and underlined.

DQ2.5-glia-alpha 1a: PFPQP**Q**LPY.

DQ2.5-glia-alpha 1b: PYPQP**Q**LPY.

DQ2.5-glia-alpha 2: PQP**Q**LPYPQ (the most common epitope).

DQ2.5-glia-alpha 33-mer: LQLQPFPQP**Q**LPYPQP**Q**LPYPQP**Q**LPYPQPQPF (six overlapping epitopes; also shorter versions [19-mer; 26-mer] of this 33-mer are known).

DQ2.5-glia-gamma 1: P**Q**QSFPQ**QQ** (seven other gamma-epitopes are distinguished but these appear less responsive).

DQ2.5-glia-gamma 26-mer: FLQP**Q**QPFP**Q**QP**Q**QPYP**Q**QP**Q**QPFPQ (five overlapping epitopes).

DQ2.5-glia-omega 1: PFPQP**Q**QPF.

DQ2.5-glia-omega 2: PQP**Q**QPFPW.

DQ8-glia-alpha 1: **Q**GSFQPSQ**Q**.

16.4 Wheat Breeding

16.4.1 Aims

Wheat is a self-pollinating crop. Farmers can cultivate it by sowing seed material saved from the preceding year. Initially, cultivation practice applied mixtures of tetraploid and hexaploid genotypes. Through conscious or unconscious selection of spontaneous mutants, landraces gradually adapted to local environmental conditions, but there is a limit to the improvement of end-use quality that can be achieved in that way. Since the early twentieth century, genetics has been used in professional wheat breeding, including pure line selection and targeted breeding. Breeders are always interested in new genetic variation. This can be achieved through introgression (trait transfer through hybridization followed by back-crossing) from other wheat species, but this process also introduces many undesired traits that subsequently have to be selected against. Alternatively, genetic variation can be induced within a cultivar through the application of mutagenic chemicals or ionizing irradiation (mutation breeding).

In wheat breeding, the main focus was traditionally directed on yield and quality trait improvements. This has culminated in the 1960s in the Green Revolution through the introduction of dwarf genes. The dwarf genes (actually, dwarfing mutations) reduce energy investment in vegetative growth (stalks), and increase grain yield. These new varieties were highly recognized worldwide. Today's breeding aims still include yield (especially starch quantity) and gluten and starch quality (for improved milling and baking quality). Adaptations to the biotic and abiotic environment, such as disease resistance genes, are receiving increasingly more attention because of the spread of major diseases and the threats of climate change, e.g. increased drought.

16.4.2 Coeliac-Safe Wheat

Recently, a new goal has appeared for wheat breeding: removal of coeliac immuno-genicity. This means the selection and development of wheat lines with fewer gluten genes (especially gliadin genes) and/or with gluten genes without intact immunogenic coeliac epitopes. Maintenance of food-industrial quality (milling and baking/bread quality) and good field performance are prerequisites.

Both gliadins and glutenins contain immunogenic epitopes within their protein sequences that trigger CD, but the gliadins contain the highest numbers of the immunogenic epitopes including the most dominant ones. On a positive note, the gliadins are only of secondary importance for food-industrial quality compared with the glutenins. They may be partly omitted or replaced by other proteins to a certain extent while retaining food technological properties (Van den Broeck et al. 2011). However, only a few natural alpha-gliadin genes, notably those on chromosome 6B, are free from immunogenic epitopes. Also, although wheat varieties and species have been identified with reduced immunogenicity, the reduction is largely insufficient to be safe to coeliac patients. As the amount of gluten in wheat flour is ~7%, equalling 70,000 ppm (pers. comm. Johan de Meester, Cargill) and as these gluten proteins are encoded by large gene families located at different sites in the wheat genome, no classical breeding or food processing strategies have been developed yet that produce wheat-based food products approaching safety for coeliac patients at all (i.e., containing less than 20 ppm gluten). However, this may change with recently developed advanced breeding technologies (Jouanin et al. 2018a).

16.4.3 Removing or Silencing Gluten Genes

Using chemical treatments such as ethyl-methane sulfonate (EMS) or ionizing irra-diation, random mutations can be generated in the plant genome. Radiation-induced wheat deletion lines have been produced in the 1960s that lack complete sets of genes, e.g., the alpha-gliadin locus on chromosome 6D with many coeliac-immunogenic epitopes (Van den Broeck et al. 2009). Such lines could be used as a start of a targeted breeding program, but crossing with other deletion lines often results in lethality making deletion lines less useful for this purpose. The strategy of combining dele-tions has, however, successfully been used in the development of ultra-low gluten barley (Tanner et al. 2016), supported by the fact that barley is a diploid crop species and gluten in barley is not relevant for the quality of beer. EMS mutation breeding can generate large numbers of random mutations and could be applied to mutate gliadin genes, but it would be very resource-intensive to trace and combine mutations in multiple genes, from many plants, into one single, coeliac-safe and well-performing wheat plant (Jouanin et al. 2018a). A clear need for a more sophisticated approaches remains.

Two modern biotechnological approaches have recently been developed that may provide a tool towards producing wheat that is safe for CD patients: RNA interference (RNAi) and CRISPR/Cas9 gene editing. RNA interference is, as the term implies, a system to interfere with the production of, in this case, gluten proteins through their RNA transcripts, even though the DNA still contains the intact genes. A single RNAi construct can be designed for a conserved region that is common for many gluten genes. With expression of such a construct in transformed wheat lines, up to 92% reduction of the gliadins, and a 10–100 fold reduction of epitopes as detected in T-cell tests were achieved (Gil-Humanes et al. 2010). Similarly, the expression of twenty α-gliadin genes was decreased although the expression of other storage proteins increased (Becker et al. 2012). In another approach, induced expression of the DEMETER gene, preventing changes in DNA methylation, repressed gliadin as well as glutenin gene expression in the endosperm (Wen et al. 2012). Some of the wheat lines with reduced immunogenicity but with the baking quality largely intact (Gil-Humanes et al. 2014) are sufficiently low in gliadins that food challenge trials with consumers are being planned.

RNAi requires stable genetic modification (GM) of the construct into wheat to silence the gliadins. The resulting transgenic lines therefore face expensive and time-consuming food-safety assessments for regulatory approval and consumer's acceptance, notably in the EU because of current GMO legislation.

16.5 Gene Editing

16.5.1 CRISPR/Cas9 Gene Editing

CRISPR/Cas9 ('clustered regularly interspaced short palindromic repeats and associated protein 9') has the potential to simultaneously and precisely modify multiple gliadin-encoded epitopes and/or delete (some of the) gliadin genes, while potentially maintaining food-technological quality. In CRISPR/Cas9 gene editing, a single guide RNA plus a Cas9 endonuclease is brought into embryogenic cells of a wheat cultivar. The guide RNA directs the endonuclease to the target DNA sites in gliadin genes, where it creates a double-strand break. This triggers the native DNA repair system of the cell. All living cells have extensive DNA repair mechanisms and these are continuously repairing the many spontaneous and induced mutations in the DNA. This repair system is highly accurate, but can sometimes make mistakes, and double-strand breaks are particularly difficult to repair. Mistakes may then result in small deletions of one or a few nucleotides at the site of the break in the gliadin gene. However, in wheat, where alpha-gliadin genes are tandemly repeated, simultaneous double-strand breaks may occur in consecutive genes, and this could result in deletions of DNA fragments carrying one or more gliadin genes (Fig. 16.1). A great advantage of the CRISPR/Cas9 system is that it can be targeted simultaneously at multiple gene sequences. Important to note is that such mutations are identical

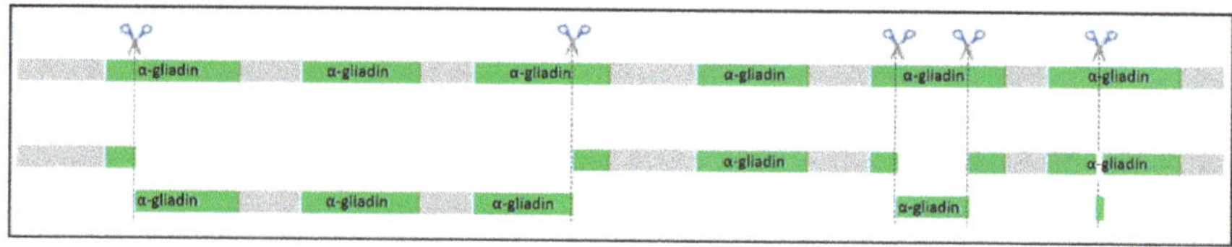

Fig. 16.1 Representation of a part of a single alpha-gliadin *Gli-2* locus on wheat chromosome 6 with the various mutagenic actions that can be induced by CRISPR/Cas9 (taken from Jouanin et al. 2019). Note that after generation of a double-strand break, the DNA repair mechanism may occasionally result in different mutation types, ranging from a point mutation, to an insertion or a deletion of nucleotides, to a deletion of a gene fragment, and a deletion of a (part of a) gene locus. A mosaic of mutations may occur in a single regenerated gene-edited plant. Mutations in genes on different chromosomes will segregate in later generations

to naturally random-occurring mutations upon double-strand breaks, but in case of CRISPR/Cas9 these mutations are at desired locations.

The next step in this process involves the regeneration of plants from the gene-edited cells. These plants will be self-pollinated to enable the production and further testing of seeds with the induced (desired) targeted mutations in the gliadins. The breeder is especially interested in plants that have desired mutations but do no longer contain the construct in its genome. According to Mendel's laws, and in case a single CRISPR/Cas9 construct is present, self-pollination will result in progeny of which one quarter will be free of the CRISPR/Cas9 construct.

Recently two proof-of-concept studies have been carried out in wheat. Sánchez-Léon et al. (2018) targeted two conserved sites next to the epitope-containing region in 45 alpha-gliadin genes of a single wheat line. In this study, 47 plants of the progeny were confirmed as being gene-edited. In one of the resulting lines, up to 35 of the 45 genes appeared mutated, with small or larger deletions around the target sites. This line showed a 85% reduction of total gluten protein as measured with the R5 gluten quantification assay (R-Biopharm, Darmstadt), which is approved for the detection and quantification of small amounts of gluten in gluten-free products. In the other study, Jouanin et al. (2019) confirmed 117 gene-edited plants after simultaneously targeted multiple sites in alpha-gliadin and gamma-gliadin genes; indeed, mutations in both gene families in the same plant were generated.

Although targeted to gliadin genes, not all potential sites in these genes will be mutated in a single line as most repair actions will not result in a mutation. As a consequence, the progeny will consist of a population of plants that each may contain a mosaic of edited and unaffected genes, and perhaps loss of some genes. The edits on different chromosomes will segregate in next self-pollinated generations. To limit the number of individual plants to be screened, a rigorous reduction through quality-directed selection steps is required (Fig. 16.2). This should result in a workable selection program, which aims at maintaining only the few most promising plants (genotypes) for multiplication, cultivation and eventual application in coeliac-safe(r) food (Jouanin et al. 2020).

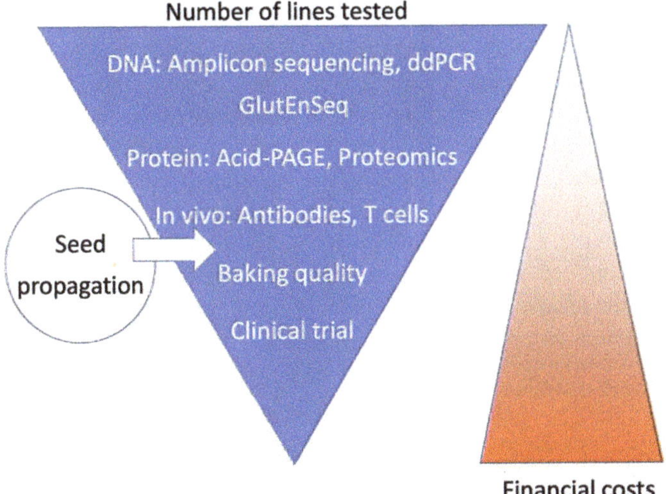

Fig. 16.2 Schematic overview of the sequential steps in the screening pipeline (from Jouanin et al. 2020). The first steps, screening at the DNA and protein level, may have to handle large numbers of plants and should therefore preferentially include fast, high throughput, and cheap methods. Subsequently, more detailed but time-consuming and expensive methods can be employed to precisely characterize a limited number of promising lines at the immunological level. The last steps include screening of food-technological quality and coeliac safety. These require sufficient amounts of grains, thus a cultivation step on the field

The use of CRISPR/Cas9-mediated gene editing in breeding thus includes two major aspects: first, a precisely defined target DNA sequence; and second, the right choice of screening technologies at the DNA and protein level, supplemented later on with screening at the phenotypic and industrial quality level, to select the most promising plant lines.

16.6   Perspectives of Coeliac-Safe(r) Wheat

16.6.1   The Gluten-Free Market

The development of safe-gluten wheat for coeliac disease patients is still a long way to go, as this really must have the equivalent of 20 ppm gluten or less to be legally approved and labelled as 'gluten-free'. However, 'gluten-free' is worldwide becoming a billion dollar/euro market, including, next to coeliac patients, also people that want to reduce their gluten intake for supposed health reasons, or people with self-diagnosed NCW/GS. This market goes beyond the coeliac disease patients who need a strict gluten-free diet.

Products for coeliac disease patients must be labelled as 'gluten-free' implying the maximum threshold of 20 ppm gluten. Another legally accepted category is 'low in gluten' with a threshold of 100 ppm, for products aiming at less-sensitive people. In addition, vague marketing terms are also sometimes used, for instance the company Arcadia Biosciences now sells a "Reduced Gluten GoodWheat" variety with 65% less gluten for non-coeliac or non-wheat allergic consumers who only want to reduce the amount of gluten in their diet. Depending on the level of reduced coeliac immunogenicity that can be reached by gene editing, wheat lines and the products thereof may therefore become categorized as 'gluten-free' (when these contain an equivalent of the legally accepted <20 ppm gluten), 'low in gluten' (<100 ppm), or 'reduced in gluten' (an undefined category).

16.6.2 Gene Editing Products as Non-GM

Although the CRISPR/Cas9 edited lines produced in the pilot studies mentioned in this paper (Sanchez-Leon et al. 2018; Jouanin et al. 2020) are not yet safe for CD patients, they demonstrate the power of gene editing for modifying tens of gliadin genes at once in bread wheat. Technically, gene editing requires foreign DNA to be introduced in the plant genome, but only as a transient step in the targeted gene editing. After self-pollination, part of the offspring will not inherit the construct while they do inherit edited gliadin genes. As such, applying CRISPR/Cas9 as a biological mutagenic agent is basically identical to chemical (EMS) or physical (ionizing radiation) mutagenesis. Therefore, in many countries worldwide, gene-edited plants, as a product, are considered as not GM, because its production follows the conventional breeding rules, including those of mutation breeding. However, the EU regulation is different as it follows the process instead of considering the end product. On July 25th 2018 the European Court of Justice ruled that (according to Directive 2001/18/EC) plants produced through gene editing as a mutation technique, are not exempted from GM regulation as long as it has not been "conventionally used" in "a number of applications" and have "a long safety record". This decision will have serious consequences for coeliac patients within the EU regarding the availability of gene-edited safe foods (Jouanin et al. 2018b). However, for the (coeliac) patients and their societies, the methods used to produce such safe products are irrelevant. Their concern relates to proper testing and labelling.

16.6.3 Labelling

Regarding labelling, the R5 assay is especially developed to quantify the gluten content, especially in gluten-free products, for food labelling purposes to inform coeliac patients at their purchase. The threshold of 20 ppm gluten is determined by the Codex Alimentarius and legally approved internationally worldwide. However, the R5 assay will be useless for coeliac-safe food products derived from gene-edited wheat that still contain gluten, but these lack the coeliac-immunogenic epitopes. Such new products will require a re-evaluation of the current legislation and a matching labelling strategy, in accordance with the total quantity and quality (severity) of coeliac-immunogenic gluten epitope sequences in the product.

16.6.4 Production Chain

Gene-edited wheat lines should be produced in a production chain that is entirely and guaranteed separated from regular-wheat production chains. This implies separate facilities on the farms regarding the field for cultivation, the sowing seed batches, the harvest machineries, the transport and storage facilities, but also at the processing factories, regarding the production lines, packaging and labelling, and transport to the retail. All these critical points should be under strict control.

16.7 Conclusions

Wheat is a healthy and nutritious crop. It contains two (in durum and pasta wheat) or three (in bread wheat) different genomes, making the crop genetically highly complex. Bread wheat contains about one hundred gluten genes, many of which are expressed into proteins representing several classes of gluten. Proteins of several classes have immunogenic properties to which sensitive individuals may develop diseases, of which coeliac disease is the most common and best studied. Immuno-genic responses have been linked to specific gluten fragments, mainly occurring in the gliadin proteins. The genome complexity of wheat prevents the application of conventional breeding to generate completely coeliac-safe wheat. Pilot studies clearly demonstrate the feasibility of CRISPR/Cas9 gene editing technology for the produc-tion of coeliac-safe(r) wheat plants and food products. If such products are becoming realistic for the gluten-free consumer's market, critical control points (CCPs) in the entire production chain should be clearly defined and strictly monitored, as well as the legal aspects regarding 'gluten-free' labelling be re-evaluated. On behalf of the concerning consumers, in Europe the current GMO status resulting from the use of gene editing in breeding new coeliac-safe wheat varieties should be reconsidered in agreement with the non-GMO status of such products elsewhere worldwide.

Acknowledgements The writing of this article was partially made possible by the Well on Wheat (WoW) project (https://www.um-eatwell.nl/wow/), an international research project under the umbrella of the Health Grain Forum and ICC-Vienna, addressing the effect of wheat consumption and processing on food composition changes and the related effects on gastrointestinal function and symptoms in humans. The WoW project is partially funded by the Dutch Topsector AgriFood project TKI 1601P01. The Centre of Genetic Resources, The Netherlands (CGN) is thanked for providing office space and facilities to LG.

References

Altenbach SB, Chang H, Simon-Buss A et al (2020) Exploiting the reference genome sequence of hexaploid wheat: a proteomic study of flour proteins from the cultivar Chinese Spring. Funct Integr Genomics 20:1–16. https://doi.org/10.1007/s10142-019-00694-z

Atchison J, Head L, Gates A (2010) Wheat as food, wheat as industrial substance; comparative geographies of transformation and mobility. Geoforum 41:236–246. https://doi.org/10.1016/j.geoforum.2009.09.006

Becker D, Wieser H, Koehler P, Folck A, Mühling KH, Zörb C (2012) Protein composition and techno-functional properties of transgenic wheat with reduced alpha-gliadin content obtained by RNA interference. J Appl Bot Food Qual 85:23–33

Directive 2001/18/EC of the European Parliament and of the Council of 12 March 2001 on the deliberate release into the environment of genetically modified organisms and repealing Council Directive 90/220/EEC—Commission Declaration. Off J L 106:0001–0039 (17/04/2001)

Gil-Humanes J, Pistón F, Tollefsen S, Sollid LM, Barro F (2010) Effective shutdown in the expression of celiac disease-related wheat gliadin T-cell epitopes by RNA interference. Proc Natl Acad Sci USA 107:17023–17028

Gil-Humanes J, Piston F, Altamirano-Fortoul R, Real A et al (2014) Reduced-gliadin wheat bread: An alternative to the gluten-free diet for consumers suffering gluten-related pathologies. PLoS ONE 9:e90898. https://doi.org/10.1371/journal.pone.0090898

Gilissen LJWJ, van der Meer IM, Smulders MJM (2014) Reducing the incidence of allergy and intolerance to cereals. J Cereal Sci 59:337–353. https://doi.org/10.1016/j.jcs.2014.01.005

Gilissen LJWJ, van den Broeck HC (2018) Breeding for healthier wheat. Cereal Foods World 63:132–136. https://doi.org/10.1094/CFW-63-4-0132

Gilissen LJWJ, Smulders MJM (2020) Biotechnological strategies for the treatment of gluten intolerance. In: Rossi M (ed) Gluten quantity and quality in wheat and in wheat-derived products. Elsevier, Amsterdam (in press)

Goryunova SV, Salentijn EMJ, Chikida NN, Kochieva EZ, van der Meer IM, Gilissen LJWJ, Smulders MJM (2012) Expansion of the gamma-gliadin gene family in Aegilops and Triticum. BMC Evol Biol 12:215. https://doi.org/10.1186/1471-2148-12-215

Huo N, Zhu T, Altenbach S, Dong L, Wang Y, Mohr T et al (2018a) Dynamic evolution of α-gliadin prolamin gene family in homeologous genomes of hexaploid wheat. Sci Rep 8:5181. https://doi.org/10.1038/s41598-018-23570-5

Huo N, Zhang S, Zhu T, Dong L, Wang Y, Mohr T et al (2018b) Gene duplication and evolution dynamics in the homeologous regions harboring multiple prolamin and resistance gene families in hexaploid wheat. Front Plant Sci 9:673. https://doi.org/10.3389/fpls.2018.00673

Igrejas G, Branlard G (2020) The importance of wheat. In: Igrejas et al (eds) Wheat quality for improving processing and human health (Chap 1). Springer Nature, Switzerland AG, pp 1–8. https://doi.org/10.1007/978-3-030-34163-3

Jouanin A, Boyd LA, Visser RGF, Smulders MJM (2018a) Development of wheat with hypoimmunogenic gluten obstructed by the gene editing policy in Europe. Front Plant Sci 9:1523. https://doi.org/10.3389/fpls.2018.01523

Jouanin A, Gilissen LJWJ, Boyd LA, Cockram J, Leigh FJ, Wallington EJ, van den Broeck HC, van der Meer IM, Schaart JG, Visser RGF, Smulders MJM (2018b) Food processing and breeding strategies for coeliac-safe and healthy wheat products. Food Res Int 110:11–21. https://doi.org/10.1016/j.foodres.2017.04.025

Jouanin A, Schaart JG, Boyd LA, Cockram J, Leigh FJ, Bates R, Wallington EJ, Visser RGF, Smulders MJM (2019) Outlook for coeliac disease patients: towards bread wheat with hypoimmunogenic gluten by gene editing of α- and γ-gliadin gene families. BMC Plant Biol 19:333. https://doi.org/10.1186/s12870-019-1889-5

Jouanin A, Gilissen LJWJ, Schaart JG, Leigh FJ, Cockram J, Wallington EJ, Boyd LA, Van Den Broeck HC, Van der Meer IM, America AHP, Visser RGF, Smulders MJM (2020) CRISPR/Cas9 gene editing of gluten in wheat to reduce gluten content and exposure—reviewing methods to screen for coeliac safety. Front Nutr 7:51. https://doi.org/10.3389/fnut.2020.00051

Kasarda DD (2013) Can an increase in celiac disease be attributed to an increase in the gluten content of wheat as a consequence of wheat breeding? J Agric Food Chem 61:1155–1159. https://doi.org/10.1021/jf305122s

Matsuoka Y (2011) Evolution of polyploid Triticum wheats under cultivation: the role of domestication, natural hybridization and alloploid speciation in their diversification. Plant Cell Physiol 52:750–764. https://doi.org/10.1093/pcp/pcr/pcr018

Rustgi S, Shewry P, Brouns F (2020) Health hazards associated with wheat and gluten consumption in susceptible individuals and status of research on dietary therapies. In: Igrejas G, Ikeda T, Guzmán C (eds) Wheat quality for improving processing and human health. Springer, Cham, Switzerland, pp 471–515. https://doi.org/10.1007/978-3-030-34163-3_20

Salentijn EM, Mitea DC, Goryunova SV, van der Meer IM, Padioleau I, Gilissen LJWJ, et al. (2012) Celiac disease T-cell epitopes from gamma-gliadins: immunoreactivity depends on the genome of origin, transcript frequency, and flanking protein variation. BMC Genom 13:277. https://doi.org/10.1186/1471-2164-13-277

Sanchez-Leon S, Gil-Humanes J, Ozuna CV, Gimenez MJ, Sousa C, Voytas DF, Barro F (2018) Low-gluten, nontransgenic wheat engineered with CRISPR/Cas9. Plant Biotechnol J 16:902–910. https://doi.org/10.1111/pbi.12837

Scherf KA, Catassi C, Chirdo FG, Ciclitira PJ, Feighery C, Gianfrani C, Koning F, Lundin KEA, Schuppan D, Smulders MJM, Tranquet O, Troncone R, Koehler P (2020) Recent progress and recommendations on celiac disease from the working group on prolamin analysis and toxicity. Front Nutr 7:29. https://doi.org/10.3389/fnut.2020.00029

Shewry PR (2019) What is gluten—why is it special. Front Nutr 6:101. https://doi.org/10.3389/fnut.2019.00101

Sollid LM, Qiao SW, Anderson RP, Gianfrani C, Koning F (2012) Nomenclature and listing of celiac disease relevant gluten T-cell epitopes restricted by HLA-DQ molecules. Immunogenetics 64:455–460. https://doi.org/10.1007/s00251-012-0599-z

Sollid LM, Tye-Din JA, Qiao SW, Anderson RP, Gianfrani C, Koning F (2020) Update 2020: nomenclature and listing of celiac disease–relevant gluten 1324 epitopes recognized by CD4+ T cells. Immunogenetics 72:85–88. https://doi.org/10.1007/s00251-019-01141-w

Tanner GJ, Blundell MJ, Colgrave ML, Howitt CA (2016) Creation of the first ultra-low gluten barley (Hordeum vulgare L.) for coeliac and gluten-intolerant populations. Plant Biotechnol J 14:1139–1150. https://doi.org/10.1111/pbi.12482

Tye-Din JA, Stewart JA, Dromey JA, Beissbarth T, van Heel DA, Tatham A, et al. (2010) Comprehensive, quantitative mapping of T cell epitopes in gluten in celiac disease. Science Transl Med 2(41):41ra51–41ra51. https://doi.org/10.1126/scitranslmed.3001012

Van den Broeck HC, van Herpen TWJM, Schuit C, Salentijn EMJ, Dekking L, Bosch D, Hamer RJ, Smulders MJM, Gilissen LJWJ, van der Meer IM (2009) Removing celiac disease-related gluten proteins from bread wheat while retaining technological properties: a study with Chinese Spring deletion lines. BMC Plant Biol 9:41. https://doi.org/10.1186/1471-2229-9-41

Van den Broeck HC, Gilissen LJWJ, Smulders MJM, van der Meer IM, Hamer RJ (2011) Dough quality of bread wheat lacking alpha-gliadins with celiac disease epitopes and addition of celiac-safe avenins to improve dough quality. J Cereal Sci 53:206–216. https://doi.org/10.1016/j.jcs.2010.12.004

Wen S, Wen N, Pang J, Langen G, Brew-Appiah RA, Mejias JH, Osorio C, Yang M, Gemini R, Moehs CP, Zemetra RS (2012) Structural genes of wheat and barley 5-methylcytosine DNA glycosylases and their potential applications for human health. Proc Natl Acad Sci USA 109:20543–20548

Luud J. W. J. Gilissen (Ph.D. 1978 Radboud University Nijmegen, The Netherlands) works on the health aspects of cereal and grain foods (with focus on oats and whole grains). He co-initiated the Allergy Consortium Wageningen (2001–present), the Dutch Coeliac Disease Consortium (2004–2013), the Dutch Oat Chain (2007–present), and the Well-on-Wheat project (2016–present) concerning non-celiac wheat sensitivity. He is (co)author of about 120 peer-reviewed research papers and is co-editor of two books (H-index 31).

Marinus J. M. (René) Smulders is group leader and manager in Plant Breeding, Wageningen University and Research, The Netherlands. He studies genetic mapping of traits in crops with polyploid genomes, such as rose and wheat. He participates in societal discussions on opportunities of new plant breeding techniques, including gene editing, for generating crop varieties that will contribute to addressing the challenges of agriculture in the twenty-first century, and that have direct benefits for consumers, including people with coeliac disease. He is (co)author of 180 peer-reviewed research papers.

Chapter 17
Near-Isogenic Lines as Powerful Tools to Evaluate the Effect of Individual Phytochemicals on Health and Chronic Diseases

Binning Wu, Jairam K. P. Vanamala, Surinder Chopra, and Lavanya Reddivari

Abstract The growing epidemic of chronic diseases in the twenty-first century has emphasized the importance of developing staple crops with enhanced nutritional value and health benefits. Plant-based diets are associated with reduced risk of many chronic diseases, which are partially due to the health-promoting effects of enriched phytochemicals. Despite the great effort put into understanding the effectiveness of phytochemicals through epidemiological and human intervention studies, the results are inconclusive. Phytochemicals used in these studies, and other cell or animal-based models, are often purified compounds and not the whole-food matrix. Emerging evidence suggests that the food matrix can significantly influence phytochemical stability, bioavailability as well as bioactivity. However, the studies where whole foods are used could encounter difficulties in dissecting out contributions of specific phytochemicals from a mixture of nutrients. Near-isogenic lines (NILs), whose genetic backgrounds differ only in one or a few genes offer feasibility to study for developing whole foods that vary in a specific class of phytochemicals, making them powerful tools. This chapter focuses on reviewing the importance of using NILs in studying the effect of individual phytochemicals in the whole-food matrix to provide evidence-based staple crops to counter the growing epidemic of chronic diseases globally.

Keywords Near-isogenic lines · Phytochemicals · Plant foods · Chronic diseases

B. Wu · S. Chopra
Department of Plant Science, The Pennsylvania State University, University Park, PA, USA

Interdisciplinary Graduate Program in Plant Biology, The Pennsylvania State University, University Park, PA, USA

J. K. P. Vanamala
Department of Food Science, The Pennsylvania State University, University Park, PA, USA

B. Wu · L. Reddivari (✉)
Department of Food Science, Purdue University, West Lafayette, IN, USA
e-mail: lreddiva@purdue.edu

© The Author(s), under exclusive license to Springer Nature Switzerland AG 2021
A. Ricroch et al. (eds.), *Plant Biotechnology*,
https://doi.org/10.1007/978-3-030-68345-0_17

17.1 Plant-Based Diets and Human Health

The rising incidence of chronic and non-communicable diseases will be a significant challenge that poses a major threat to human health in the twenty-first century. Prevalence of obesity, type 2 diabetes, cardiovascular disease (CVD), various cancers, and other inflammation-associated diseases are driven by increasing urbanization, including shifts in lifestyle and dietary patterns over the past five decades (Bauer et al. 2014).

Over the past decades, a substantial scientific effort was directed towards understanding the diet-health relationship as well as identifying critical dietary components that can counter chronic diseases. Accumulating epidemiological evidence suggests that diets rich in fruit and vegetables can reduce the risk of cancers and many chronic diseases. Furthermore, recent studies took into account the food nutritional composition to answer the specific question of which nutrient/non-nutrient within the food matrix could be potentially health-beneficial. In this manner, different putative bioactive compounds were found to be responsible for explaining the inverse association between consumption of plant-based diet and disease incidence. An example of non-nutrient bioactive compounds is carotenoids, fat-soluble yellow/orange pigments. Combined results from several epidemiological studies have revealed the influence of carotenoids intake on the risk of CVDs. We measured α- and β- carotene concentrations in ~570 (α-carotene in 565 and β-carotene in 572) nondiabetic Mexican American (MA) children, aged 6–17 years. We recently reported that serum carotenoid concentrations were under strong additive genetic influences based on variance components analyses and that the common genetic factors may influence β-carotene and obesity and lipid traits in MA Children (Farook et al. 2017).

The idea of consuming plant bioactive compounds to boost health is widely advocated not only because these compounds possess antioxidant properties, but also because they are naturally occurring compounds universally distributed in edible plants, making them as safer options compared to therapeutic drugs. Plants are a rich source of secondary metabolites and have been the foundation of many traditional medicine systems since ancient time. These secondary metabolites are not essential for plant growth and development but could exhibit a diverse spectrum of biological functions and participate in many important bioactivities such as signal exchange, detoxification, seed germination, and pollination. Based on their chemical structures and biosynthetic origins, phytochemicals can be classified as carotenoids, phenolics, alkaloids, nitrogen-containing compounds, and organosulfur compounds (Fig. 17.1). The mechanistic basis underlying the protective effects of phytochemicals against inflammatory diseases can be summarized as (1) Antioxidant and free radical scavenging activities; (2) Modulation of cellular activities of inflammation-related cells; (3) Modulation of pro-inflammatory enzymes; (4) Interference with other elements from the inflammation cascade (Bellik et al. 2013). For example, the activity of antioxidant enzymes such as superoxide dismutase, catalase, glutathione peroxidase, and glutathione reductase can be elevated by a flavonoid known as quercetin. Other phytochemicals such as curcumin, resveratrol, and epigallocatechin

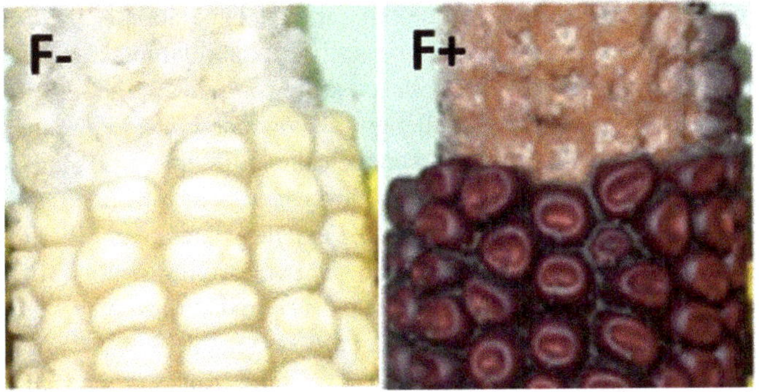

Fig. 17.1   Maize near-isogenic lines *P1-ww* (F−) and *P1-rr* (F+) differ in phlobaphenes accumulation (Wu et al. 2020)

can suppress pro-oxidant enzyme activity. Cyclooxygenase (COX) and lipoxygenase (LOX) involved in eicosanoids biosynthesis are highly implicated in intestinal inflammation. Flavonoids such as rutin, luteolin, and apigenin were found to suppress COX activity whereas quercetin and kaempferol were found to suppress LOX activity.

Though there is a growing body of evidence on the positive effects exerted by phytochemicals to modulate host redox status and immune response in the context of inflammation, relationships between phytochemicals and other disease-triggering factors are also being examined. Accumulating studies have attributed the protective effects of flavonoids partially to their prebiotic potential that helps to shape the intestinal microbial community. For instance, in vitro studies have shown that addition of anthocyanin-rich extracts or purified anthocyanin compounds could elevate the growth of *Lactobacillus* spp. and *Bifidobacterium* spp. (Cassidy and Minihane 2017). These genera are known as short-chain fatty acid (SCFA) producers and are considered beneficial bacteria to be used in probiotic preparations.

17.2   Current Challenges in Assessing Health-Promoting Properties of Phytochemicals

Current evidence that supports the protective roles of plant-based diet against chronic diseases was primarily based on epidemiological data encompassing retrospective case–control studies and prospective cohort studies. Albeit these studies were conducted on a large scale where the lifestyle or dietary habits of individuals were monitored over a long time period, the validation of both types of studies is heavily dependent on the use of questionnaires or biomarkers to measure the intake of certain food types. This is a challenging task for both the volunteers and the researchers, as the former are required to provide accurate recall of their diet whereas the latter need to carefully interpret the results for the biomarkers. The data are likely to provide only patchy information that is not representative of habitual intake.

Besides the difficulties that come with scientific approaches, the selection of food material for the study is equally challenging. Evidence collected from epidemiological studies provides a rationale to conduct human intervention studies to make health claims for functional food products. Nevertheless, data from human intervention studies have been scarce and parts of the existing data are conflicting. Tomato lycopene is an example, where the U.S. Food and Drug Administration (FDA) placed concerns on its role in the reduction of prostate cancer risk due to contradicting results from multiple epidemiological studies. The FDA evaluated 13 observational studies conducted in different countries (US, Canada, China, England, and Greece) with different sources of tomato exposure (raw tomato, cooked tomato, tomato sauce, and tomato juice). With all the studies reviewed by the FDA, only five studies reported an inverse correlation between tomato consumption and risk of prostate cancer.

Difficulties in the interpretation of epidemiological results can also be seen in other phytochemicals. For example, a study on more than 11,000 male physicians taking 50 mg β-carotene supplement daily for over 11 years showed no significant changes in terms of CVD incidence compared to the placebo-receiving group (Hennekens et al. 1996).

Purified phytochemicals, as opposed to their naturally occurring forms, are commonly used in human intervention studies. Nevertheless, the question remains whether some isolated phytochemicals with altered chemical structures will confer the same health benefits as they do in their original forms and their native food matrix. Matrix-bound phytochemicals are found to exert higher and prolonged antioxidant activity than their free forms, as the digestion processes which involve massive enzymatic reactions lead to a continuous release of these health beneficial compounds from the linking cell wall polysaccharides to regulate their bioaccessibility. Though it is suggested to utilize phytochemicals in original forms to retain their efficacy, whole-food studies have the downside of being extremely difficult to pinpoint contributions of any single class of compounds from a mixed nutritional background. Firstly, our food is chemically complex and consists of macro and micronutrients as well as non-nutrients. Besides phytochemicals, various substances such as resistant starch and dietary fiber are also potentially health-beneficial. Moreover, food microstructure can greatly influence the bioavailability of phytochemicals. For instance, phenolic acids exhibited higher bioaccessibility in water-based fruit-juice blends than in milk-based fruit-juice blends (Rodríguez-Roque et al. 2015).

17.3 The Importance of Near-Isogenic Lines

Utilization of near-isogenic lines (NILs) could be an excellent alternative to overcome the limitations of using whole-food or isolated compounds in assessing the health-beneficial effects of phytochemicals. NILs are homozygous lines with a fixed genetic background that differs from each other by only one or a few loci. The development of NILs usually begins with a cross of a donor parent and a recurrent parent, followed by repeated backcross breeding with the recurrent parent and extensive genotyping. NILs can be used in developing food materials that differ only in the content of certain phytochemicals while maintaining their natural forms tethered to the whole-food matrix.

17.4 The Development of Near-Isogenic Lines

NILs can be developed through either non-transgenic breeding methods or transgenic methods. In this part, both types of methods will be discussed on a case-by-case basis.

17.4.1 *Isogenic Food Materials Developed Through Selective Breeding*

To develop anthocyanin contrasting maize NILs, researchers selected genotypes with a functional (high anthocyanin accumulation) and non-functional (devoid of anthocyanin) *r1* locus respectively and backcrossed them to W22 background to achieve homozygous genotypes except for *r1* constitution. The two genotypes were further crossed with a commercial line Dekalb 300 for high kernel production, and the F1 progeny seeds were used to develop anthocyanin-rich and anthocyanin-free NILs (Toufektsian et al. 2008).

A more detailed and complex case regarding the dissection of purple pericarp trait was reported in wheat, where scientists developed six NILs carrying various combinations of a gene called *purple pericarp* (*pp*) (Gordeeva et al. 2015). Three genes implicated in this process were: *Pp3* (located on chromosome 2A), *Pp-A1* (7A), and *Pp-D1* (7D). Parental materials used were two wheat NILs i:S29*Pp-A1Pp-D1Pp3* (dark purple pericarp, dark red coleoptile) and i:S29*Pp-A1pp-D1pp3* (colorless pericarp, light red coleoptile), both of which were developed in spring bread wheat 'Saratovskaya 29' background with 'Janetzkis Probat' being the donor of *pp-A1* and 'Purple' or 'Purple Feed' being the donor of *Pp3* and *Pp-D1*. Three wheat NILs were developed through the process: i:S29*Pp-A1Pp-D1pp3*^{PF} (colorless pericarp, dark red coleoptile), i:S29*Pp-A1pp-D1Pp3*^{PF} (light purple pericarp, colorless coleoptile) and i:S29*pp-A1pp-D1Pp3*^{PF} (both pericarp and coleoptile are colorless).

17.4.2 Isogenic Food Materials Developed Through Genetic Engineering

In cases where desirable phytochemicals cannot be selected from existing germplasms, transgenic approaches can be used to develop monogenic mutants with increased nutritional value. A famous attempt for this is Golden Rice. To tackle a worldwide health problem of vitamin A deficiency, scientists introduced β-carotene biosynthetic pathway into rice endosperm to develop β-carotene-expressing orange rice known as Golden Rice. To reconstitute an active β-carotene biosynthesis pathway in rice endosperm, the transformation process included four genes sourced from daffodil or maize and bacteria *Erwinia uredovora*, to be expressed in endosperm.

Besides β-carotene, anthocyanin remains a popular candidate to confer better dietary benefits in food crops. Tomato contains high amounts of carotenoids (majorly lycopene) but suboptimal levels of flavonoids, which justify the introduction of anthocyanin to offer better health benefits. To produce tomato fruit with intense anthocyanin expression in both peel and inner flesh, scientists expressed genes of two transcription factors, snapdragon sourced Delila (*Del*) and Roseal (*Ros1*) in Micro-Tom, resulting in different transgenic lines with fruit coloration ranging from light to dark purple. The dark purple phenotype (*Del/Ros1*N) with the greatest anthocyanin accumulation was crossed with a commercial tomato line Money Maker and maintained through five generations in this background (Butelli et al. 2008).

17.5 Application of Near-Isogenic Lines in Diet Experiments

Toufektsian et al. (2008) aimed at investigating the protective effect of dietary anthocyanins (ACN) against ischemia–reperfusion injury in rat. They utilized two maize NILs differed only in *r1* locus to formulate ACN-rich and ACN-free diet for animal feeding. In this study, 62 male Wistar rats were randomly assigned to two treatment groups and subjected to 8 weeks of feeding trial. Ex vivo heart perfusion assay showed that rat fed with ACN-rich diet had significantly reduced ischemic zone compared to mice fed with ACN-free diet (Toufektsian et al. 2008).

Wu et al. (2020) utilized two maize NILs differed in *p1* locus, flavan-4-ols-rich red corn (F+) and flavan-4-ols-free (F−) white corn (Fig. 17.1), in combination with a carboxymethylcellulose (CMC)-induced low-grade colonic inflammation model to investigate the disease preventive effect of flavan-4-ols. This study reported that a nine-week feeding of F+ diet to mice was effective in decreasing the mRNA expression level of a pro-inflammatory cytokine IL-6. Mucus is considered as the first line of defense against enteric pathogens such as invasive bacteria, colonic inflammation is always associated with impaired mucus barrier characterize by attenuated mucus thickness. Under CMC exposure, F+ diet was found to greatly restored mucus thickness compared to F− diet, indicating the role of flavan-4-ols in maintaining the intestinal barrier function (Wu et al. 2020).

Butelli et al. (2008) did a mice pilot study using two tomato NILs MT and *Del/Ros1*N to explore the health-promoting property of anthocyanins in a cancer context. They supplemented standard mice pellet diet with 10% MT or *Del/Ros1*N and provided diets to cancer-prone $Trp53^{-/-}$ knockout mice for over 200 days to compared life expectancy between treatment groups. This study reported that mice fed MT diet had an average life span of 145.9 d and a maximum life span of 211 d, comparable to those fed with the control diet. However, the average life span of mice consuming *Del/Ros1*N diet was increased to 182.2 d with a maximum life span of 260 d, which could be due to the capacities of anthocyanins to modulate host redox status and to interact with immune signaling pathways (Butelli et al. 2008).

Rising incidence of chronic disease is calling for safer and efficient dietary strategies, NILs with contrasting phytochemical profiles can be of great importance to address the precise question regarding the diet-health relationship. This area opens plenty of opportunities for plant scientists and nutritionists to work in concert and endeavor towards the development of evidence-based functional foods and novel prevention/treatment options for disease.

References

Bauer UE, Briss PA, Goodman RA, Bowman BA (2014) Prevention of chronic disease in the 21st century: elimination of the leading preventable causes of premature death and disability in the USA. Lancet 384:45–52. https://doi.org/10.1016/S0140-6736(14)60648-6

Bellik Y, Boukraâ L, Alzahrani HA et al (2013) Molecular mechanism underlying anti-inflammatory and anti-Allergic activities of phytochemicals: an update. Molecules 18:322–353. https://doi.org/10.3390/molecules18010322

Butelli E, Titta L, Giorgio M et al (2008) Enrichment of tomato fruit with health-promoting anthocyanins by expression of select transcription factors. Nat Biotechnol 26:1301–1308. https://doi.org/10.1038/nbt.1506

Cassidy A, Minihane AM (2017) The role of metabolism (and the microbiome) in defining the clinical efficacy of dietary flavonoids. Am J Clin Nutr 105:10–22. https://doi.org/10.3945/ajcn.116.136051

Farook VS, Reddivari L, Mummidi S et al (2017) Genetics of serum carotenoid concentrations and their correlation with obesity-related traits in Mexican American children. Am J Clin Nutr. https://doi.org/10.3945/ajcn.116.144006

Gordeeva EI, Shoeva OY, Khlestkina EK (2015) Marker-assisted development of bread wheat near-isogenic lines carrying various combinations of purple pericarp (Pp) alleles. Euphytica. https://doi.org/10.1007/s10681-014-1317-8

Hennekens CH, Buring JE, Manson JE et al (1996) Lack of effect of long-term supplementation with beta carotene on the incidence of malignant neoplasms and cardiovascular disease. N Engl J Med 334:1145–1149. https://doi.org/10.1056/NEJM199605023341801

Rodríguez-Roque MJ, de Ancos B, Sánchez-Moreno C et al (2015) Impact of food matrix and processing on the in vitro bioaccessibility of vitamin C, phenolic compounds, and hydrophilic antioxidant activity from fruit juice-based beverages. J Funct Foods. https://doi.org/10.1016/j.jff.2015.01.020

Toufektsian M-C, de Lorgeril M, Nagy N et al (2008) Chronic Dietary Intake of Plant-Derived Anthocyanins Protects the Rat Heart against Ischemia-Reperfusion Injury. J Nutr 138:747–752. https://doi.org/10.1093/jn/138.4.747

Wu B, Bhatnagar R, Indukuri VV et al (2020) Intestinal Mucosal Barrier Function Restoration in Mice by Maize Diet Containing Enriched Flavan-4-Ols. Nutrients 12:896. https://doi.org/10.3390/nu12040896

Binning Wu received her B.S. degree in Agronomy from China Agricultural University in 2016 and started her Ph.D. study in the Plant Biology Program at The Pennsylvania State University under the mentorship of Dr. Lavanya Reddivari and Dr. Surinder Chopra. She joined Dr. Reddivari's group in the Food Science Department at Purdue University as a Visiting Scholar in January 2019 to continue her Ph.D. research focusing on the health-promoting effects of dietary flavonoids.

Jairam K. P. Vanamala Ph.D. Associate Professor (2013–2018), Penn State Food Science department and USDA Food Quality and Gut Microbiome Grant Panel Manager (2018–2019), focuses on food, gut bacteria, and intestinal epithelial/immune cell interaction in health and disease, particularly colon cancer development and obesity. He employs cell culture, animal (mice, rat, rabbit and pig), and human studies in his research. Dr. Vanamala secured grants from USDA, foundations (AICR), as well as industry and co-authored several peer-reviewed publications.

Surinder Chopra has a Ph.D. from Vrije University Brussels, Belgium and a post-doctoral fellowship with Thomas Peterson, Iowa State University. He served the ICRISAT, India and currently directs a cereal molecular genetics and epigenetics research program focusing on plant development, plant-pathogen and plant–insect interactions. These basic and applied research projects have been supported by NSF, USDA NIFA, and USAID grant programs. He mentors post-doctoral fellows, graduate and undergraduate students and teaches plant genetics and crop biotechnology courses.

Lavanya Reddivari Ph.D. is an Assistant Professor in the Purdue Food Science department focuses on plant food bioactive components and gut bacterial metabolism in health and low-grade inflammation-driven chronic diseases. The goal is to develop food-based strategies to counter diseases such as ulcerative colitis and colon cancer. Dr. Reddivari has co-authored several peer-reviewed publications in the area of food and health and received grants from the USDA-NIFA to study the interaction of food, gut bacteria and the host in health and disease.

Part VI
Contributions of Genome Editing to Agriculture

Chapter 18
Policies and Governance for Plant Genome Editing

Joachim Schiemann, Frank Hartung, Jochen Menz, Thorben Sprink, and Ralf Wilhelm

Abstract Genome editing and modification techniques enable breeders to create single point mutations and to insert or delete new DNA sequences at a specific location in the plant genome thus for the first time making possible the precise modulation of traits of interest with unprecedented control and efficiency. The advent of genome editing has evoked enthusiasm but also controversy, creating regulatory and governance challenges worldwide. Plant genome editing could play a key role in developing crops that will contribute significantly to attaining multiple Sustainable Development Goals provided that accompanying the rapid scientific progress also policy and governance problems will be solved. Today, several countries, most of which located in the Americas, have adapted legislations to these technologies or released guidelines supporting the use of genome editing. Other countries are debating the path to come either because there is no clarity on the legal classification or due consensus is hampered by a renewed GMO debate. In recent years (2017–2020), eight countries have introduced guidelines clarifying the legal status of genome edited products and many of those are actively committed to international harmonization of their policies. In this chapter which is mainly based on a recent up-to-date review published by Menz et al. (Front Plant Sci 11(588027), 2020) we provide an overview on the current and potentially future regulatory environment for genome edited plants at national and international level.

Keywords Genome editing · Regulation · Legislation · Policies

18.1 Outdated Laws for New Techniques Provide Room for Uncertainty

Most national and international legislations do not explicitly refer to products of genome editing due to its novelty and diversity of products. Most regulations of biotechnology applications in breeding refer to the use and commercialization of

J. Schiemann (✉) · F. Hartung · J. Menz · T. Sprink · R. Wilhelm
Institute for Biosafety in Plant Biotechnology, Julius Kuehn-Institute, Quedlinburg, Germany
e-mail: joachim.schiemann@t-online.de

conventional genetically modified organisms (GMOs) and products thereof. Thus, for conventional GMOs the legal status is clear and often in line or similar to the definition given in the Cartagena Protocol on Biosafety—an international agreement which aims to ensure the safe handling, transport and use of so called "living modified organisms (LMOs)". The protocol defines an LMO as '[…]*any living organism that possesses a novel combination of genetic material obtained through the use of modern biotechnology*' (Secretariat of the Convention on Biological Diversity 2000) and many national legislations use *somewhat similar* definitions for a GMO. These definitions originate in the years before 2000 when modern biotechnology was mostly considered as insertion or deletion of recombinant DNA in organisms beyond the species border and when most genome editing methods were neither known nor applied. Current biosafety regulations governing import, cultivation and use of GMOs for food and feed were established to meet concerns that genetic engineering potentially generates unforeseen risks for human and animal health and the environment. However, a generic risk caused by the technology itself has not been proven since more than 30 years of biosafety research (European Commission 2001–2010, 2020; Nicolia et al. 2014; Leopoldina and Akademieunion 2019). In most countries, a GMO intended for cultivation and/or import for food and feed production needs to pass a rigorous safety assessment to be approved. Since breeders or importers have to provide extensive data for the safety assessment, GMO approvals are time and cost intensive in many countries. In the European Union, the approval of a genetically modified (GM) crop costs 11–17 million Euro and takes on average 6 years (EuropaBio 2019). Since most European member states restricted or banned cultivation of GM plants on their territories due to, inter alia, biosafety concerns of stakeholder groups, only Spain and Portugal still grow one GM cultivar (ISAAA 2018). Already since 2011, breeding companies and research institutions requested advice from national competent authorities whether and how genome edited plants are regulated as GMOs resulting in the same costly legal provisions that apply to GMOs. The intended commercial release of canola modified for herbicide-tolerance using oligonucleotide directed mutagenesis (ODM) by the company Cibus caused legal controversy in many countries and provoked controversial discussions on the regulation of genome edited products especially in the EU (for details see Sprink et al. 2016). The Court of Justice of the European Union (CJEU) provided regulatory clarity in July 2018—with an unexpected outcome for European breeders and scientists working with genome editing methods. The CJEU decided that organisms obtained by the new techniques of genome editing are GMOs within the meaning of the Directive 2001/18/EC on the release of GMOs into the environment, and they are subject to the obligations in the legal framework laid down by the GMO Directive. Exemptions from the obligations only apply to organisms derived by random mutagenesis as they show a history of safe use before the law came into force in 2001. In contrast to the EU, Canada regulated the ODM-modified canola like any other crop plant considering the novelty of its trait without focusing on the technique used. The heterogeneity of the regulatory environment for genome edited plants will undoubtedly raise concerns and problems when more and more genome editing products will reach the market.

Genome editing employs variants of site-directed nuclease (SDN) technologies and oligonucleotide-directed mutagenesis (ODM). Genome editing by using SDNs can be categorized in three types:

1. the induction of single point mutations or InDels (SDN-1),
2. short insertions or editing of a few base pairs by an external DNA template sequence (SDN-2) and
3. the insertion of longer strands of allochtonous (transgenes) or autochtonous (cisgenes) sequences (SDN-3).

The SDN-1/2/3 terminology has been adopted by many countries to legally categorize SDN applications.

18.2   The Global Regulatory Status of Genome Edited Plants in 2020

Although scientists and breeders in numerous countries are working on the development of genome edited crops for several years, only a few countries defined their regulatory environment for those plants between 2014 and 2016 (see Sprink et al. 2016 and Ishii and Araki 2017). However, in the last three years more and more countries amended their current biotechnology regulations or clarified the legislations' interpretation with regards to genome editing and products thereof.

The first countries that released advices, opinions or regulations on genome editing are located on the American continent, namely Argentina, Chile, the United States and Canada. Brazil, Colombia and Paraguay enacted normative resolutions on genome editing after Argentina and Chile had released their resolutions. Some of these legislations clearly define which techniques of genome editing will or will not result in the creation of a GMO. Consequently, local breeders and producers gain clarity beforehand. Since countries with product-oriented regulatory concepts and long-lasting GMO cultivation like Canada or the United States did not change their regulatory environment, genome edited plants could be placed on the market without specific regulatory burdens. New Zealand developed a new framework already in 2014 but had to stick with its outdated GMO regulations after a high court decision similar to the CJEU ruling in Europe. Nevertheless, controversial discussions are ongoing both in the EU and in New Zealand as more and more agricultural trading partners promote genome editing and products thereof such as Israel, Japan, and Australia, declaring not to specifically regulate plants derived by some techniques of genome editing.

In many other countries the legal status of genome editing is not decided yet or still under discussion. Examples are Norway, Switzerland, Russia, and India. Several African countries are currently debating the future regulatory environment for genome edited plants. South Africa and Sudan are cultivating GM crops, and South Africa is already discussing genome editing and related regulations. Recently, Burkina Faso, Nigeria and Ghana started cropping GM plants, while Uganda still

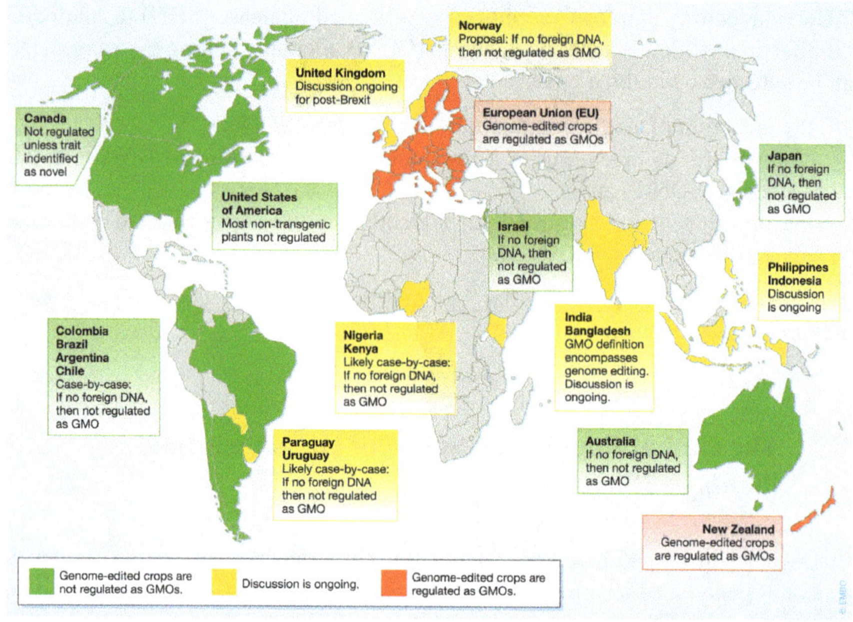

Fig. 18.1 Current state of genome-editing legislation. Taken from Schmidt et al. (2020)

debates the establishment of a GMO legislation without explicitly naming genome editing. The recent declaration of the Africa Biennial Biosciences Communication (ABBC) Symposium stated that regulatory frameworks should facilitate the access to genome editing and awareness on genome editing should be created among African policy and decision-makers (ABBC 2019). Recently, India has released a draft document on genome edited organisms suggesting a tiered risk approach for the regulation of genome edited products which is open for discussion with stakeholders.

In the following, we discuss policies and governance developments related to genome edited plants for prominent countries based on available/accessible information in the recent time period (see Fig. 18.1). We sorted the countries by their progress (decided, undecided yet, and in discussion) to integrate genome editing and resulting products into their regulatory environment.

18.2.1 Latin American States

Since 2015, Latin American countries established explicit regulations for genome editing. Argentina released a resolution in 2015, Chile followed in 2017. In 2018 and 2019, Brazil, Colombia and Paraguay released their resolutions, and the Central American countries Guatemala, Honduras and El Salvador introduced a common biotech policy. Furthermore, Uruguay and the Dominican Republic argued within the WTO for a genome editing policy based on scientific consensus. The policies in Latin America extended the existing GMO regulatory frameworks with resolutions to clarify the legal status of genome edited organisms. Common for all policies is that an interested party can consult or notify a named agency, e.g. CONABIA in Argentina, in a mandatory or voluntary procedure. Within a given time (20–120 days) the agency determines case-by-case whether a modified plant falls under the country's GMO/LMO definition and whether the obligations for a GMO apply. Insertions and/or deletions or base substitutions in the plant genome made by SDN-1, SDN-2 or ODM do not result in a GMO, provided that no residues of foreign recombinant DNA are detected in the final plant. Evidence for the absence of transgene residues which might have been introduced during the modification process must be provided (Whelan and Lema 2015). Brazil and Paraguay especially name the techniques that do not result in the formation of a GMO. Since in most cases recombinant DNA sequences of foreign origin are introduced into the plant genome by SDN-3, the resulting plants are categorized as GMOs. An exception might be the complete replacement of an allele (allele swap).

18.2.2   United States of America

The US regulatory system has not been changed with the emergence of genome editing and products thereof. In fact, genome editing was incorporated in a lasting discussion of renewing the biotech regulation. In 2015, the Administration of President Obama issued a memorandum directing the Environmental Protection Agency (EPA), the Food and Drug Administration (FDA), and the U.S. Department of Agriculture (USDA) to clarify the roles and responsibilities of these agencies regulating biotechnology products under the Coordinated Framework for the Regulation of Biotechnology). Plans were published to modernize the Coordinated Framework for the regulation of biotechnology (EPA 2017), providing information about the types of biotechnology product areas regulated by each competent authority (i.e., EPA, FDA, or USDA). The US Secretary of Agriculture reconfirmed in March 2018 that USDA's Animal and Plant Health Inspection Service (APHIS), which is the primary regulatory authority for plant products in the USA, does not regulate or has any plans to regulate plants that otherwise could have been developed through traditional breeding techniques under §7 CFR part 340 which is the operative statue governing regulation of genetic engineering within APHIS. Also in October 2018, the FDA committed in context of its Plant and Animal Biotechnology Innovation Action Plan (FDA 2018) to pursue advances in policy priorities in order to establish a science- and risk-based approach for product developers and to remove barriers for future innovation in plant and animal biotechnology. Under the Administration of President Trump, at June 11, 2019, an Executive Order on Modernizing the Regulatory Framework for Agriculture Biotechnology Products (The White House 2019) described reforms to promote agricultural innovation and to streamline current regulations for biotechnology and emerging technologies. In line with this order APHIS proposed a revision of the

§7 CFR part 340 (USDA APHIS 2019). The new proposal was a product of several feedback phases with stakeholders asking for public comments on the proposed revisions. In May 2020, the proposal was implemented in the new SECURE rule (USDA APHIS 2020) as a comprehensive revision of APHIS' biotechnology regulations and will be fully implemented by 2021. The revised framework is meant to provide clear, predictable and efficient regulatory pathways for applicants, when the plant products are unlikely to pose a plant pest risk. In effect, the new rule exempts categories of products developed through genome editing under most conditions from obligations under §7 CFR part 340 when changes in the plant genome are:

1. deletion(s) of any size;
2. targeted substitutions of a single base pair; or
3. solely introductions from sequences derived from the plant's natural gene pool or edits from sequences which are known to correspond to the plant's natural gene pool.

Although not explicitly named, the classification resembles mostly the classification by SDN-1/-2/-3. In addition, APHIS may exempt product-depending on the individual case.

In a new exemptions and confirmation process applicants may request confirmation that their products are exempted and APHIS will provide a written confirmation within 120 days. This replaces the previously offered voluntary consultation process "Am I regulated according to 7 CFR part 340?" (AIR) which allowed interested parties to determine the regulatory status according to the product's plant pest character prior to marketing in the US. In almost all responses from APHIS, one advises the interested party to consult also with FDA and EPA but both agencies do not offer a comparable service to APHIS so far. Since FDA is responsible for the biosafety evaluation of food additives, it evaluates the safety of plant derived foods and feed products through a consultation procedure. FDA firms up on a product-based regulation by comparing substantial equivalences of a novel food product to a known comparator (CAST 2018). EPA's responsibilities are related to products generating pesticides (e.g. Bt-Toxins) or to food containing pesticide residues. As many genome edited products will not produce pesticides per se it is expected that EPA will only play a minor role in the evaluation of most genome edited crops in the US.

18.2.3 Canada

Like in the USA, Canada's regulatory system has not been changed with the emergence of genome editing, but due to its product-oriented policy the system is flexible and able to cope with all plants, irrespective of the breeding method used (Smyth 2017). All plant products, whether obtained through biotechnology (e.g. transgenesis or genome editing) or conventional breeding including undirected mutagenesis, are subject to the same oversight in the regulatory framework for plants with novel traits (PNTs). The Canadian legislation is based on the novelty of product characteristics, considering all products case-by-case. Plant products classified as PNT are subject to extended oversight and are tested for allergenicity, toxicity and impact on non-target organisms. In the product-based legislation there is no clear definition of novelty, but a rule of thumb of about 20% difference in the respective trait(s) to a reference product has been established (Smyth 2017). The Canadian Food Inspection Agency (CFIA) offers guidance to determine novelty and when to notify the agency.

18.2.4 *Israel*

In March 2017, Israel reconfirmed the statement from 2016 that plants modified by genome editing are not subjected to the Seed Act (Genetically Modified Plants and Organisms) from 2005 and will not be considered as GMO. The National Committee for Transgenic Plants published its decision that genome edited plants do not fall under the regulation when only small deletions or sequence edits occurred (USDA FAS 2018b). Interested parties introducing new cultivars to Israel need to demonstrate that no foreign DNA was incorporated in the organism's genome.

18.2.5 *Japan*

In relation to its integrated innovation strategy, the Japanese cabinet decided in June 2018 (Japanese Cabinet Office 2018) that handling, cultivation and release of genome edited organisms should be clarified under the Cartagena Act and the Food Sanitation Act by the end of March 2019. Thereupon, an expert panel of the Japanese Ministry of the Environment (MOE) suggested tha organisms derived by SDN-1 should not be specifically regulated (USDA FAS 2018a; Igarashi and Hatta 2018). In December 2018, the MOE released a practical guide addressed to interested parties to resolve which genome editing technologies result in the production of an LMO according to the Cartagena Act and which information should be provided to the respective ministries/authorities (MoE Japan 2018). In March 2019, the MOE clarified its genome editing policy: Organisms modified by insertion of *extracellularly processed* nucleic acids generally result in LMOs and the obligations of the Japanese Cartagena Act apply. Exemptions can be made when the absence (e.g. by outcrossing) of inserted nucleic acids or its replicated products in the genome was confirmed (USDA FAS 2019b). Accordingly, organisms derived by SDN-1 are exempted from regulation when no foreign sequences (e.g. coding for the SDN) remain in the genome. For SDN-2 and ODM where extracellularly processed nucleic acid sequences pose as template for the recombination, their absence has to be demonstrated on a case-by-case basis. SDN-3 applications are generally regarded as LMO although it is not clarified how allele swaps will be regulated. Irrespective of the applied method, in every case the competent authority has to be informed on the edited sequences in the organism.

For food derived from genome edited organisms, the regulations of the Ministry of Health, Labour and Welfare (MHLW) apply. In September 2019 the Councilor for Environmental Health and Food Safety released handling procedures for the marketing of food and feed products and additives, entirely or partially derived from genome editing or from crossbred genome edited progeny (MHLW Japan 2019). Before placing on the market, producers need to consult the MHLW about the regulatory status of the respective product. The MHLW determines case-by-case whether a specific safety assessment is required (as for GM food) or a notification is sufficient. A notification requires information on the editing technique, the genes targeted for modification, year and month of marketing and other details from developers or in case of imported products from importers. After its publication by the MHLW marketing will be enabled; if food is processed from notified genome edited products a separate notification is not required. For food, SDN-1 and SDN-2 derived products harboring genomic substitutions or indels of one to several bases are considered as similar to conventional products and in most cases a notification is sufficient; while for food derived from SDN-3 containing foreign genes a safety assessment is mandatory. It remains unclear which method will be accepted to prove the absence of transgenes or extracellularly processed nucleic acid sequences in the final product.

18.2.6 Australia

In 2016, the Australian government initiated the third review of their National Gene Technology Scheme to clarify the scope of regulation in light of the rapid technical progress, involving national and international stakeholders. The final report was released in October 2018 (Australian Government—The Department of Health 2018). In order to facilitate the flexibility of the regulatory scheme, several recommendations proposed a reorganization of the legislation and the establishment of a risk tiering. It was proposed to ensure a regulatory level proportionate to risk and to avoid over-regulation or under-regulation, respectively. Based on the identification of new risks or the history of safe use, allocation of organisms between categories should be ensured with appropriate flexibility. Furthermore, the existing legal process-based trigger and the subsequent risk assessment were maintained in the review (Australian Government—The Department of Health 2018). Instead of reorganizing the entire legislation as proposed, the Australian government published a first set of updated amendments of the Gene Technology Scheme in April 2019: Genome editing is defined as gene technology; organisms modified with SDN-1 are exempted from the obligations of the regulation as a nucleic acid template was not added; ODM, SDN-2, and SDN-3 derived products are regulated as GMO (Mallapaty 2019; Thygesen 2019). The majority of amendments in the Australian GMO regulation were implemented in October 2019 (Australian Federal Executive Council 2019). For products from genome editing designated for human consumption, different rules will apply and are still under discussion (see New Zealand).

18.2.7 New Zealand

Already in 2013, New Zealand's environmental protection authority decided on the regulatory status of genome edited plants (Environmental Protection Authority 2013) Plants edited by SDN-1 were considered as closely related to plants treated by undirected mutagenesis and were exempted from New Zealand's GMO regulation in the Hazardous Substances and New Organisms (HSNO) act (Ministry for the Environment New Zealand 1998). This interpretation was overruled by a High Court decision in 2014 (Kershen 2015). The act was updated by clarifying that all mutagenesis techniques established after 1996 result in GMOs. Since then, genome edited plants are regulated as GMOs in New Zealand and the respective biosafety regulations of the HSNO act apply (Kershen 2015).

For developing food and feed standards, New Zealand and Australia jointly operate the statutory authority FSANZ (Food Standards Australia New Zealand). The standards are published ("Food Standards Code") and apply to food produced for sale in, or imported into Australia and New Zealand. In 2018, FSANZ started a stakeholder consultation to determine whether such imports need a pre-market assessment and approval as it is established for conventional GMOs (FSANZ 2018a). A preliminary report was released quoting and summarizing the responses to key questions (FSANZ 2018b). Many stakeholders drew attention to differences in definitions of genome editing between the laws in New Zealand and Australia and the common Food Standards Code. Furthermore, stakeholders suggest harmonization of regulatory approaches for genome editing, both domestically and internationally, as the way to facilitate trade and certainty while providing the agricultural sector and consumers access to innovative products (FSANZ 2018b). In 2019, New Zealand's Royal Society released a critical opinion on the current situation for genome editing in New Zealand and proposed options to change current legal obligations proportionate to risk and in accordance with the Australian genome editing regulations (Royal Society/Te Aparangi 2019).

18.2.8 European Union

Plant genome editing policies and governance have been discussed lively in Europe (Eriksson 2018) and there was hope for evidence-based regulatory developments. In contrast, in July 2018 the European Court of Justice (CJEU) ruled that products resulting from targeted mutagenesis methods are regulated under the provisions of the Directive 2001/18/EC for the deliberate release of GMOs (Court of Justice of the European Union 2018). Due to the ruling previous legal interpretations or decisions of EU competent authorities became obsolete and had to be retracted. The EU is now challenged to enforce the judgment and the Member States are obliged to monitor compliance with regards to (unauthorized) genome edited plants and products thereof which identification and differentiation is hardly possible (Grohmann et al. 2019). The European Network of GMO Laboratories (ENGL) emphasized the difficulties and technical limits in identifying genome edited plants, concluding that the enforcement of the current European GMO legislation is challenging (ENGL 2019). The validation of an event-specific detection method and its implementation for market control is not feasible for genome edited plant products carrying a DNA alteration that is not unique.

On request of the European Council the European Commission initiated a study published in April 2021 regarding the status of "novel genomic techniques" (European Commission, 2021). Further initiative by the Commission to update the legislation may follow.

In their statement "Towards a scientifically justified, differentiated regulation of genome edited plants in the EU" (Leopoldina and Akademieunion 2019) highly recognized German academies recommend the following:

1. "Amendment of European genetic engineering legislation: In a first step, the European genetic engineering legislation should be amended. This should include a revision of the GMO definition or the associated exemptions within the current legislative period of the European Parliament in order to exempt genome edited organisms from the scope of genetic engineering legislation if no foreign genetic information is inserted and/or if there is a combination of genetic material that could also result naturally or through traditional breeding methods. ... This would also align European legislation with the regulation of some of the EU's major trading partners in the agricultural sector.

2. A fundamentally new legal framework: Beyond the short-term amendment of current genetic engineering legislation, a second step should comprise developing a fundamentally new legal framework that is detached from the previous, process-based regulatory approach to genetic modification. This longer-term action is the logical next step from a scientific point of view. The current process-centric approach cannot be scientifically justified. However, it is also unwarranted for regulation to distinguish between breeding methods with and without transgenic DNA. Risks to humans, nature, and the environment can only arise from the plant (or its new traits) and the way in which it is used, but not from the process on which the genetic modification is based. A new legal framework must therefore link the requirement of authorisation, registration, or declaration to resulting traits. The requirement, nature and scope of a science-based risk assessment should be determined on the basis of the innovative nature of the product or trait concerned."

18.2.9 China

China massively invests in genome editing research and is a leading country regarding genome editing publications (Cohen and Desai 2019). A growing number of published researches with genome edited crops grown in field trials demonstrate the ease to test genome edited crops under field conditions in China. Discussion on risk analysis of genome editing products has been initiated in China since 2015 and a working group within the National Biosafety Committee (NBC) was established in September 2016 to provide technical assistance on how to regulate new techniques including genome editing. Respective regulations have not been issued yet (Gao et al. 2018), but the Chinese government closely monitors foreign policies on genome editing (USDA FAS 2019a). Due to China's strong investment in genome editing, one can expect genome editing-friendly policies and governance in the coming years.

18.2.10 Russian Federation

Since 2016, the Russian Federation law prohibits the cultivation of GM plants and the breeding of GM animals on the territory of the Russian Federation, except for the cultivation and breeding of plants and animals required for scientific or research purposes. The apparent anti-GMO attitude of the Russian government changed in April 2019, when the Russian Ministry of Education and Science issued the decree No. 479 to reduce the deficits in Russian biotechnology and to address future development in genetic technologies including gene editing (Russian Governmen 2019). With the decree a billion dollar research program was initiated (Dobrovidova 2019). Besides animal and medicinal biotechnology, the financing will support the improvement of genome editing in plants essential for the Russian agricultural production. The decree defines plant products derived by *"some types"* of genome editing as being equivalent to those derived by conventional plant breeding. Moreover, hindering regulations were named and recommendations for improvements provided. Therefore, an update of the Russian policies and governance that promotes genome editing is expected in the coming years.

18.2.11 India

In January 2020, the Indian Department of Biotechnology drafted genome editing guidelines (Indian Ministry of Science and Technology 2020). The guidelines propose a tiered regulatory approval process based on categorization in regulatory groups depending on the genome editing type explored. Group 1 combines plants which genomes harbor one or a few base pair edits or deletions based on SDN-1 or ODM, whereas plants which harbor a few or several base pair edits based on SDN-2 using a template are belonging to Group 2. The distinction between *a few* and *several* is not conclusive in the draft. Risk assessment in Group 1 and Group 2 is performed on a case-by-case basis and requires confirming the targeted edit, ruling out biologically significant off-targets as well as testing for the efficacy of the traits and for their equivalence to reference varieties except the edited trait. Group 3 resembles plants with large DNA changes and insertion of foreign DNA for which the same stringent risk assessment as for conventional transgenic plants applies.

18.2.12 Switzerland

In November 2018, the Swiss Federal Council released plans to modify the current gene technology regulations in order to adapt them to the latest developments in genome editing (Generalsekretariat UVEK 2018). According to a study from the Federal Department of the Environment, Transport, Energy and Communications (DETEC) and the Federal Department of Economic Affairs, Education and Research (EAER), the current Swiss GMO regulation from 2004 is inadequate to cope with plants derived from genome editing (Transkript 2018). Switzerland proclaimed a moratorium on GMO cultivation until the end of 2021 and it is unclear if the moratorium encompasses also plants modified by genome editing. Amended regulations are planned to include a categorization of products and technologies into different risk classes. A first outcome of the debate was expected by the end of 2019, but an official statement is pending (Hardegger 2019). It has to be noted that Switzerland is not part of the EU, but surrounded by EU member states and that new regulations might be affected by transboundary trade issues with the EU.

18.2.13 Norway

Currently, the Norwegian GMO authorization process is entangled with the European authorization procedure (Eriksson et al. 2017). In 2018, the Norwegian Biotechnology Advisory Board proposed a re-evaluation of the Norwegian regulatory framework for GMOs (Norwegian Biotechnology Advisory Board 2018). Currently, conventional GMOs and genome editing products are categorized into four risk tiers based on differences in genetic modification technology, organism, potential of invasiveness and social parameters (Bioteknologirådet 2018). Relevant criteria are stability and thus heritability of a genetic modification, whether the change could have been induced by conventional breeding techniques, and whether the change crosses species boundaries. For an organism or product categorized in the lowest level, a notification of competent authorities (and their response) may be sufficient. At higher categories, organisms would require approval before release and subjected to more stringent risk management requirements. An official statement by the Norwegian government how genome edited plants will be regulated is still missing. Since Norway is an associated country to the EU with regards to trade and travel, uncoordinated policies and governance might be problematic.

18.2.14 Developments in Global Organizations

The Organization for Economic Co-operation and Development (OECD) recognizes the increasing impact of emerging new breeding technologies such as genome editing on global economies. Consequently, an OECD Conference on Genome Editing: Applications in Agriculture was held in June 2018 bringing together relevant stakeholders from more than 35 countries (*OECD Review of Fisheries: Policies and Summary Statistics 2017* 2017; Friedrichs et al. 2019). Participants discussed that regulatory approaches for genome editing should be determined to achieve policy objectives considering both precaution and innovation through better communication between all stakeholders. Different legal systems should understand their respective regulatory and policy approaches to genome editing and a common understanding should be obligatory (Friedrichs et al. 2019).

In November 2018, the delegations of Australia, Argentina, Brazil, Canada, the Dominican Republic, Guatemala, Honduras, Paraguay, the United States of America and Uruguay signed an international statement on agricultural applications of precision biotechnology in the WTO Committee on Sanitary and Phytosanitary Measures (CSPM). The delegations agreed to engage for the exploration of science-based opportunities for regulatory frameworks and the avoidance of trade barriers for products derived from genome editing (Commitee on Sanitary and Phytosanitary Measures 2018). In their declaration the states affirmed that cultivars derived from genome editing should be regulated alike conventional cultivars due to their high similarity. Deregulation of genome editing techniques offers new opportunities for SMEs and national research institutions. Thus, a harmonization at national and international level should be ensured to exploit the full potential of genome editing. Furthermore, within the CSPM the United States with support from Argentina and Paraguay raised specific trade concerns (STC 452) about restrictions from the European Union resulting from the implementation of the CJEU Ruling in July 2018 (Commitee on Sanitary and Phytosanitary Measures 2019), leading to unjustified barriers to trade of genome editing products.

References

ABBC (2019) ABBC 2019 declaration pretoria. Accessed 06 Sept 2019. https://abbcsymposium. org/blog?slug=pretoria-abbc-r2019-declaration

Australian Federal Executive Council (2019) Gene technology amendment (2019 Measures No. 1) regulations 2019

Australian Government—The Department of Health (2018) The third review of the National Gene Technology Scheme October 2018 final report.

Bioteknologirådet (2018) Forslag til oppmyking av regelverket for utsetting av genmodifiserte organismer

CAST (2018) Regulatory barriers to the development of innovative agricultural biotechnology by small businesses and universities

Cohen J, Desai N (2019) With its CRISPR revolution, China becomes a world leader in genome editing. Accessed 02 Sept 2019. https://www.sciencemag.org/news/2019/08/its-crispr-revolution-china-becomes-world-leader-genome-editing

Commitee on Sanitary and Phytosanitary Measures (2018) International statement on agricultural applications of precision biotechnology: communication from Argentina, Australia, Brazil, Canada, The Dominican Republic, Guatemala, Honduras, Paraguay, The United States of America and Uruguray

Commitee on Sanitary and Phytosanitary Measures (2019) Specific trade concerns 2018: G/SPS/GEN/204/Rev.19

Court of Justice of the European Union (2018) Organisms obtained by mutagenesis are GMOs and are, in principle, subject to the obligations laid down by the GMO directive: judgment in Case C-528/16 Confédération paysanne and Others v Premier ministre and Ministre de l'Agriculture, de l'Agroalimentaire et de la Forêt. Press release no 111/18.

Dobrovidova O (2019) Russia joins in global gene-editing bonanza. Nature 569:319–320. https://doi.org/10.1038/d41586-019-01519-6

ENGL (2019) Detection of food and feed plant products obtained by new mutagenesis techniques: report endorsed by the ENGL Steering Committee Publication date 26 Mar 2019

Environmental Protection Authority (2013) Determination of whether or not any organism is a new organism under section 26 of the Hazardous Substances and New Organisms (HSNO) Act 1996: APP201381, April 19. Accessed 05 Sept 2019.

EPA (2017) Modernizing the regulatory system for biotechnology products: final version of the 2017 update to the coordinated framework for the regulation of biotechnology

Eriksson D (2018) The Swedish policy approach to directed mutagenesis in a European context. Physiol Plant 164:385–395. https://doi.org/10.1111/ppl.12740

Eriksson D, Brinch-Pedersen H, Chawade A, Holme IB, Hvoslef-Eide TAK, Ritala A et al (2017) Scandinavian perspectives on plant gene technology: applications, policies and progress. Physiol Plant 162:219–238. https://doi.org/10.1111/ppl.12661

EuropaBio (2019) Pricing innovation out of the EU: counting the costs of GMO Authorisations

European Commission (2001–2010) EUR 24473—a decade of EU-funded GMO research

European Commission (2020) Decree amending the list of techniques for obtaining genetically modified organisms traditionally used without any noted drawbacks with regard to public health or the environment: communication from the Commission—TRIS/(2020) 01601. Accessed 26 May 2020. https://ec.europa.eu/growth/tools-databases/tris/en/index.cfm/search/?trisaction=search.detail&year=2020&num=280&mLang=EN

European Commission (2021) Study on the status of new genomic techniques under Union law and in light of the Court of Justice ruling in Case C-528/16. https://ec.europa.eu/food/system/files/2021-04/gmo_mod-bio_ngt_exec-sum_en.pdf

FDA (2018) FDA's plant and animal biotechnology innovation action plan

Friedrichs S, Takasu Y, Kearns P, Dagallier B, Oshima R, Schofield J et al (2019) Policy considerations regarding genome editing. Trends Biotechnol 37:1029–1032. https://doi.org/10.1016/j.tibtech.2019.05.005

FSANZ (2018a) Consultation paper: food derived using new breeding techniques

FSANZ (2018b) Preliminary report: review of food derived using new breeding techniques—consultation outcomes

Gao W, Xu W-T, Huang K-L, Guo M-z, Luo Y-B (2018) Risk analysis for genome editing-derived food safety in China. Food Control 84:128–137. https://doi.org/10.1016/j.foodcont.2017.07.032

Generalsekretariat UVEK (2018). Neue gentechnische Verfahren: bundesrat prüft Anpassung der rechtlichen Regelung. Accessed 09 May 2019. https://www.uvek.admin.ch/uvek/de/home/uvek/medien/medienmitteilungen.msg-id-73173.html

Grohmann L, Keilwagen J, Duensing N, Dagand E, Hartung F, Wilhelm R et al (2019) Detection and identification of genome editing in plants: challenges and opportunities. Front Plant Sci 10:236. https://doi.org/10.3389/fpls.2019.00236

Hardegger A (2019). Neue Gentech-Verfahren sollen in der Schweiz liberaler reguliert werden als in der EU. Accessed 25 Jan 2019. https://www.nzz.ch/schweiz/crisprcas-bundesrat-strebt-libera lere-regelung-an-als-die-eu-ld.1452558

Indian Ministry of Science and Technology (2020) Draft document on genome edited organisms: regulatory framework and guidelines for risk assessment

ISAAA (2018) Global status of commercialized biotech/GM crops in 2018: biotech crops continue to help meet the challeges of increased population and climate change: executive summary. ISAAA Brief 54

Ishii T, Araki M (2017) A future scenario of the global regulatory landscape regarding genome-edited crops. GM Crops Food 8:44–56. https://doi.org/10.1080/21645698.2016.1261787

Igarashi K, Hatta K (2018) Env Ministry committee proposes deregulating some genetically edited organisms. Mainichi Japan

Japanese Cabinet Office (2018) Integrated innovation strategy: provisional translation

Kershen DL (2015) Sustainability council of New Zealand trust v. The environmental protection authority: gene editing technologies and the law. GM Crops Food 6:216–222. https://doi.org/10. 1080/21645698.2015.1122859

Leopoldina DFG, Akademieunion (2019) Wege zu einer wissenschaftlich begründeten, differen- zierten Regulierung genomeditierter Planzen in der EU: Stellungnahme = Towards a scientifically justified, differentiated regulation of genome edited plants in the EU. Halle (Saale), Berlin, Mainz: Deutsche Akademie der Naturforscher Leopoldina e.V; Deutsche Forschungsgemeinschaft; Union der Deutschen Akademien der Wissenschaften e.V

Mallapaty S (2019) Australian gene-editing rules adopt 'middle ground.' Nature. https://doi.org/ 10.1038/d41586-019-01282-8

Menz J, Modrzejewski D, Hartung F, Wilhelm R, Sprink T (2020) Genome edited crops touch the market: a view on the global development and regulatory environment. Front Plant Sci 11(588027)

MHLW Japan (2019) Food hygiene handling procedures for food and additives derived from genome editing technology

Ministry for the Environment New Zealand (1998) Hazardous substances and new organisms (Organisms not genetically modified) regulations 1998: HSNO regulation

MoE Japan (2018) Decision guidance: to genome editing technologies users. https://www.env.go. jp/press/2_2_%20genome%20editing_En.pdf

Nicolia A, Manzo A, Veronesi F, Rosellini D (2014) An overview of the last 10 years of genet- ically engineered crop safety research. Crit Rev Biotechnol 34:77–88. https://doi.org/10.3109/ 07388551.2013.823595

Norwegian Biotechnology Advisory Board (2018) A forward-looking regulatory framework for GMO

OECD Review of Fisheries: Policies and Summary Statistics 2017 (2017)

Royal Society/Te Aparangi (ed) (2019) Gene editing: legal and regulatory implications

Russian Government (2019) On approval of the federal scientific and technical program for the development of genetic technologies for 2019–2027: resolution No. 479

Schmidt SM, Belisle M, Frommer WB (2020) The evolving landscape around genome editing in agriculture. EMBO Rep 21:e50680

Secretariat of the Convention on Biological Diversity (2000) Cartagena protocol on biosafety to the convention on biological diversity: text and annexes. Secretariat of the Convention on Biological Diversity, Montreal

Smyth SJ (2017) Canadian regulatory perspectives on genome engineered crops. GM Crops Food 8:35–43. https://doi.org/10.1080/21645698.2016.1257468

Sprink T, Eriksson D, Schiemann J, Hartung F (2016) Regulatory hurdles for genome editing: process- vs product-based approaches in different regulatory contexts. Plant Cell Rep 35:1493– 1506. https://doi.org/10.1007/s00299-016-1990-2

The White House (2019) Executive order on modernizing the regulatory framework for agricultural biotechnology products: land & agriculture. Accessed 12 June 2019. https://www.whitehouse. gov/presidential-actions/executive-order-modernizing-regulatory-framework-agricultural-bio technology-products/

Thygesen P (2019) Clarifying the regulation of genome editing in Australia: situation for genetically modified organisms. Transgenic Res 28:151–159. https://doi.org/10.1007/s11248-019-00151-4

Transkript (2018) Bundesrat setzt Genscheren auf Agenda. Accessed 10 Oct 2019. https://transkript.de/meldungen-des-tages/detail/bundesrat-setzt-genscheren-auf-agenda.html

USDA APHIS (2019) Proposed rules for 7 CFR parts 340 and 372 movement of certain genetically engineered organisms

USDA APHIS (2020) 7 CFR parts 330, 340, and 372: RIN 0579–AE47

USDA FAS (2018a) Japan discusses genome editing technology: GAIN report number: JA8048

USDA FAS (2018b) Israel agricultural biotechnology annual 2018: GAIN report number: IS18011

USDA FAS (2019a) Regulatory process getting more unpredictable, additional requirements on trials and data for approvals: GAIN report number: CH 18085

USDA FAS (2019b) Environment ministry finalizes policy for regulating genome editing: GAIN report number: JA9024

Whelan AI, Lema MA (2015) Regulatory framework for gene editing and other new breeding techniques (NBTs) in Argentina. GM Crops Food 6:253–265. https://doi.org/10.1080/21645698.2015.1114698

Joachim Schiemann Prof. Dr. Joachim Schiemann has been active for 25 years in the field of GMO biosafety research as well as risk assessment, risk management and risk communication of GM plants. Until his retirement in September 2016 he was head of the Institute for Biosafety in Plant Biotechnology at the Julius Kuehn-Institute (JKI), Federal Research Centre for Cultivated Plants. Since 2006 he is Honorary Professor at University of Lüneburg. From 2003 to 2009 he was member of the Panel on Genetically Modified Organisms of the European Food Safety Authority (EFSA), from 2002 to 2012 member of the Executive Committee of the International Society for Biosafety Research (ISBR), and from 2004 to 2008 ISBR President.

Frank Hartung Dr. Frank Hartung has been active for 25 years in the field of DNA repair and recombination in plants. Since 2010 he is working on risk assessment of GM plants in the Institute for Biosafety in Plant Biotechnology (SB) at the Julius Kuehn-Institute (JKI), Federal Research Centre for Cultivated Plants. Since 2018 he is deputy head of the Institute SB. Based on his long and outstanding experience on DNA repair and recombination, he is from 2011 on involved in the application and risk evaluation of New Plant Breeding Techniques, especially Genome Editing. In 2012, he was working as external expert for the European Food Safety Authority (EFSA). Since 2017 he is co-chair of the European Plant Science Organisation (EPSO) working group on agricultural techniques.

Jochen Menz With a keen interest in the development of genome editing, Dr. Jochen Menz joined the group of Dr. Thorben Sprink in the Institute for Biosafety in Plant Biotechnology at the Julius Kuehn-Institute (JKI), Federal Research Centre for Cultivated Plants. In a post-doctoral position in the current H2020 project CHIC, he investigated root chicory (*Cichorium intybus*) as a new multipurpose crop for the European bioeconomy with a focus on genome editing and its assessment of biosafety and international regulation. Jochen Menz is an agricultural biologist who has focused his studies mainly on agricultural biotechnology. In 2016, he completed his doctorate at the Institute of Crop Science at the University of Hohenheim in Stuttgart.

Thorben Sprink Dr. Thorben Sprink is currently a senior scientist at the Julius Kuehn-Institute (JKI), Federal Research Centre for Cultivated Plants, and leader of the working group genome editing and synthetic biology in the Institute for Biosafety in Plant Biotechnology. Thorben Sprink is active in genome editing research incl. biosafety, risk and technology assessment since its beginning and he is additionally active in the field of GMO biosafety research as well as risk assessment, risk management and risk communication of conventional GM plants. He is a work package leader in the current H2020 project CHIC in which he is investigating Chicory as a new multipurpose crop for the European bioeconomy with a focus on genome editing and its assessment and regulation. Since 2015 he is giving lectures at the Universities of Hannover and Goettingen.

Ralf Wilhelm Since 2017, Dr. Ralf Wilhelm is head of the Institute for Biosafety in Plant Biotechnology in the Julius Kuehn-Institute (JKI)—Federal Research Centre for Cultivated Plants, in Quedlinburg, Germany. He has been working as senior scientist in biosafety research of genetically modified plants since 2001. His expertise spans from methods for biosafety assessment and management in plants to related regulatory issues. From 2001 to 2006 he was part time consultant for biotechnology. He is board member of the International Society for Biosafety Research (ISBR), and co-chair of the Agricultural Technology Working Group of the European Plant Science Organisation (EPSO).

Chapter 19
Exploring the Roots of the Old GMO Narrative and Why Young People Have Started to Ask Critical Questions

Philipp Aerni

Abstract The history of modern plant breeding is implicitly present in everything we cultivate and eat today. Therefore, the strategy in retail marketing to advertise premium organic products as 'natural' and therefore 'safe', as opposed to products from 'agro-industry' and its 'genetically modified' (GM) products, is highly misleading. After all, almost all food products are a product of culture, not nature, and the organic industry constitutes an important part of agro-industry as well. Yet, it is the radical simplification of 'good' versus 'bad' agriculture that makes the narrative so popular, no matter if it is fiction or fact. It provides a normative orientation without the need to delve deeper into the subject. The consequences of this rather shallow debate on sustainable agriculture has led to real consequences in the form of incoherent and burdensome regulation designed to prevent the use of genetically modified (GM) crops in agriculture. The same narrative is now being extended to the latest breeding techniques associated with CRISPR Cas9 and other gene-editing tools. They tend to be labelled as GMO 2.0 by stakeholders who oppose agricultural biotechnology in general. This label was also implicitly embraced by the High Court of New Zealand as well as the European Court of Justice (ECJ) in their decisions to subject the latest gene-editing techniques to GMO regulation, no matter whether the end product is transgenic or not. Especially for New Zealand, the decision runs against the country's success story as a global powerhouse of agricultural innovation. This chapter argues that a different regulatory environment is only possible if the old GMO narrative loses its credibility with the next generation of concerned citizens. In view of the current global crises related to climate change and COVID-19, many of them find it increasingly irresponsible to discard an important platform technology such as gene-editing just because it is 'new'. If they do not receive convincing answers to their critical questions, they may start to sort out fiction from fact on their own and integrate it into a counter-narrative that is not just more meaningful for their generation but also more effective in enabling sustainable change.

P. Aerni (✉)
Center for Corporate Responsibility and Sustainability at the University of Zurich, Zurich, Switzerland
e-mail: philipp.aerni@ccrs.uzh.ch

A. Ricroch et al. (eds.), *Plant Biotechnology*,
https://doi.org/10.1007/978-3-030-68345-0_19

277

19.1 Introduction

Arguments against the use of modern biotechnology in agriculture tend to be linked to a wider historical account on the impact of agroindustry and modern technology on farm practices and farm income, consumer health and habits, and the environment. In this chapter we scrutinize narratives of decline as well as narratives of salvation concerning modern agriculture and explain why the narrative of decline tends to appeal more to affluent consumer societies.

Agricultural biotechnology and, of latest, gene editing, may just offer some additional tools to make breeding more precise and efficient, yet their fate is decided in a political arena that is not so much concerned about the technology itself but normative views of what should be part of sustainable agriculture and what should not (Aerni 2011).

The normative nature of the discussion often prevents an outcome-oriented view that is guided by questions about the institutional framework conditions necessary to ensure the safe and equitable use of gene-editing techniques and how to harness them effectively to address the urgent global sustainability challenges in agriculture. As long as the discussion takes place on a highly polarized political meta-level, dominated by bipolar framings such as farmer families being exploited by agroindustry, traditional farming being destroyed by modern agriculture, the 'purity' of organic farming being contaminated by agricultural biotechnology, etc., collective action across political factions is unlikely to take place.

The narrative in favor or against the use of agricultural biotechnology often starts with claims about the positive and negative impact of the so-called Green Revolution. The Green Revolution started after World War II as a public sector-driven initiative to ensure food security and to contain communism in non-aligned states in the Global South (Kingsbury 2009). It eventually lost public support after the end of the Cold War in the 1990s, when political priorities shifted from national security concerns to public concerns about the use of taxpayer's money for intensive agriculture and consumer concerns about food safety (Aerni et al. 2012). Despite its positive effects for global food security, the protagonists of the Green Revolution may have cared too little and too late about its negative side effects for the environment and economic inclusiveness. However, one has to bear in mind that the purpose of the Green Revolution was to increase global food security. Its protagonists never claimed that it would also be able to effectively address concerns related to social justice and environmental sustainability (Kingsbury 2009).

The end of the Green Revolution coincided to some extent with the rise of the "Gene-revolution", which was initiated with the successful commercialization of the first genetically modified (GM) crops in the 1990s. Even though the Gene Revolution was already driven by private sector investments by then, there was still a widespread confidence expressed by global food security experts that governments would commit themselves to invest in a "Doubly Green Revolution" to boost global food supply while reducing dependence on environmentally harmful inputs (Conway 1999).

Yet, eventually this narrative stumbled over the fact that the Gene Revolution lacked the generous financial support from the public sector once the Cold War ended. As a result, agricultural technology became largely associated with the rise of the global profit-oriented life-science company epitomized by 'Monsanto'.

However, the public sector initially supported the use of genetic engineering in agriculture in the 1970s and 1980s very much. At that time, university-based spin-off firms developed and, later on, made use of the new technique to create crops with a built-in resistance to pests and other traits that could address environmental concerns. Their innovations were widely seen as an opportunity to move away from the wide and indiscriminate use of toxic agrochemicals. But, once it became clear that ownership of the technology ended up mostly in the hands of Monsanto through the company's large investments in in-house R&D as well as its acquisition of numerous small start-up companies that pioneered the technology, public skepticism started to grow and the call for regulation of the technology became louder (Schurman et al. 2003). Ironically, it was then the costly and time-consuming regulatory approval process in European response to public opposition that increased industrial concentration further. After all, only the large and established companies had the means to cope with the burdensome regulation (Bonny 2017). Negative public perception also led many donor countries to stop supporting public sector agricultural research to improve orphan crops in developing countries if such research did not explicitly exclude the use of agricultural biotechnology (Aerni 2006; Juma 2016). Many of these projects subsequently obtained financial support from the Bill Gates Foundation and became more embedded in public–private partnerships in which industry played a constructive role by offering its proprietary technologies and its know-how for free, provided that the main beneficiaries are small-scale farmers. Lack of access to technology in developing countries can therefore hardly be attributed to the proprietary nature of the technology but is rather related to the regulatory environment in which such projects had to operate (Fukuda-Parr 2007; Aerni 2018).

The rise of gene-editing techniques over the past decade has led, once again, to the creation of many successful new spin-off firms. They do no more embrace the initial narrative of the Green Revolution, which mostly relied on boosting agricultural productivity through the widespread adoption of input-intensive high yielding varieties (HYVs). Instead, they focus on value-added traits that increase the quality rather than the quantity of the food product. As of concerns related to the interference with nature, company representatives argue that, compared to classical mutagenesis,[1] gene-editing is a less-invasive and more precise breeding technique. Moreover, according to them, there is no reason why gene-edited crops should be incompatible with sustainable agro-ecological practices (Clark 2016), also in view of the fact that it is not the gene-editing tool kit that is costly, but the approval process. In fact,

---

[1] Classical mutagenesis a breeding techniques that involves genetic engineering but is not regulated as such. Since the mid-twentieth century, more than 3200 mutant plant varieties, produced by radiation and chemical mutagenesis or by somaclonal variation, found their way to the food market (Pathirana 2011).

they could address certain persistent environmental problems in organic agriculture (Ritchie 2017).

Yet, such issues are not part of a public debate that tends to divide the world into good and bad agriculture, into agriculture 'with nature' and agriculture 'against nature'. This public understanding reveals a lack of basic knowledge about the nature of the 'human being', which is characterized by its lack of specialization to survive in a particular ecosystem. Humans create innovative tools and practices to make up for these natural deficiencies. This equips them to survive in almost any ecosystem on this planet (Gehlen 1988). Innovation is also the way humankind responses to arising scarcities resulting from population growth and increasing affluence (Boserup 1965). From this perspective, human evolution must be understood as a continuous adaptation process in which nature has been converted into culture. The result is what we tend to call the anthropocene, an empirical fact that tends to be falsely interpreted in normative terms assuming that the human motive behind it was 'greed' when in fact it may have been 'concern for the survival of the community' (Crutzen 2006).

These fundamental empirical insights derived from philosophical anthropology are not taught anymore in our basic education system, because every complex issue related to human–environment interactions is explained in normative terms. The GMO narrative as it is embraced in the public debate as well as by the education system, implies, that agricultural modernization should have never taken place because it would be based on exploitation and environmental destruction (Aerni et al. 2009). Facts that contradict the rather one-sided story are simply ignored. Yet, the story can be convincingly applied to any new developments in the field of agricultural biotechnology. This explains why gene-editing is conveniently labelled as GMO 2.0 by its opponents.[2] This may not be surprising in Europe, where governments tend to belief that direct payments rather than innovation will render agriculture more sustainable. But, how was it possible that a country like New Zealand, which managed to sustainably transform and diversify its export-oriented agricultural economy through entrepreneurship and innovation has turned against the commercial use of gene-editing in agriculture, despite its considerable potential to reduce greenhouse gas emissions in its large livestock industry?

In this chapter we argue that the regulatory decisions in Europe and New Zealand tend to strangle innovation by regulation in the field of agricultural biotechnology not just domestically but also in countries that depend on aid and trade with these economies. The trend may only be reversed, if the normative baseline assumptions of the popular narrative against agricultural biotechnology are increasingly questioned in the light of new product developments and new empirical insights. But merely presenting facts in response to false claims about the fatal health and environmental consequences resulting from the cultivation and consumption of GMOs will not do. Instead, there is a need for a new storyline with a strong ethical component that focuses on the socioeconomic concerns related to inclusiveness. This new storyline must start with the main paradigm shift behind the Sustainable Development Goals (SDGs) approved by the UN General Assembly in 2015 and designed to enable a

---

[2] https://www.foeeurope.org/gmo20.

global sustainable transformation by 2030. The SDGs are ambitious in the sense that they aim to address poverty, economic and social inequality and environmental challenges simultaneously through a global partnership (SDG 17) that helps to enable inclusive and sustainable economic change through investments in entrepreneurship and innovation (SDG 8). Moreover, almost every SDGs is directly or indirectly linked to the future of food and agriculture. In this context, agricultural biotechnology in general and gene-editing in particular, have a great potential to become tools of economic empowerment and sustainable change, especially where the SDG-related challenges are greatest. This potential can only be realized, if the platform technology is sufficiently user-friendly and accessible to also enable low income countries, where agriculture hugely matters, to make use of it. And that is the point where especially Europe is slowly losing its moral highground because access to the technology has so far been prevented, not because it is in the "hands of industry", but because Europe managed to export its costly and dysfunctional biosafety regulation to low-income countries that highly depend on European aid and trade. In fact, all of the patents on genetic engineering owned by Monsanto have expired and the new crucial patents on gene-editing are held by universities that tend to pursue a much more open approach that encourages sharing rather than exclusivity.

Interestingly, the call to reconsider the role of gene-editing and look at its potential to address the UN SDGs on a case-by-case basis comes increasingly from young green politicians and early-career scientists in the fields of biology and environmental sciences. They dare to challenge the cherished views in public that were largely shaped by a prior generation that sticks to the belief that recent global sustainability challenges can be addressed with an approach that reflects the mindset of the Cold War of the twentieth century rather than the global knowledge economy of the twenty-first century. It is also these young critical thinkers who have the moral authority to shift the global narrative on agricultural biotechnology since they are unencumbered by the baggage of vested interests. If they succeed, then the content of our education system as well as of our public debates on agricultural biotechnology will also shift. Empirical facts are then likely to gain more weight in the highly normative nature of the debate.

19.2 Historical Context

Science-based plant breeding started with the discovery and application of Gregor Mendel's work on genetics. As a gifted breeder and trained mathematician, the Augustinian Monk Mendel was able to conduct his controlled trials with pea plants in his monastery garden in the mid nineteenth century without any time pressure. He was merely driven by the desire to test his hypothesis of recessive and dominant alleles and had no interest in creating commercial value from it. One could call it therefore an irony that his discovery of the laws of inheritance eventually enabled the emergence of a profitable modern seed industry in the twentieth century (Kingsbury 2009).

It was only four decades after Mendel's original publication in 1866, when scientists in the United States and the United Kingdom paid attention to his scientific breakthrough. They were able to validate the principles of Mendelian genetics in plant breeding and their governments started to promote its application in public agricultural research institutes and universities. However, Germany and France did not see any need to embrace Mendelian genetics in plant breeding at that time. Both countries already had a flourishing seed industry dominated by commercially-oriented plant breeders relying on the mass selection of landraces. They argued that Mendelian genetics would first have to prove its superiority to other established methods of breeding (Kingsbury 2009). Even though it proved very useful as a predictive tool in breeding, the science of genetics was indeed still far from being properly understood. Moreover, there were already concerns voiced in academia that the discovery of science-based breeding will lead to a shift of power from the farmer, who also used to be the plant breeder, to the scientist and his institution, with the result of making farmers more dependent on external input (Kingsbury 2009). Biodynamic agriculture, developed in the 1920s by the founder of the anthroposophy, Rudolf Steiner, can be considered as one of the counter-movements that looked for an alternative.

Unlike widely held contemporary beliefs in public about corporate control of seeds, science-based plant breeding was initially promoted and funded by the public sector whereas private breeders largely continued to rely on traditional breeding approaches; despite its drawbacks. Only when the hybrid vigor effect was first demonstrated in F1 hybrid corn in the 1920s in the United States, the seed industry started to see the business case in Mendelian genetics. The phenomenon of hybrid vigor was observed as a result of a controlled cross-pollination of inbred strains. It manifests itself in desirable physiological traits of plants (uniformity, stability, high yield). But these traits appear consistently in the phenotype only in the first generation. First generation hybrid vigor was therefore a sort of natural intellectual property rights protection. Farmers were prepared to buy seeds from the company rather than re-use the seeds from their harvest or get them for free from a public sector breeding institution because the resulting yield increases combined with the labor saving effects more than made up for the price of the seed (Duvick 2001).

19.2.1 The Green Revolution

Science-based plant breeding based on Mendelian genetics combined with the innovative breeding approaches applied by Norman Borlaug to increase the harvest index and photo insensitivity of wheat in Mexico in the 1940s, were crucial in the creation of high yielding varieties (HYV). They performed well in favorable conditions, but did not have the hybrid vigor effect. Their widespread adoption in Latin America and Asia led to the so-called 'Green Revolution', which helped to increase the yields and improve food security in many parts of the Global South during the Cold War period (Kingsbury 2009).

The 'Green Revolution' was a public sector initiative launched by the US government with additional support from the Rockefeller Foundation. The motivation to improve global food security was, however, not merely a humanitarian cause, but also driven by the desire to contain communism in non-aligned developing countries. It was reasoned that improved food security in these countries would prevent them from embracing communism (Anderson et al. 1991).

The Green Revolution was also accompanied by the creation of international agricultural research centers, the so called CGIARs in Southeast Asia, the Middle East, Africa and Latin America, designed to do research on the improvement the yields of regionally relevant food crops such as rice, wheat and potato.

Overall, the investment in the Green Revolution paid off, since average yields increased substantially in developing countries which participated in this global initiative, while those countries, which embraced socialism or communism proved to be more likely to be affected by famine and starvation (Aerni 2011). However, one must also take into account that the Green Revolution was a US-supported public sector initiative which was deliberately not left to the markets. Agribusiness nevertheless benefited from the research insights gained in the CGIAR centers and the fact that HYVs were highly dependent on external input, which required governments in the Global South to buy irrigation equipment, fertilizer and agrochemicals from agribusiness in bulk. Subsequently, governments sold it at a subsidized price to its farmers (Murray 1994). The result was often a misuse as well as an overuse of natural resources as well as health problems in the agricultural labor force as a result of mostly unprotected exposure to agrochemicals (Pingali 2012). Adoption rates of input-intensive HYVs were high in regions with a high share of capital-intensive farms, but very low in more marginal regions characterized by low-input, rainfed, semi-subsistence farming. In other words, farmers in marginal areas with little access to markets did not benefit from any yield increases, particularly in Africa (Byerlee and Morris 1993).

19.2.2 The End of the Cold War and Its Negative Impact on International Agricultural Research

The United States and Europe became less interested in funding agricultural research in developing countries after the end of the Cold War. Food security ceased to be a major international concern for donor countries and its policy makers started to pay more attention to domestic consumer and tax payer concerns about food safety and sustainability in agriculture (Aerni et al. 2012). As a consequence, more money in agriculture was spent on direct support schemes for farmers that were considered non-trade distorting and therefore compatible with the agreements of the World Trade Organisation (WTO) that was created in 1995. It asked member states to de-couple agricultural subsidies from agricultural production, while offering them also ways to address non-trade concerns such as food safety and the environment

(Aerni 2009). The resulting shift in government support in OECD countries from production-oriented subsidies to the implementation of the concept of multifunctional agriculture that is to compensate farmers for their contribution to the public good caused a major power shift in the global agro-food value chain. While agribusiness was the great beneficiary from public support schemes of agriculture during the Cold War, it was now the global retailers who found themselves as the powerful gatekeepers that defined the terms of sustainable agriculture (Freidberg 2007). Even though they are not directly involved in the daily business of agriculture, they became the standard-setting power for 'Good Agricultural Practices', the so-called GlobalGAP standard.[3] In addition, retailers embraced the normative nature of the public debate on sustainable agriculture by advertising 'the goodness' of premium fair trade and organic labels (Freidberg 2007). However, based on empirical evidence, this normative claims are increasingly questioned: a sustainability performance assessment comparing business-to-consumer (B2C) standards, including organic and fair trade with business-to-business (B2B), that aim to ensure the sustainability across agribusiness value chains, revealed, that B2B standards tend to outperform the supermarket B2C labels because of their continuous commitment to self-improvement (Giovannucci et al. 2014).

While retailers were happy to respond to consumer fears about GM food by labeling products with GMO-free even when they are not (Gabrielczyk 2019), they were certainly not at the root of public resistance against GMOs in the 1990s. Back then, it was not just health concerns but also socioeconomic concerns that were voiced frequently by opponents, who denounced the concentration of ownership of the technology with a few large corporations that increasingly dominated the seed industry.

19.2.3 *The Legacy of Monsanto*

In fact, there was one company eager to take the lead in transforming global agribusiness into a life science industry. Its name was Monsanto and the controversy surrounding it is still dominating the debate in the media today, even though the company stumbled a long time ago and was bought in 2016 by the German company Bayer, a global leader in agrochemicals and pharmaceuticals.

The rise of Monsanto in the 1990s was due to its early investments in agricultural biotechnology. When most agro-chemical companies still relied on the assumption that they could sell agro-chemicals in bulk to boost yields during the Cold War period, Monsanto saw a great potential in genetic engineering to offer solutions to farmers that would prove superior to traditional means of plant protection and better for the

---

[3] Farmers that aim to sell to European retailers have to meet this standard in one form or another, if they want to sell to European retailers. The problem with such private standards is that they do not have to comply with the WTO standards of non-discrimination and that the costs of compliance are borne exclusively by the farmers (Freidberg 2007; Aerni 2018).

environment. The first GM products the company offered were corn with have a built-in pesticide resistance and soybean with herbicide tolerance, which allowed for no-tilling practices reducing soil erosion as well as greenhouse gas emissions. Overall, farmers liked its products because they made farming less labor-intensive and yields more predictable (Brookes and Barfoot 2009). The company also generated large profits thanks to its focus on simple traits that met a huge demand in farming and crops that reach a global market. Its success was however also accompanied by a sort of missionary zeal, a strong belief that the company is a force of good in agriculture (Perlak et al. 2001). It believed in the technology to the extent, that it discarded any public concerns as a sort of ignorance that would eventually give way to more enlightened views.

There was little Monsanto had to fear in the 1990s since the United States Government had already decided that GM crops will be regulated under the existing web of federal statuary authority and regulation related to environmental protection and public health. Moreover, the company owned most of the relevant patents and its competitors in the pesticide industry started to face a public reputation problem (Aerni and Rieder 2001).

Yet, the tides turned eventually with the growing power of Monsanto within agro-industry and the increasing public resistance against GMOs in Europe. As in the United States, the company wanted to reap the fruits of its investment in agricultural biotechnology research by pushing for a fast approval of its genetically modified (GM) crops in Europe and elsewhere assuming that skepticism must be merely a result of ignorance. What it failed to take into account was the fact, that GMOs became increasingly associated with reckless corporate arrogance, disregard for public concerns and US imperialism (Aerni and Rieder 2001; Lamphere and East 2017).

The example of Monsanto shows how a company can lose its license to operate if the public distrusts it because of the conviction that it abuses its dominant position to generate profits at the expense of exposing people to unknown long-term risk concerning public health and the environment. The company and its technology eventually became linked to various undesirable things such as the killings of beautiful monarch butterflies, the creation of superweeds, infertility in humans, cancer tumors in rats, and the commercial sale of sterile seeds to naïve farmers (Aerni 2018; Smyth et al. 2014).

Even though, there is no doubt that the large-scale cultivation of GM crops has caused environmental problems as well, these problems were not related to the technology itself but due to unsustainable agricultural practices (NAS 2016). Sterile cotton seeds, often associated with 'terminator technology', have never been sold, yet the story continues to pop up over and over again in public debates, especially in relation to farmers committing suicide in India (Lynas 2013). In return, other more commendable issues related to Monsanto's investments in data-driven precision agriculture designed to improve soil health and reduce water consumption and greenhouse gas emissions, failed to be noticed by the public at all because it does not fit the reductionist frame of good and evil forces in the battle for sustainable agriculture (Pham and Stack 2018).

While Monsanto tended to abuse its dominant monopoly position, bullied foreign regulatory agencies and belittled justified public concern about unintended side effects resulting from a relatively new technology in the 1990s, there was also a lot of opportunism involved in joining the movement against Monsanto. The organic farming industry saw it as an opportunity to present itself as the 'pure' and 'innocent' form of agriculture and many policy makers endorsed preventive regulation of the "disliked" GMOs in the hope of enhancing their chances for re-election. Even teachers in high schools were happy to show their students award-winning documentary movies such as 'The World According to Monsanto'[4] because it would not cost them much time to prepare and become informed about the facts. Such documentaries help create a simple narrative in which 'good' NGOs fight on behalf of the public interest against 'evil' agribusiness in pursuit of private profits. Such highly emotional narratives make it clear on which side you should stand. They also allow teachers to present themselves as "critical thinkers" to their students (Aerni and Oser 2011; Aerni 2018).

19.3 Gene-Editing as GMO 2.0?

The pro- and contra arguments in favor or against GMOs in today's public debate are very similar to the ones heard in the 1990s, despite the fact that the technology has continuously evolved since the commercial launch of the first GM crops in the United States in the 1990s. Advanced gene-editing techniques such as CRISPR-Cas9 have demonstrated their potential to render crops more productive but also to contribute to climate change adaptation and mitigation, animal welfare, food waste reduction and improved nutrition (Francisco Ribeiro and Camargo Rodriguez 2020; Ricroch 2019). It made agricultural biotechnology also more precise, less invasive to the more affordable. Moreover, advanced gene-editing also enables breeders to better address undesirable off-target effects in the plant DNA which are already much lower compared to the ones observed when less precise mutagenesis techniques are applied in plant breeding (Lee et al. 2020; Liu et al. 2020a). Finally, computational approaches have been applied to further minimize potential off-target effects of the CRISPR Cas9 mediated System in Plant Breeding (Liu et al. 2020b). Despite the great potential of gene-editing, it tends to be portrayed in the public debate as GMO 2.0 in order to justify the demand to regulate it in the same way like GMOs. This is convenient because the negative narrative of the story of GMOs can just be extended to the new technologies without the need to study the subject further. This seems possible because the public debate is not about facts but about normative viewpoints, moral concerns and precautionary tales.

---

[4] The documentary directed by Marie-Monique Robin was released in 2008 and won the Rachel Carson Prize in 2009 (see https://topdocumentaryfilms.com/the-world-according-to-monsanto/).

19.3.1 The Popular Established Narrative of Precaution

Like certain traditions that are passed on from one generation to the next, narratives that transport shared values, beliefs and emotions help us navigate through a complex world by interpreting new events in a pre-established mental frame (Olson and Witt 2019). They help people to say something meaningful about a complex issue they are not really familiar with. For example, by raising a concern about a new technology regarding its possible long-term impact, one neither needs to know anything about the technology itself nor about the probability and the consequences of a possible long-term impact. It is a normative view meant to express personal values that are expected to be respected. At the same time, being concerned that short-term business-oriented activities may have long-term negative consequences for society and the environment makes one appear to be informed, responsible and far-sighted. How can someone possibly challenge such a narrative of precaution unless he or she is paid by a lobbying organization or simply very naïve? In other words, very general normative statements are a way of self-protection, they help one to avoid the risk of appearing ignorant and immunize against potential criticism and further inquiry (Aerni and Grün 2011).

The narrative of precaution has also been used to warn of the consequences of the 'Anthropocene', a term that was originally coined by the Noble Prize Winner Paul Crutzen to describe a new epoch in the history of the earth that is characterized by the deep impact of human interventions on the earth system (Crutzen 2006). In the educational context, the Anthropocene is connected to a normative story of human-induced environmental decline (Steffen et al. 2015). It builds implicitly upon the myth that humankind was once in harmony with nature but then decided to subject and exploit it at the expense of the environment and the natural resources on which we depend. The popularity of this storyline may also explain the enormous commercial success of Yuval Harari's book 'Homo sapiens: A brief history of humankind' (Harari 2015).

Opponents of GMOs are aware of the power of the narrative of precaution. They also know that their public legitimacy is largely derived from the role they play in such narratives: the defenders of 'the poor and the environment' against 'corporate interests'. This also explains why they have an active interest in extending it to gene-editing (Bain et al. 2019).

This may be convenient for opponents of agricultural biotechnology to preserve their popular role as representatives of the public interest in the public drama on GMOs, and GMO 2.0 in the sequel. But will the people be prepared to buy it? This probably depends on the degree of curiosity they show about the wide range of techniques and applications associated with gene-editing and how they could be applied to address global sustainability challenges.

19.3.2 Gene-Editing: What It Does, How It Evolves and Who Owns It

Gene-editing techniques enable researchers to edit genes at almost any specific location of the genome of a living organism. In this context, they make use of restriction enzymes (nucleases) to delete specific portions of a gene or to insert other DNA exploiting the cell's natural DNA repair mechanisms (Sprink et al. 2020). While genome editing, broadly speaking, encompasses several technologies, the precision, ease, and flexibility of the gene-editing system called CRISPR Cas9 (an abbreviation for "clustered regularly interspaced short palindromic repeats") has been unmatched so far (Cong et al. 2013). Several universities such as Vilnius University, University of California Berkeley, the Broad Institute and the University of Vienna contributed to its successful application and publication in 2012.

In the meantime, researchers also make use of genome-editing systems to edit RNA, regulate gene expression, and change single nucleotides within the genome (Cox et al. 2017; Qi et al. 2013; Wang et al. 2020). The CRISPR tool box of precision breeding has expanded significantly through the discovery of other useful additional RNA-guided nucleases in bacterial genomes (in addition to Cas9) (Zhang et al. 2019), base editing, and prime editing. The tools offer complementary strengths and weaknesses to edit almost any target site (Veillet et al. 2020).

19.3.3 Improved Access to Advanced Biological Discovery Platforms

Advances in large-scale germplasm-characterization has led to the creation of Genetic sequence databases (GSD) that provide free and unrestricted access to data archives for scientists anywhere. The largest database available is the collaborative framework of the International Nucleotide Sequence Database Collaboration (INSDC).[5]

As a result, gene banks are being increasingly converted into biological discovery platforms, enhancing the importance and value of crop diversity as a source of haplotypes encoding desirable traits in plant breeding. This will contribute to a deep collection of germplasm characterized at sequencing and agronomic levels for identification of marker-trait associations and superior haplotypes (Varshney et al. 2020). Improved and easier access to such knowledge-based platforms combined with investments in basic gene-editing equipment in plant research laboratories will also allow plant breeders in developing countries to leap-frog more conventional breeding approaches that proved more expensive and time-consuming, as well as less effective (Gaffney et al. 2020).

---

[5] https://www.insdc.org/.

19.3.4 Gene Editing for the Genetic Improvement of Orphan Crops

Advanced precision breeding in collaboration with local research institutes and other domestic stakeholders in low income countries have already resulted in great advances in the genetic improvement of orphan crops that are of high relevance for global food security. Gene-editing techniques have been used to make such crops more resistant to abiotic and biotic stress factors, more nutrient efficient, better protected against post-harvest losses as well as more valuable from a commercial point of view by inserting certain quality traits (Bart and Taylor 2017; Bull et al. 2018; Oliva et al. 2019; Bellis et al 2020). This may not just contribute to improved food security and less malnutrition but also promote sustainable intensification in tropical agriculture resulting in yield increases, less greenhouse gas emissions, less deforestation, and less pressure on the use of natural resources (Willet et al. 2019).

These are all very promising developments that indicate the great potential of gene-editing to contribute to the UN Sustainable Development Goals (UN SDGs) in general and climate change mitigation and adaptation in particular (Tylecote 2019).

19.4 The Challenge of Regulation

When gene-editing was first applied to plant breeding in 2013, the question was already in the room whether the public at large and consumers in particular would accept food derived from such a technology (Zhang et al. 2019). There was a general agreement that minimal point mutations created by gene-editing would face less public resistance than gene-editing techniques that involve the insertion of foreign DNA, especially if the genetic make-up of the resulting product remains indistinguishable from its conventional counterpart and offers a concrete health or sustainability benefit.

For that reason, a distinction was made between three types of alterations based on the degree of invasiveness of the applied Site Directed Nuclease (SDN).[6] SDN-1 type comprises interventions that merely introduces base-pair changes or small insertions/deletions without addition of foreign DNA. SDN-2 types makes use of small DNA templates to generate directed sequence changes by homologous recombination[7] involving specific nucleotide substitutions. SDN-3 inserts larger DNA elements of foreign origin using an approach similar to SDN-2. Yet, unlike SDN-2, SDN-3 is clearly regarded as transgenic because it introduces significant pieces of foreign DNA (Agapito-Tenfen et al. 2018).

---

[6] SDNs produce a sequence-specific DNA break that is repaired by the plant's natural DNA repair mechanisms; as the repair is inherently imperfect, it results in target-site variants.

[7] Homologous recombination (HR) is the genetic consequence of physical exchange between two aligned identical DNA regions on two separate chromosomes or on the same chromosome.

19.4.1   United States: Prior Experience with GMO Determines Regulation of Gene-Editing

In the United States, where GM crops have been grown and consumed for almost two decades, government authorities did not see any need for the creation of any additional regulatory body for the assessment and approval of gene-edited crops. In 2004 the US Food and Drug Administration (FDA) published the opinion that CRISPR-edited mushroom could enter the market without oversight since it could be attributed to the SDN 1 type of gene-editing (Waltz 2016). This made the product the first CRISPR-edited organism to receive such market authorization (Wang et al. 2019).

Later on, additional CRISPR-edited products such as false flax with enhanced omega3 oil, soy bean that produce trans-fat free soy oil, drought-resistant soybean, and the TALEN-edited potatoes that produce less acrylamide (a known carcinogen) have been approved for commercialization (Wang et al. 2019; Urnov et al. 2018). All these products have also in common that they provide clear benefits for human health and the environment (Urnov et al. 2018).

Finally, in May 2020, the US government published its new regulatory policy in the Federal Register that largely exempts SDN 1 type of gene-edited crops from regulatory oversight easing their path to the market. However, moving a gene between species or rewiring its metabolism will still be subject to regulatory review.[8] Yet, another regulatory clarification may make it easier to create minor variations designed to tailor GM crops to different climates. The changes will go into effect in 2021 (Stockstad 2020).

19.4.2   EU: Stuck in the Past

The situation in Europe evolved into the opposite direction. The decision of European Court of Justice in July 2018 to subject gene-edited crops to the same preventive regulations that govern conventional GM organisms, is widely considered to be a rule that is not compatible with the basic philosophy behind the precautionary principle (Aerni 2019) and not grounded in scientific principle (Agapito-Tenfen et al. 2018). It is also not clear how the ruling can be enforced in practice given the dependence of Europe on the import of soybean from countries where SDN 1 type of gene-editing is not subject to burdensome regulation. On the other hand, it signals to investors and

---

[8] While a trigger mechanism (SDN-1, beyond SDN-1) determines whether a submitted product requires regulatory overview or not, the US regulation could still be called product rather than process-based regulation. After all, the initial product-based approach to GM crop regulation proposed by the White House Office of Science and Technology Policy (OSTP) in 1986 also included an initial test to determine the regulatory pathway based on the concept of substantial equivalence (Aerni and Rieder 2001). Substantial equivalence is the initial step designed to test if there are toxicological and nutritional differences in the new food compared to a conventional counterpart. If no such differences are found, they are declared as 'substantially equivalent'.

skilled researchers that Europe has ceased to be place to invest in cutting-edge plant research with a great potential for sustainable agriculture as well as climate change mitigation and adaptation (Bierbaum et al. 2020).

While it is still unclear how many countries outside Europe and the United States will eventually regulate gene-edited crops, there is a clear trend in favor of the regulatory approach pursued by the United States. Important agricultural economies such as Brazil, Argentina, Chile and Colombia in Latin America as well as Israel, Australia and Japan have already decided that SDN 1 type of gene-edited will be exempted from gene regulation (Schmidt et al. 2020).

19.4.3 New Zealand: A Powerhouse of Agricultural Innovation Strangled by Regulation

There is however one exemption: New Zealand. In 2014, the High Court of New Zealand decided that gene-editing has to be subject to the country's very burdensome GMO regulation based on the Hazardous Substances and New Organisms (HSNO) Act dating back to 1996. It was pointed out that the Act is not well drafted and should be revised. Law makers addressed the issue by clarifying that conventional mutagenesis techniques are exempted while, gene editing, no matter how similar they are to mutagenesis, must be regulated as GMOs.

This was celebrated by the "Sustainability Council of New Zealand" which originally challenged the New Zealand Government in the High Court for considering the SDN 1 type of gene-editing to be more similar to mutagenesis than transgenic crops and therefore not subject to the HSNO Act.

The question is, however, how sustainable is the "Sustainability Council of New Zealand" really? In 2019, an expert panel set up by Royal Society Te Apārangi of New Zealand argued that the HSNO Act of 1996 needs to be overhauled to provide a better basis for assessing the risks and opportunities of particular applications of gene editing.[9] In this context, it should be taken into account that the body of knowledge and experience in the area of agricultural biotechnology is today much larger than in 1996 and the urgent need to address global environmental challenges, such as climate change, is much greater than in the twentieth century. Therefore, gene-editing should not be discarded as one of many options to address such challenges.

The pressure for New Zealand increased when Australia decided in April 2019 that SDN-1 type of gene-editing will not be subject to stringent GMO regulation because it does not involve any transfer of genetic material and has a potential to address agronomic and environmental problems in agriculture (Mallapaty 2019). This paves the way for the approval Australia's innovative gene-edited barely which has shown to increase yield, quality and nutrient efficiency.[10]

---

[9] 2019 ROYAL SOCIETY TE APĀRANGI Report on Gene-editing (see https://www.royalsociety. org.nz/assets/Uploads/Gene-Editing-FINAL-COMPILATION-compressed.pdf).

[10] See https://www.farmweekly.com.au/story/6867769/gene-editing-creates-superior-barley-trait/.

New Zealand faces similar environmental challenges like Australia. Its greenhouse gas emissions per capita are high compared to its small population size,[11] and have risen significantly in the past few decades mainly because of emission-intensive activities in agriculture and forestry, two of the main pillars of the New Zealand economy.

At the same time, New Zealand passed a zero-carbon bill to reduce its greenhouse gas emissions to a near-neutral level by 2050. In particular, it aims to cut methane emissions mostly produced in the livestock industry by 10% by 2030, and up to 47% by 2050.[12] Gene-editing may only be one of several options to reduce greenhouse gases in agriculture in general and methane emissions in the livestock industry in particular. Nevertheless, it could play a significant role in helping New Zealand to meet its ambitious targets, especially in view of the fact that its Crown Research Institutes do cutting-edge research in agricultural biotechnology.[13]

In this context, the Green Party of New Zealand was challenged for issuing a statement in August 2020 that "science should guide the policies and decisions of the members of the parliament". Andrew Hoggard, the leader of the Federated Farmers of New Zealand challenged them by wondering why they would then not listen to the science principle when it comes to the potential of agricultural biotechnology to address climate change.[14] The Federated Farmers played a supportive role in the 1980s when the government decided to reform agricultural policy. As a result of these policy reforms, New Zealand developed a more open and competitive farming system in which the government assumed the role of a coach that supported farmers with biosecurity measures, marketing and large-scale investments in agricultural research and development (R&D). It also made New Zealand one of the most advanced innovation-oriented agricultural economies in the world and its semi-privatized Crown Research Institutes became a leader not just in agricultural innovation but also basic agricultural research. New Zealand in fact re-invented the original spirit of the land-grant college system in the United States in the nineteenth century and tailored it to the challenges and opportunities of rural development of the twenty-first century. This allowed it, in the long-term to reduce poverty in rural areas, and make agriculture more sustainable not through a system of conditional direct payments like in Europe, but through investments in innovations that also generated positive externalities for society and the environment (Aerni 2009).

---

[11] 17.2 tonnes (in carbon dioxide equivalent) per person.

[12] See https://www.dw.com/en/climate-change-new-zealand-passes-zero-carbon-law/a-51145459.

[13] AgResearch has developed a genetically modified high metabolisable energy (HME) ryegrass with a high potential to reduce methane emissions in dairy farming by 30% while also making it more productive and less dependent on irrigation 50%. It has to be field-trialed in the United States since New Zealand law does not allow for it (see https://www.nzherald.co.nz/the-country/news/article.cfm?c_id=16&objectid=12262826).

Principal scientist Greg Bryan has also found it can store more energy for better animal growth, be more resistant to drought and produce up to 23% less methane from the dairy livestock it feeds.

[14] https://www.scoop.co.nz/stories/PO2008/S00172/refreshing-to-hear-the-green-party-wants-to-embrace-science-feds.htm.

In view of this history, the regulatory approach to agricultural biotechnology runs completely against the country's own success story in making agriculture more competitive, diverse and sustainable through innovation. At the beginning, it may also have made sense to refrain from producing GM crops since one of its major importers were countries with a de-facto ban on GMOs, mostly located in Europe. But, the situation today is quite different: the three largest importers of agricultural products from New Zealand are China, Australia and the USA. They all grow GM crops and tend to favor regulation that exempt SDN-1 type of gene-editing from GMO regulation (even though the situation in China is still uncertain) (Fritsche et al. 2018).

19.4.4 The Impact of Regulation on Innovation

The highly preventive GMO regulatory frameworks and the subsequent court decisions in the European Union and Zealand to subject gene-editing techniques to the GMO regulatory frameworks developed twenty years ago, has an impact on investment in agricultural biotechnology in these regions and the migration of skilled researchers in the field. Many gene-edited crops developed in laboratories in Europe and New Zealand must rely on field testing abroad (Jorasch 2020). Moreover, many of the innovative global agribusiness companies with roots in New Zealand and Europe tend to build up more research facilities in countries where the research infrastructure has a similar quality but the regulatory environment is more friendly toward the actual use of the technology (Graff and Hamdan-Livramento 2019). The trend manifests itself also in the filing of patents related to gene-editing in general and gene-edited plants and animals in particular. The United States and China are once again emerging as the two rivals in the accumulation of applied knowledge, assessed by the number of approved patents in the field of gene-editing. In return, European countries lag strongly behind and New Zealand does not even show up in the landscape of intellectual property rights (IP) granted to inventions related to gene-editing (Martin-Laffon et al. 2019). This may also explain why most of the new companies that are emerging in the field of gene-editing are found in the United States and China too.

19.5 Challenged by the Young Generation

In view of the regulatory deadlock that strangles innovation in plant breeding and prevents the emergence of solution-oriented start-up companies in the industry, established stakeholders in Europe and New Zealand have come under increasing scrutiny by new players. There are movements of young plant scientists who desire to make a difference through their research and therefore increasingly dislike the fact that their research on gene-editing in the laboratory is never meant to be used in agriculture. In response to the ruling of the ECJ in July 2018, young researchers in the Netherlands created the "Gene Sprout Initiative",[15] which is now spreading across Europe. It encourages more bold thinking on the creation of a sustainable European bioeconomy that does not discard the potential of gene-editing just because it is new. In New Zealand, a letter signed by 155 young students and researchers in the biological and environmental sciences to the Green Party in November 2019 has urged it to rethink its stance on the regulation of agricultural biotechnology. It was a reminder to the public in New Zealand, that incumbents and their defensive views of sustainable development are increasingly challenged as irresponsible by a younger generation.[16] May this be an indication that young people increasingly refuse to buy the defensive narrative of their earlier mentors who frame agricultural biotechnology only as a threat to sustainable development while disregarding its opportunities?

19.5.1 If You Want to be Rebel, Just Act the Way We Did, When We Were Young!

Children are known to rebel against the cherished views of their parents. This is however difficult if parents argue that they themselves used to be rebels themselves challenging the establishment. This was the case with those who belonged to the 68 generation. They have largely shaped the view that technology and innovation are a threat to sustainability that needs to be regulated with a 'better safe than sorry' regulatory approach (Aerni and Grün 2011). They proved very successful in passing on their 'subversive' values to their offspring by pointing at their creative protest actions against nuclear power and the Vietnam war and the music that challenged the taste of their parents (Grau 2018). Yet, today, these progressive narratives of the 1970s have become the conservative narrative of the establishment today, designed to prevent rather than enable change. Therefore, they do not resonate strongly anymore with the millennials. After all, they do not provide any answers to cope with challenges of the twenty-first century. This may explain the generational conflict of the Green Party in Germany. A faction of the Green Party, mostly represented by the younger generation, challenged the entrenched views of the established members against agricultural biotechnology by pointing at the potential of gene-editing in addressing certain sustainability problems (Zinkant 2020). Their main argument was that creating taboos in the sustainability discourse does not lead anywhere. Rather than just say 'no' to gene-editing, they want to learn more about the technology on a case-by-case basis and dislike the de-facto censorship on the debate within the Party. For the time being, the incumbents in the Party were able to smother the little fire. But on the long run, it is the younger generation that is likely to prevail.

---

[15] https://www.genesproutinitiative.com/.

[16] https://www.agscience.org.nz/young-scientists-letter-to-a-divided-green-party-calls-for-a-rev iew-of-our-gm-law-to-help-tackle-climate-crisis/.

19.6 The Landscape of Intellectual Property Has Changed and No One Seems to Have Noticed

Even though the younger generations generally recognize the importance of innovation for sustainable change, they are very concerned about the potential negative impact of Intellectual Property (IP) protected technologies on fair access and benefit sharing, especially when it involves patents granted to large companies. The argument is that the patent holding company would take advantage of the temporary monopoly right granted by the patent either to exclude others from use, or to extract undue rents from users who depend on their innovation. Public resentment is particularly high when the IP protected innovations are broadly used in research and have the potential to contribute to global food security or public health.

But is the issue really black and white in the sense that innovating companies generate profits at the expense of people? While there are certainly those who continue to defend the classic IP system based on exclusivity, the IP system has substantially evolved in direction of openness, transparency and sharing because circumstances in which the innovation processes takes place have changed significantly too. Companies operate in a particular innovation ecosystem that requires them to share their IP protected technologies in return for granting access to a much broader knowledge platform (Graff and Hamdan-Livramento 2019).

This is particularly true for the field of agricultural biotechnology. First of all, most of the crucial patents are held by universities and can be easily accessed thanks to third party licensing schemes. For example, as the primary source of exchange of plasmids containing ZFN, TALEN and CRISPR, Addgene has become one of the most important third party licensing intermediaries enabling a better separation university's broader mission and pursuit of commercial application (Graff and Sherkow 2020).

During the COVID-19 crisis, Addgene played a crucial role in facilitating the open sharing of plasmids and resources for COVID-19 research (Tsang and LaManna 2020). It ensures access to all researchers under specific terms that contain restrictions against commercial users. However, the IP protected core CRISPR inventions, mostly held by universities, have been offered non-exclusively to a range of companies that specialize in selling research equipment and standardized reagents (CRISPR products for use in research labs, including academic and commercial R&D).

As for gene-editing applied to crop breeding, the Broad Institute and Corteva Agriscience allow the CGIAR centers free access to their gene-editing technologies for breeding crops to be used by small-holder agriculture (Srivastava 2019). A potential licensing fee will only be considered, if the resulting products reach a certain threshold of commercial success. In addition, Broad and Corteva have developed a joint licensing framework that brings together patents that both parties have come to control and offer them as non-exclusive commercial licenses to any interested users of CRISPR technology in crop agriculture, including direct competitors of Corteva in its major markets (Graff and Sherkow 2020).

Second, thanks to the fact that there is now a wide range of gene-editing techniques available (using different types of nuclease), a return to a de-facto monopoly power, which Monsanto had earlier due to its ownership of the crucial set of patents necessary to apply genetic engineering to plant breeding, is not going to happen anymore (Veillet et al. 2020; Graff and Sherkow 2020). The more flexible licensing approaches also result in greater competition and innovation as well as more ways to customize and tailor products to particular customer needs.

Third, the IP Landscape looks completely different compared to a few decades ago, when Monsanto still dominated the field. Most important biotech patents held by Monsanto have expired and initiatives have been launched to make cheaper agro-biogenerics more widely available despite regulatory hurdles (Jefferson et al. 2015). Despite a thicket of complex legal and regulatory hurdles, agrobiogenerics offer an opportunity for new companies to enter the market with alternative products that are more affordable.

Finally, a vast body of knowledge in the public domain, based on publications in scientific journals and the open access to the description of existing and expired patents, have made possible advances in gene-editing techniques in the first place (Graff and Sherkow 2020).

19.6.1 Harnessing the Technology as a Tool of Empowerment

All these trends, combined with open-access biological discovery platforms and the rapidly decreasing costs of gene-editing tool kits may offer a great opportunity for developing countries to make use of the new tools of precision breeding to address their priority concerns in agriculture. This already happened before with tissue culture technology designed to clone clean planting material of vegetatively propagated crops. Once this technology became cheaper and more user-friendly, low-cost tissue culture laboratories have been set up in the rural areas in Africa and Latin America with the purpose of empowering farmers to create better solutions for themselves. The projects showed that tissue culture is a useful technique that enables farmer groups to make better use of their local knowledge of clean planting material of crops such as cassava or banana. Once they are trained on how to manage, operate and use low-cost tissue culture laboratories, they learn how to multiply clean planting material in vitro with the purpose of subsequently selling the raised clean stakes to other farmers. This became in many cases a flourishing business with local knowledge content, produced by locals and for locals. It shows clearly, that advanced breeding technologies must not stand in contradiction with local traditional knowledge (Aerni 2006; Jacobsen et al. 2019).

In return, access to biological resources may ironically become highly restricted due to the inconsistent and burdensome requirements to access genetic resources (GR) under the Nagoya Protocol[17] on Access and Benefit Sharing (ABS) of the Convention on Biological Diversity (CBD), which has entered into force in October 2014.

19.6.2 The Need to Rethink Access and Benefit-Sharing (ABS)

According to the CBD, states exercise sovereign rights over domestic Genetic Resources (GR) and traditional knowledge associated with it, as well as its subsequent uses that must be based on prior-informed consent including Material Transfer Agreements. Even though the Nagoya Protocol aims to create conditions that encourage research on the conservation and sustainable use of biological diversity (Article 8a), it makes no exceptions for academic research or conservation-related research under the term 'utilization' (Article 2c) (Prathapan et al. 2018).

It is the concrete implementation of the Nagoya Protocol that raises particular concerns related to access and inclusiveness. The Regulation (EU) No 511/2014 of the European Parliament and of the Council on compliance measures for users from the Nagoya Protocol[18] is a good case to illustrate how good intentions lead to an outcome that prevents rather than enables knowledge sharing in plant breeding. The regulation has expanded the 'where' (extending to GR kept outside their country of origin), the 'when' (extending retroactively to GR accessed prior to the CBD entering into force), the 'what' (extending the GR obligations to commercial materials made with GR and derivatives), the 'who' (extending the obligations from parties directly accessing a GR to parties using commercial products made with GR) (Herrlinger and Kock 2016). Even though this broad definition of sovereign rights over genetic resources will substantially affect the right to access, transfer and use GR, many tropical countries with biodiversity hotspots have welcomed these strict requirements and adopted similar national regulation to protect the sovereign rights of *their* GR (Herrlinger and Kock 2016).

Yet, the belief that genetic resources are comparable to oil or mineral mines where the resource has a commercial value on its own, turned out to be wrong. Moreover, there are alternatives to do the research in megabiodiversity hotspots. There are the numerous ex situ collections that can be freely accessed, including the growing number of genetic sequences database (GSD). Moreover, the commercially interesting biodiversity of microorganisms is mostly located in temperate zones and not in the areas with megabiodiversity. The claim of ownership of microorganisms through the exertion of sovereign rights over genetic resources is also highly problematic from a public health point of view. After all, pathogens know no national borders

---

[17] https://www.cbd.int/abs/.

[18] https://ec.europa.eu/environment/nature/biodiversity/international/abs/legislation_en.htm.

and their spread must be contained through partnerships in non-commercial research. Long negotiations related to ABS requirements to obtain specimens in such research partnerships may not be in the public interest because they slow down the response to public health threats, manifested in the risk of infectious diseases spreading in populations that have no immunity (Overmann and Scholz 2017). COVID-19 has made the world again aware of this problem.

19.6.3 More Creative Solutions to Protect the Public Interest While Enabling Innovation

Unlike the Nagoya Protocol, which regards GR as being of proprietary nature, the so-called 'Plant Treaty' (FAO Treaty for Plant Genetic Resources in Food and Agriculture) adopted in 2001, facilitates access to the genetic materials of important food crops in a multilateral system for research, breeding and training. In accepting the terms and conditions—including benefit-sharing—of a standard material transfer agreement (SMTA) prior informed consent (PIC) is automatically granted. Since the members of the treaty are member states only and is limited to a few important crops, it does not create predictable benefits for breeders, its basic principle of 'free access but not access for free' could serve as a template for a comprehensive reform of the IP system in food and agriculture (Herrlinger and Kock 2016). It would rely on a subscription-based 'flat rate' model to access GR that is based on clearly defined rights and obligations on the terms of knowledge sharing while still ensuring that investors in innovation can generate the necessary returns to reimburse their fixed costs invested in R&D. The revenues necessary to operate and maintain such club good could be derived from a value-added tax on seeds, considering that the cost of seed makes up on average less than 2% of the total expenditure for external input paid by farmers. After all, the value-added of improved seed may manifest itself only in the consumed end product (e.g. improved taste, lower price, less stains).

This model has been successfully embraced in the music industry once peer-to-peer file sharing undermined the prior proprietary business model. The search for a new business model in the music industry was necessary because the information (music) became increasingly de-coupled from the carrier of information (music playing device) due to digitalization. The same is now becoming a reality in plant breeding with the growing reliance on digital sequences information, because the information carrier (genetic resource) is increasingly de-couple from the information (genetic sequence).

The digital revolution is also likely to transform the business model in the seed industry and a well-designed subscription-model that includes all actors interested in creating commercial value out of GR and all products, which are used in the field of food and agriculture could have a great potential (Metzger and Zech 2020; Zech 2015).

19.7 Concluding Remarks

Over the past three decades, agricultural biotechnology research has extended beyond input-trait genetically-modified (GM) products and expanded into the commercialization of output-trait GM products. Thanks to new plant breeding techniques (NPBTs), such as genome editing, the breeding process has become faster, less invasive and more precise. The required toolkit to do gene-editing is also much more accessible and affordable because the approach to intellectual property rights (IPR) protection has moved from a focus on creating profits by excluding others to generating value beyond the owner of the IP through a focus on sharing under mutually agreed terms. Moreover, advances in the creation of large biological discovery platforms provide access to a much larger ex-situ germplasm pool and genetic sequences information for breeders enabling them to tailor breeding to locally relevant circumstances to a much greater extent.

The application of gene-editing can, but does not have to involve the transfer for genes; and that has become a crucial criteria on how to regulate the technology. Currently, only the EU and New Zealand regulate gene-editing techniques like GMOs. Low-income countries that depend on aid from and trade access especially in Europe, are however likely to follow this highly restrictive approach eventually. In the EU as well as New Zealand, opponents of GMOs were successful in convincing the courts that gene-editing techniques would merely represent a new generation of GMOs (GMO 2.0). This was in a way convenient because it allowed the opponents to adopt a prevailing negative narrative on agricultural biotechnology that is believed to represent the norms and values of society, and with it, help preserve the bipolar baseline assumptions that shape the public debate on sustainable agriculture. In this polarized debate, which largely dispenses with the historical context and the possibility to combine different approaches depending on the local circumstances, moving away from agricultural trade and advanced agricultural technologies is the only way to preserve food sovereignty, healthy food, and our natural resource base. This narrative implies that everything used to be sustainable prior to making agriculture a business. In this chapter, we illustrated that such baseline assumptions do clearly not correspond to historical evidence. Throughout history, institutional and technological change in agriculture were largely a response to a growing population that could not be fed anymore with strictly traditional agricultural practices. Such changes are never 'risk-free', but have led to new challenges that once again have to be addressed by trying different approaches and combinations. But what is the alternative to this process of trial and error? Relying on approaches that make the affluent minority feel good about itself but do neither address the scarcity problems nor the environmental challenges resulting from poverty in the majority world in low income countries?

The popular bipolar narrative may provide meaning and orientation at low cost. It allows people to side with the 'good' and condemn the 'bad' in agriculture without the need to search for additional information. But, this has also unintended consequences, because ultimately, it will be the poor and the environment that will suffer most from the exclusion from agricultural trade and new technologies.

A growing share of the younger generations seem to grasp this danger. They are very much concerned about the failure of the United Nations to achieve the ambitious Sustainable Development Goals (SDGs) by 2030, mainly because of the infertile and rather uninformed bipolar debate on sustainable agriculture. In Europe, millennials have become impatient with environmental policies that make use of the precautionary principle to postpone action rather than enable it. This manifested itself recently within the Green Party of Germany where the youth faction demanded from the incumbents to have a serious debate on gene-editing and explore its potential on a case-by-case basis. In New Zealand, young researchers in the biological and environmental sciences were wondering in an open letter to the Green Party if they really care about future generations.

Even though there is no evidence that these seeds of resistance against the prevailing narrative of GMOs in agriculture will sprout more widely, they are signs of hope for a more constructive debate in future that will also result in a more solution-oriented type of regulation on gene-editing.

Acknowledgements I would like to thank Isabelle Schluep for the thorough proofread and critical feedback.

References

Aerni P (2006) Mobilizing science and technology for development: the case of the Cassava Biotechnology Network (CBN). AgBioforum 9(1):1–14

Aerni P (2009) What is sustainable agriculture? Empirical evidence of diverging views in Switzerland and New Zealand. Ecol Econ 68(6):1872–1882

Aerni P (2011) Food sovereignty and its discontents. ATDF J 8(1–2):23–40

Aerni P (2018) The use and abuse of the term 'GMO' in the 'Common Weal Rhetoric' against the application of modern biotechnology in agriculture. In: James H (ed) Ethical tensions from new technology: the case of agricultural biotechnology. CABI Publishing, pp 39–52

Aerni P (2019) Politicizing the precautionary principle: why disregarding facts should not pass for farsightedness. Front Plant Sci 10:1053

Aerni P, Grün K-J (eds) (2011) Moral und Angst. Vandenhoeck & Ruprecht Verlag, Göttingen

Aerni P, Oser F (eds) (2011) Forschung verändert Schule. Seismo Verlag, Zürich (Jan 2011)

Aerni P, Rieder P (2001) 'Public policy responses to biotechnology'. Chapter BG6.58.9.2 in Our fragile world: challenges and opportunities for sustainable development. Encyclopaedia of Life Support Systems (ELOSS), UNESCO, Paris

Aerni P, Rae A, Lehmann B (2009) Nostalgia versus pragmatism? How attitudes and interests shape the term sustainable agriculture in Switzerland and New Zealand. Food Policy 34(2):227–235

Aerni P, Karapinar B, Häberli C (2012) Rethinking sustainable agriculture. In: Cottier T, Delimatsis P (eds) The prospects of international trade regulation: from fragmentation to coherence. Cambridge University Press, Cambridge, UK, pp 169–210

Agapito-Tenfen SZ, Okoli AS, Bernstein MJ, Wikmark OG, Myhr AI (2018) Revisiting risk governance of GM plants: the need to consider new and emerging gene-editing techniques. Front Plant Sci 9:1874

Anderson RS, Levy E, Morrison BM (1991) Rice science and development politics: research strategies and IRRI's technologies confront Asian diversity (1950–1980). Clarendon Press

Bain C, Lindberg S, Selfa T (2019) Emerging sociotechnical imaginaries for gene edited crops for foods in the United States: implications for governance. Agric Human Values 1–15

Bart RS, Taylor NJ (2017) New opportunities and challenges to engineer disease resistance in cassava, a staple food of African small-holder farmers. PLoS Pathog 13(5):e1006287

Bellis ES, Kelly EA, Lorts CM, Gao H, DeLeo VL, Rouhan G et al (2020) Genomics of sorghum local adaptation to a parasitic plant. Proc Natl Acad Sci 117(8):4243–4251

Bierbaum R, Leonard SA, Rejeski D, Whaley C, Barra RO, Libre C (2020) Novel entities and technologies: environmental benefits and risks. Environ Sci Policy 105:134–143

Bonny S (2017) Corporate concentration and technological change in the global seed industry. Sustainability 9(9):1632

Boserup E (1965) The conditions of agricultural growth: the economics of agrarian change under population pressure. George, Allan & Unwin.

Brookes G, Barfoot P (2009) Global impact of biotech crops: income and production effects 1996–2007. AgBioforum 12(2):184–208

Bull SE, Seung D, Chanez C, Mehta D, Kuon JE, Truernit E et al (2018) Accelerated ex situ breeding of GBSS-and PTST1-edited cassava for modified starch. Sci Adv 4(9):EAAT6086

Byerlee D, Morris M (1993) Research for marginal environments: are we underinvested? Food Policy 18(5):381–393

Clark S (2016) Sustainable agriculture–beyond organic farming. MDPI AG

Cong L, Ran FA, Cox D, Lin S, Barretto R et al (2013) Multiplex genome engineering using CRISPR/Cas systems. Science 339:819–823

Conway G (1999) The doubly green revolution: food for all in the twenty-first century. Cornell University Press

Cox DBT, Gootenberg JS, Abudayyeh OO, Franklin B, Kellner MJ et al (2017) RNA editing with CRISPR-Cas13. Science 358:1019–1027

Crutzen PJ (2006) The "anthropocene". In Earth system science in the anthropocene. Springer, Berlin, Heidelberg, pp 13–18

Duvick DN (2001) Biotechnology in the 1930s: the development of hybrid maize. Nat Rev Genet 2(1):69–74

Francisco Ribeiro P, Camargo Rodriguez AV (2020) Emerging advanced technologies to mitigate the impact of climate change in Africa. Plants 9(3):381

Freidberg S (2007) Supermarkets and imperial knowledge. Cult Geogr 14(3):321–342

Fritsche S, Poovaiah C, MacRae E, Thorlby G (2018) A New Zealand perspective on the application and regulation of gene editing. Front Plant Sci 9:1323

Fukuda-Parr S (ed) (2007) The gene revolution: GM crops and unequal development. Earthscan

Graff GD, Sherkow JS (2020) Models of technology transfer for genome-editing technologies. Ann Rev Genomics Human Genet 21

Graff GD, Hamdan-Livramento I (2019) The global roots of innovation in plant biotechnology. Econ Res Working Paper (59)

Grau A (2018) Die Rechtfertigungsideologie enthemmter Konsumkinder. Cicero Magazin für Politische Kultur, 06 Jan 2018 (https://www.cicero.de/kultur/68-68er-gesellschaft--hedonismus-marx-marcuse)

Gabrielczyk T (2019) Gentechtäuschung mit Siegel? Transkript 25(3):47–54

Gaffney J, Tibebu R, Bart R, Beyene G, Girma D, Kane NA et al (2020) Open access to genetic sequence data maximizes value to scientists, farmers, and society. Glob Food Secur 26:100411

Gehlen A (1988) Man: his nature and place in the world (first published in German in 1940). Columbia University Press

Giovannucci D, von Hagen O, Wozniak J (2014) Corporate social responsibility and the role of voluntary sustainability standards. In: Schmitz-Hoffmann C et al (eds) Voluntary standard systems. Springer, Berlin, Heidelberg, pp 359–384

Harari YN (2015) *Homo sapiens*: a brief history of humankind. Random House

Herrlinger C, Kock M (2016) Biodiversity laws: an emerging regulation on genetic resources or 'IP on Life' through the backdoor. Biosci Law Rev 13(4):119–131

Jacobsen K, Omondi BA, Almekinders C, Alvarez E, Blomme G, Dita M et al (2019) Seed degeneration of banana planting materials: strategies for improved farmer access to healthy seed. Plant Pathol 68(2):207–228

Jefferson DJ, Graff GD, Chi-Ham CL, Bennett AB (2015) The emergence of agbiogenerics. Nat Biotechnol 33(8):819–823

Jorasch P (2020) Will the EU stay out of step with science and the rest of the world on plant breeding innovation? Plant Cell Rep 39(1):163–167

Juma C (2016) Innovation and its enemies: why people resist new technologies. Oxford University Press

Kingsbury N (2009) Hybrid: the history and science of plant breeding. Chicago University Press

Lamphere JA, East EA (2017) Monsanto's biotechnology politics: discourses of legitimation. Environ Commun 11(1):75–89

Lee JH, Mazarei M, Pfotenhauer AC, Dorrough AB, Poindexter MR, Hewezi et al (2020) Epigenetic footprints of CRISPR/Cas9-mediated genome editing in plants. Front Plant Sci 10:1720

Liu Q, Yang X, Tzin V, Peng Y, Romeis J, Li Y (2020a) Plant breeding involving genetic engineering does not result in unacceptable unintended effects in rice relative to conventional cross-breeding. Plant J https://doi.org/10.1111/tpj.14895

Liu G, Zhang Y, Zhang T (2020b) Computational approaches for effective CRISPR guide RNA design and evaluation. Comput Struct Biotechnol J 18:35–44

Lynas, M. (2013) Terminator seeds will not usher in an agricultural judgement day. The Conversation, December 24, 2013 (available online: https://theconversation.com/terminator-seeds-will-not-usher-in-an-agricultural-judgement-day-21686)

Mallapaty S (2019) Australian gene-editing rules adopt 'middle ground'. Nature 23 Apr 2019 (available online: https://www.nature.com/articles/d41586-019-01282-8)

Martin-Laffon J, Kuntz M, Ricroch AE (2019) Worldwide CRISPR patent landscape shows strong geographical biases. Nat Biotechnol 37:613–620

Metzger A, Zech H (2020) A comprehensive approach to plant variety rights and patents in the field of innovative plants. Forthcoming in Christine Godt/Matthias Lamping (eds), In Honour of Hanns Ullrich (tbc), Springer. Available at SSRN: https://ssrn.com/abstract=3675534, https://papers.ssrn.com/sol3/papers.cfm?abstract_id=3675534

Murray DL (1994) Cultivating crisis: the human cost of pesticides in Latin America. University of Texas Press

National Academies of Sciences (2016) Genetically engineered crops: experiences and prospects. A report prepared by the Board on Agriculture and Natural Resources. Washington, DC (available online: https://www.nationalacademies.org/our-work/genetically-engineered-crops-past-experience-and-future-prospects)

Oliva R, Ji C, Atienza-Grande G, Huguet-Tapia JC, Perez-Quintero A, Li T et al (2019) Broad-spectrum resistance to bacterial blight in rice using genome editing. Nat Biotechnol 37(11):1344–1350

Olson ET, Witt K (2019) Narrative and persistence. Can J Philos 49(3):419–434

Overmann J, Scholz AH (2017) Microbiological research under the Nagoya Protocol: facts and fiction. Trends Microbiol 25(2):85–88

Pathirana R (2011) Plant mutation breeding in agriculture. Plant Sci Rev 6(032):107–126

Perlak FJ, Oppenhuizen M, Gustafson K, Voth R, Sivasupramaniam S, Heering D, Boyd C, Ihrig RA, Roberts JK (2001) Development and commercial use of Bollgard® cotton in the USA–early promises versus today's reality. Plant J 27(6):489–501

Pham X, Stack M (2018) How data analytics is transforming agriculture. Bus Horiz 61(1):125–133

Pingali PL (2012) Green revolution: impacts, limits, and the path ahead. Proc Natl Acad Sci USA 109:12302–12308

Prathapan KD, Pethiyagoda R, Bawa KS, Raven PH, Rajan PD (2018) When the cure kills—CBD limits biodiversity research. Science 360(6396):1405–1406

Qi LS, Larson MH, Gilbert LA, Doudna JA, Weissman JS et al (2013) Repurposing CRISPR as an RNA-guided platform for sequence-specific control of gene expression. Cell 152:1173–1183

Ricroch A (2019) Global developments of genome editing in agriculture. Transgenic Res 28(2):45–52

Ritchie H (2017) Is organic really better for the environment than conventional agriculture? Our World in Data, 17 Sept 2017 (available online: https://ourworldindata.org/is-organic-agriculture-better-for-the-environment)

Schmidt SM, Belisle M, Frommer WB (2020) The evolving landscape around genome editing in agriculture: many countries have exempted or move to exempt forms of genome editing from GMO regulation of crop plants. EMBO Reports, e50680, https://doi.org/10.15252/embr.202050680

Schurman RA, Kelso DT, Kelso DD (eds) (2003) Engineering trouble: biotechnology and its discontents. University of California Press

Smyth SJ, Phillips PW, Castle D (eds) (2014) Handbook on agriculture, biotechnology and development. Edward Elgar Publishing

Sprink T, Wilhelm RA, Spök A, Robienski J, Schleissing S, Schiemann JH (eds) (2020) Plant genome editing–policies and governance. Front Media SA

Srivastava V (2019) CRISPR applications in plant genetic engineering and biotechnology. Plant Biotechnol: Progr Genomic Era 429–459 (Springer, Singapore)

Steffen W, Richardson K, Rockström J, Cornell SE, Fetzer I, Bennett EM et al (2015) Planetary boundaries: guiding human development on a changing planet. Science 347(6223)

Stockstad E (2020) United States relaxes rules for biotech crops. Science. https://doi.org/10.1126/science.abc8305

Tsang J, LaManna CM (2020) Open sharing during COVID-19: CRISPR-based detection tools. The CRISPR J 3(3):142–145

Tylecote A (2019) Biotechnology as a new techno-economic paradigm that will help drive the world economy and mitigate climate change. Res Policy 48(4):858–868

Urnov FD, Ronald PC, Carroll D (2018) A call for science-based review of the European court's decision on gene-edited crops. Nat Biotechnol 36(9):800–802

Varshney RK, Sinha P, Singh VK, Kumar A, Zhang Q, Bennetzen JL (2020) 5Gs for crop genetic improvement. Curr Opin Plant Biol

Veillet F, Durand M, Kroj T, Cesari S, Gallois J-L (2020) Precision breeding made real with CRISPR: illustration through genetic resistance to pathogens. Plant Commun https://doi.org/10.1016/j.xplc.2020.100102

Waltz E (2016) Gene-edited CRISPR mushroom escapes US regulation. Nat News 532(7599):293

Wang T, Zhang H, Zhu H (2019) CRISPR technology is revolutionizing the improvement of tomato and other fruit crops. Hortic Res 6(1):1–13

Wang SR, Wu LY, Huang HY, Xiong W, Liu J, Wei L et al (2020) Conditional control of RNA-guided nucleic acid cleavage and gene editing. Nat Commun 11(1):1–10

Willett W, Rockström J, Loken B, Springmann M, Lang T et al (2019) Food in the Anthropocene: the EAT–lancet commission on healthy diets from sustainable food systems. The Lancet 393(10170):447–492

Zhang Y, Malzahn AA, Sretenovic S, Qi Y (2019) The emerging and uncultivated potential of CRISPR technology in plant science. Nat Plants 5(8):778–794

Zech, H. (2015). Information as Property," JIPITEC, 192, https://www.jipitec.eu/issues/jipitec-6-3-2015/4315/zech%206%20%283%29.pdf

Zinkant K (2020) Grüne Gechtechnik: Gründe fordern Umdenken. Süddeutsche Zeitung, 10.Juni (available online: https://www.sueddeutsche.de/wissen/klimawandel-landwirtschaft-gruene-gentechnik-crispr-1.4932845

Philipp Aerni is Director of the Center for Corporate Responsibility and Sustainability (CCRS) at the University of Zurich. He is an interdisciplinary social scientist with a Masters Degree in Geography and a Ph.D. in Agricultural Economics. Prior to his position at CCRS, Dr. Aerni worked at Harvard University, ETH Zurich, the University of Berne, as well as the Food and Agriculture Organisation of the United Nations (FAO). He currently teaches at ETH Zurich, the University of Zurich and the University of Basel.

Printed in the USA
CPSIA information can be obtained
at www.ICGtesting.com
LVHW021525111123
763669LV00005B/54